STRUCTURE AND PROPERTIES
OF ORIENTED POLYMERS

MATERIALS SCIENCE SERIES

Advisory Editors

LESLIE HOLLIDAY,

Brunel University,
Uxbridge, Middlesex, Great Britain

A. KELLY, Sc.D.,

Deputy Director, National Physical Laboratory,
Teddington, Middlesex, Great Britain

STRUCTURE AND PROPERTIES OF ORIENTED POLYMERS

Edited by

I. M. WARD

Department of Physics, University of Leeds, England

APPLIED SCIENCE PUBLISHERS LTD.

LONDON

APPLIED SCIENCE PUBLISHERS LTD
RIPPLE ROAD, BARKING, ESSEX, ENGLAND

ISBN 0 85334 600 3

WITH 21 TABLES AND 220 ILLUSTRATIONS

© APPLIED SCIENCE PUBLISHERS LTD 1975

Filmset by Typesetting Services Ltd., Glasgow, Scotland
Printed in Great Britain by Galliard Limited, Great Yarmouth, Norfolk, England

LIST OF CONTRIBUTORS

D. I. BOWER
   Department of Physics, University of Leeds, Leeds LS2 9JT, England

G. B. CARTER
   ICI Corporate Laboratory/Runcorn, Runcorn, Cheshire, England

M. W. DARLINGTON
   Department of Materials, Cranfield Institute of Technology, Cranfield, Bedford, England

R. A. DUCKETT
   Department of Physics, University of Leeds, Leeds LS2 9JT, England

M. J. FOLKES
   Department of Materials, Cranfield Institute of Technology, Cranfield, Bedford, England

D. W. HADLEY
   J. J. Thomson Physical Laboratory, University of Reading, Whiteknights, Reading RG6 2AF, England

L. HOLLIDAY
   Department of Polymer Science, Brunel University, Uxbridge, England

A. PETERLIN
   Institute for Materials Research, U.S. Department of Commerce, National Bureau of Standards, Washington, D.C. 20234, U.S.A.

B. E. READ
  Division of Materials Applications, National Physical Laboratory, Teddington, Middlesex, England

D. W. SAUNDERS
  Department of Materials, Cranfield Institute of Technology, Cranfield, Bedford, England

V. T. J. SCHENK
  Imperial Chemical House, Millbank, London SW1, England

G. SCHUUR
  Koninklijke/Shell Plastics Laboratorium, Delft, Broekmelenweg 20, The Netherlands

R. S. STEIN
  Polymer Research Institute, University of Massachusetts, Amherst, Massachusetts 01002, U.S.A.

A. K. VAN DER VEGT
  Koninklijke/Shell Plastics Laboratorium, Delft, Broekmelenweg 20, The Netherlands

I. M. WARD
  Department of Physics, University of Leeds, Leeds LS2 9JT, England

G. L. WILKES
  Department of Chemical Engineering, Princeton University, Princeton, New Jersey 08540, U.S.A.

PREFACE

It has become increasingly evident that there is much to be gained from a detailed understanding of the structure and properties of polymers in the oriented state. This book reflects the growth of interest in this area of polymer science and attempts to give the reader an up to date view of the present position. The individual chapters are for the most part self-contained, and cover a very wide range of topics. It is intended that each of them should serve the dual purpose of an expository introduction to the subject and a topical review of recent research.

It is inevitable that there will be differences of style and approach in the contributions from the different authors. No attempt has been made to moderate these differences, as they serve to illustrate the diversity of approaches required to give the reader a balanced view of the subject.

I should like to thank the contributors for their endeavours, and especially for their patience in accepting modifications and corrections which make for consistency in the book as a whole. I am particularly indebted to Professor Leslie Holliday who originally approached me with the proposition that such a book would be a worthwhile venture and to the publishers who have given me every assistance in making its progress as painless as possible.

University of Leeds   I. M. WARD

CONTENTS

Chapter 2
MOLECULAR ASPECTS OF ORIENTED POLYMERS
—A. PETERLIN

Chapter 3
PHYSICO-CHEMICAL APPROACHES TO THE
MEASUREMENT OF ANISOTROPY
—R. S. STEIN and G. L. WILKES

Chapter 4
ULTRA-VIOLET, VISIBLE AND INFRA-RED DICHROISM
—B. E. READ

Chapter 5
POLARISED FLUORESCENCE AND RAMAN SPECTROSCOPY
—D. I. BOWER

Chapter 6
NUCLEAR MAGNETIC RESONANCE
—M. J. FOLKES and I. M. WARD

Chapter 7
THE STIFFNESS OF POLYMERS IN RELATION TO THEIR STRUCTURE
—L. HOLLIDAY

Chapter 8
THE MACROSCOPIC MODEL APPROACH TO LOW STRAIN PROPERTIES
—D. W HADLEY and I. M. WARD

Chapter 9
SMALL STRAIN ELASTIC PROPERTIES
—D. W. HADLEY

Chapter 10
ANISOTROPIC CREEP BEHAVIOUR
—M. W. DARLINGTON and D. W. SAUNDERS

Chapter 11
ANISOTROPIC YIELD BEHAVIOUR
—R. A. DUCKETT

Chapter 12
ORIENTATION OF FILMS AND FIBRILLATION
—G. SCHUUR and A. K. VAN DER VEGT

Chapter 13
ULTRA-HIGH MODULUS ORGANIC FIBRES
—G. B. CARTER and V. T. J. SCHENK

CHAPTER 1

A GENERAL INTRODUCTION TO THE STRUCTURE AND PROPERTIES OF ORIENTED POLYMERS

L. HOLLIDAY and I. M. WARD

1.1 THE PHENOMENON OF ORIENTATION

Orientation in polymers is a phenomenon of great technical and theoretical importance. The word orientation itself conveys a number of ideas. It suggests that the structural units in the material, which in this case can refer to the polymer chains, or the segments of the polymer chains, or the crystalline regions in the polymer, are aligned to some extent. The measurement of orientation in a polymer therefore provides valuable information for an understanding of the structure and properties. It may also describe the process whereby the oriented polymer is produced as well as indicate how the physical properties are modified as a result of this process.

A theoretical introduction to the subject of orientation can usefully deal with three aspects.

(1) The process of orientation.
(2) The effects of this process of orientation on physical properties.
(3) Theoretical explanations of orientation processes, and the influence of orientation on physical properties.

1.1.1 The description of anisotropic materials

It is a safe generalisation to say that the properties of all solid materials —metals, ceramics, glasses and plastics—are dependent on their processing history. The temperature, method and speed of processing are each important variables. In any discussion of the effects of processing on properties, there is one important feature which simultaneously involves many properties, and that is whether the material is isotropic or anisotropic.

An isotropic material has the same properties in all directions. Properties such as refractive index and Young's modulus are independent of direction, and if we wish to refer the properties to a set of rectangular cartesian co-ordinates, we can rotate the axes to be in any orientation without any preference. For an anisotropic material, where the properties differ with direction, it is usually convenient to choose co-ordinate systems which coincide with axes of symmetry if this is possible. The material is then described by its properties referred to these principal directions, which affords considerable simplification.

The degree of complexity of this representation depends on the tensorial nature of the property. Properties which relate two vectors, such as dielectric constant κ, which relates electric displacement D and electric field intensity E, can be described by a symmetric second rank tensor with six independent components, $viz.$

$$D = \begin{bmatrix} \kappa_{xx} & \kappa_{xy} & \kappa_{xz} \\ \kappa_{xy} & \kappa_{yy} & \kappa_{yz} \\ \kappa_{xz} & \kappa_{yz} & \kappa_{zz} \end{bmatrix} E$$

The components of the displacement D_x, D_y, D_z are then given in terms of the components of the electric field intensity E_x, E_y, E_z by the equations $D_x = \kappa_{xx} E_x + \kappa_{xy} E_y + \kappa_{xz} E_z$
$$D_y = \kappa_{xy} E_x + \kappa_{yy} E_y + \kappa_{yz} E_z$$
$$D_z = \kappa_{xz} E_x + \kappa_{yz} E_y + \kappa_{zz} E_z$$
where κ_x, κ_{xy}, etc., are the dielectric constants.

It is always possible to rotate the system of axes until the property is described by three components, the principal components κ_1, κ_2, κ_3. The second rank tensor then reduces to

$$\begin{bmatrix} \kappa_1 & 0 & 0 \\ 0 & \kappa_2 & 0 \\ 0 & 0 & \kappa_3 \end{bmatrix}$$

In oriented polymers we are usually concerned with sheets or films where there are three axes of symmetry at right angles or with uniaxially oriented systems (either fibres or films) where there is isotropy in the plane perpendicular to the orientation direction, which defines the unique axis (the fibre axis). In the former case the system shows orthorhombic symmetry and if we choose a system of Cartesian co-ordinates x, y, z to coincide with the three principal directions in the film the second rank tensor reduces to

$$\begin{bmatrix} \kappa_1 & 0 & 0 \\ 0 & \kappa_2 & 0 \\ 0 & 0 & \kappa_3 \end{bmatrix}$$

In the uniaxially oriented case, the system is transversely isotropic and the dielectric constant tensor reduces to

$$\begin{bmatrix} \kappa_1 & 0 & 0 \\ 0 & \kappa_1 & 0 \\ 0 & 0 & \kappa_3 \end{bmatrix}$$

where we have put $\kappa_1 = \kappa_2$. For a full isotropic material $\kappa_1 = \kappa_2 = \kappa_3$, and we may rotate the x, y, z axes into any orientation.

We also have to deal with properties such as elasticity which relates the second rank tensor quantities stress and strain, and piezoelasticity which relates the second rank tensor quantity stress and the vector quantity electric polarisation. These properties are described by a fourth rank tensor and a third rank tensor respectively. It is not appropriate to attempt a summary of these more complex properties here, and the reader is referred to later chapters (*e.g.* Chapters 8 and 9) and other texts (*e.g.* Ref. 1) for further details. It is, however, worth emphasising that the properties described by high order tensors may display aniso-tropic features which are entirely absent in simpler properties such as dielectric constant or related properties such as refractive index. The classic example in crystal physics is that of cubic crystals which are optically isotropic (*i.e.* $\kappa_1 = \kappa_2 = \kappa_3$) but elastically anisotropic. A classic example in polymer physics is low density polyethylene where there appears to be a monotonic increase in orientation on stretching as monitored by refractive index measurements, but the development of mechanical anisotropy is very complicated and the Young's modulus in the stretching direction passes through a minimum value with increasing draw.

There is another phenomenon which deserves mention in passing, where the value of a given property varies monotonically along a given axis, as in a case-hardened metal. In this situation the material may be called property-graded. This may occur to a limited extent with oriented polymers as in an injection-moulded product, but it will not be con-sidered further here. Instead, it will be assumed that, *on a macroscopic scale of scrutiny*, a property has a fixed value along a given axis in space, although as will be seen later—on a molecular scale of scrutiny, the same property may alternately show high and low values along the same axis. Thus an overall measurement of Young's modulus of a fibre along its axis (representing the macroscopic scale of scrutiny) is different from the individual values obtained for the amorphous and crystalline regions along the same axis when using the technique of X-ray diffraction to measure strain (representing the molecular scale of scrutiny).

1.2 GENERAL EFFECTS OF ORIENTATION AND ANISOTROPY ON THE PROPERTIES OF MATERIALS

Many examples could be quoted to illustrate how the introduction of anisotropy during processing affects the properties of solids. Table 1 shows the effect of forging on a sample of 0.36% carbon steel.[2] The ratio of ingot diameter to forged rod diameter was 12:1. For comparison, results are also presented for a hydrostatically extruded sample of polypropylene[3] at a deformation ratio of 5.5.

In the case of the carbon steel, the forging has made the ingot markedly anisotropic as evidenced by the improvement in ductility and impact strength in the longitudinal direction. This phenomenon could be illustrated by many similar examples for metals, if necessary. Sometimes it depends on the fact that the basic structural unit of the material —be it grain or particle—is non-equiaxial. In certain cases the effect of processing is simply to line up, to a greater or lesser extent, these particles as the result of the applied stresses. Then, if the basic particle itself is anisotropic, which is a very usual situation with crystalline solids, the aggregate of partly-aligned particles will be anisotropic. This provides a method for improving the properties of solid materials.

Turning to the polymer case, there is a certain formal likeness with other materials with respect to the effect of processing on properties, but in reality the differences are greater than the similarities. Most important, the extent to which it is possible to increase stiffness, and more particularly strength, and the ease with which this can be done, find no parallel with other materials. The behaviour of polymers is unique in the magnitude of the effects obtained. Thus we see that the extruded polypropylene has increased its longitudinal yield strength and tensile strength by factors of 1.1 and 10 respectively.

There are situations where this ease of orientation in polymers is undesirable. For example, orientation occurs readily in injection moulding, and if an injection-moulded article is highly anisotropic, it is likely to

TABLE 1

EFFECT OF FORGING ON THE PROPERTIES OF STEEL AND POLYPROPYLENE

Sample	Extrusion Ratio	Longitudinal yield strength ($\times 10^7\ N\ m^{-2}$)	Longitudinal tensile strength ($\times 10^7\ N\ m^{-2}$)	Elongation at break (%)	Impact strength
Carbon steel: ingot	1	29	56	19	56 (J-Charpy)
Carbon steel: forged rod	12	34	56	27	102 (J-Charpy)
Polypropylene: billet	1	1·0	2·2	200	0·3 ft. lb.
Polypropylene: extruded rod	5·5	—[a]	22	24	14 ft. lb.

[a] No yield drop observed.

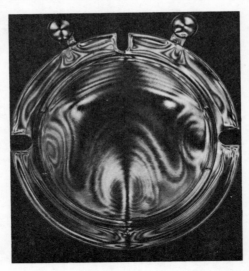

Fig. 1. Inspection of injection moulded article between crossed polaroids. (Photograph kindly supplied by M. G. Griffin, Brunel University.)

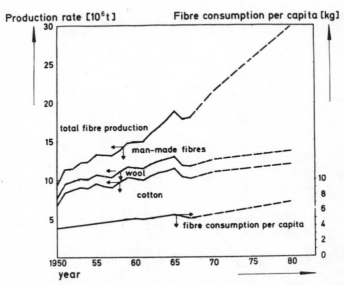

Fig. 2. Production of textile raw materials—world wide.

show regions of mechanical weakness. A simple test for anisotropy, where the object is transparent, is to examine it between crossed polaroids. Figure 1 shows how revealing this test can be.

Whilst this type of *incidental* orientation may constitute a minor disadvantage, there are whole industries which are based upon the *deliberate* exploitation of this effect. The production of synthetic fibres, and therefore of all synthetic textiles and ropes, depends upon orientation. Also the packaging industry makes extensive use of uniaxial and biaxial orientation. Recently the drawing and subsequent fibrillation of polyolefins has produced a cheap replacement for jute and sisal with profound repercussions on the economies of India and Pakistan. The extent to which synthetic fibres have supplemented natural fibres is shown in Fig. 2.

1.3 HOW ORIENTATION IS ACHIEVED IN POLYMERS

1.3.1 Orientation in natural polymers

The phenomenon of orientation in polymers is shown by the larvae of the silk moth, and by spiders. The former weave their heads about inside the cocoon, so that as they spin their silk, it is pulled and oriented. The latter, by suddenly dropping on their drag line, produce a highly oriented and very strong fibre. Robert Hooke understood something of these matters when he wrote in his 'Micrographia' in 1665 that the silk-worm 'withdraws his clue' and that fibres might perhaps be made artificially by copying the process.

An early technological development based on orientation was the manufacture of catgut (mostly in France) for uses in surgery and in musical instruments. Catgut was made by stretching the guts of certain animals to yield a coarse monofilament. Towards the end of the nineteenth century, with the development of fibres made from regenerated cellulose, the importance of orientation became more clearly appreciated, but the true significance had to wait until Staudinger's work in the 1920s. Once it was established that polymers were made up of very long molecules, a real understanding of orientation became possible, and it could be demonstrated conclusively by X-ray crystallography.

One of the most important polymers of all—cellulose—occurs naturally in a highly oriented form in the vegetable world. Wood contains 50–70% of cellulose, whilst cotton in its natural state contains 85–90%. From an X-ray examination of cellulose, it is seen to be highly crystalline, with four anhydroglucose residues in the unit cell. Each fibre is composed of elementary fibrils, approximately 50–100 Å in width, aggregated into larger fibrils about 250 Å wide. The length of the crystallites is some 300–600 Å. The result of this structural arrangement is that cellulose fibres are highly anisotropic and Young's moduli of over $5 \times 10^{10}\,\mathrm{N\,m^{-2}}$ have been reported in the fibre direction.

In contrast to the natural synthesis of cellulose, leading to an aligned crystalline arrangement of molecules, the synthesis of man-made polymers leads to a molecular arrangement which is random overall, even though in semi-crystalline polymers, such as polyethylene, there are aligned regions within the mass. For this reason a further processing step must be introduced and this will now be discussed.

1.3.2 Orientation in synthetic polymers: the drawing of thermoplastics
It has been known for many years that some thermoplastics can be oriented at room temperatures as a result of cold drawing. This is a remarkable phenomenon, in which the plastic deformation is concentrated in a small region of the specimen. The full line in Fig. 3 shows a typical stress–strain curve for a cold-drawing polymer. The stress shown is the nominal stress, related to the original cross-sectional area of the sample. The rate of strain is low, as in a normal tensile test. The behaviour at low strains is homogeneous and the stress rises steadily with increasing strain. At B the sample thins to a smaller cross-section at some point with the formation of the neck. Further extension occurs by the movement of this neck through the sample as it thins from its initial state to the final drawn state. If, now, a sample of the drawn material is taken, and this is tested in the longitudinal, or fibre direction, the upper dotted line will be obtained. In a typical case (*e.g.* low density polyethylene, nylon) this will show a several-fold increase in Young's modulus (as indicated by the greater slope) and a much greater increase in strength, the latter increasing by more than an order of magnitude. Also shown,

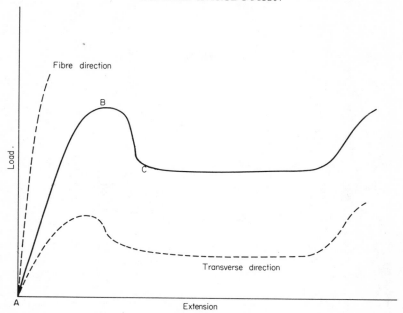

Fibre direction

Load.

B

C

Transverse direction

A

Extension

Fig. 3. Typical load extension curve for a cold drawing polymer ———. (Dashed lines ---- show load extension curve on redrawing, i.e. after first draw).

in the lower dotted line, is the result of testing the same material in the transverse direction. In most cases (low density polyethylene is an exception—see Chapter 9) the stiffness in this direction is lower than the initial stiffness or very nearly equal to it. The strength is always lower, and the energy for crack propagation falls very sharply for propagation parallel to the direction of orientation. The latter may fall by a factor of 100, which accounts for the ease of tearing in this direction.[4] In both the longitudinal and transverse directions, the elongation at break is greatly reduced.

This simple experiment of cold drawing enshrines the basic principles of fibre technology; the production of a material with high longitudinal stiffness and strength. Concomitant with these changes are corresponding changes in other properties such as refractive index (the polymer becomes birefringent), coefficient of expansion, etc.

The idealised case of cold drawing which has been discussed, is typical of the behaviour of several major thermoplastics such as polyethylene, polypropylene, nylon and polyethylene terephthalate, which readily cold draw at ambient temperatures. Other thermoplastics, which

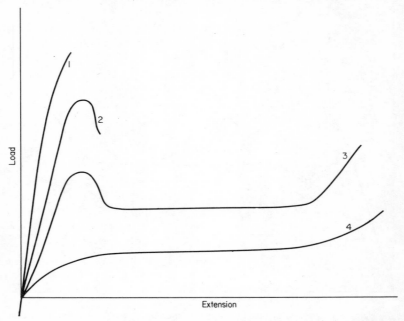

Fig. 4. *Load–extension curves for a typical polymer tested at four temperatures showing different regions of mechanical behaviour. 1. Brittle fracture. 2. Ductile failure. 3. Necking and cold drawing. 4. Rubber-like behaviour.*

are brittle at normal temperatures, can be drawn at an elevated temperature, although in most cases they do not undergo the same degree of orientation. Examples are polystyrene or polymethylmethacrylate. The polar diagrams of Young's modulus shown later (Figs. 14a and b) compare orientation effects in low density polyethylene and polymethylmethacrylate. If the load–elongation curves for a polymer at different temperatures are determined the results shown in Fig. 4 are observed. At low temperatures we see brittle fracture (Curve 1). With increasing temperature we pass through the brittle–ductile transition and Curve 2 is observed. In this temperature region, the polymer is ductile but the neck does not stabilise; cold drawing therefore cannot take place. At high temperatures (Curve 4) above the glass–rubber transition temperature T_g, the polymer behaves like a rubber and shows homogeneous deformation at all strains up to the very high strain ($\sim 500\%$) at which fracture occurs. The cold drawing behaviour of Curve 3 is shown at temperatures below T_g but above the necking rupture range of Curve 2.

Not all polymers show cold drawing; there are requirements such as a minimum molecular weight for strain hardening, and this has been extensively discussed elsewhere (see for example Ref. 5, pp. 271 and 322). It is clearly necessary for the polymer to be above its brittle–ductile transition, but this is a necessary rather than a necessary *and* *sufficient* condition for cold drawing. It should perhaps also be emphasised that there is no immediately obvious relationship between the glass transition temperature and the brittle–ductile transition temperature T_b (see for example Ref. 6).

It is possible to produce orientation at normal temperatures in a brittle thermoplastic like polystyrene, by the application of a sufficiently high hydrostatic pressure during extension.[7] The hydrostatic pressure prevents brittle failure at small strains, and can be regarded as reducing T_b below ambient temperature. Similar results may be obtained by hydrostatic extrusion, where a billet of the polymer is forced through a reducing die under the pressure of a hydraulic fluid. An example of the properties achieved in this way has been given in Section 1.2 above.

1.3.3 Orientation in thermosetting polymers
In the case of thermosets, deliberate and extensive orientation is virtually unknown. This appears to be the result of the practical difficulties involved, rather than from any theoretical obstacle. For example, it is possible that the fibre Kynol produced by the Carborundum Corporation[8] is oriented to some extent. This is produced from a melt-spun Novolak phenol–formaldehyde resin, which is later further cross-linked with formaldehyde. It is, of course, legitimate to consider carbon fibres as extreme examples of thermosets. Formed by the cyclisation and subsequent graphitisation of polyacrylonitrile (or other suitable precursors), they are highly oriented.

1.3.4 Orientation in elastomers
The remaining class of polymeric materials to be considered are the rubbers or elastomers. These are also oriented by stretching, but in this case the orientation is reversible, together with the elastic deformation.

1.3.5 Orientation in the thermoplastic elastomers: macroscopic 'single crystals'
As a postscript to this account of orientation which depends on increased chain alignment, it is necessary to mention the totally different type of orientation which can be induced in the block copolymers where

a rubbery component is copolymerised with a glassy component—the so-called thermoplastic elastomers. The styrene–butadiene–styrene (SBS) block copolymers are a typical example. When an SBS copolymer (25% weight fraction of S with a molecular weight of 10^4, the B having a molecular weight of 5.5×10^4 in the example studied by Keller et al.[9]) is extruded and annealed under carefully controlled conditions, macroscopic single crystals of unusual structure are obtained. The term single crystal refers to the regular morphology of the dispersed polystyrene domains. These are aligned as a hexagonal lattice of parallel filaments or cylinders, the long axis being in the extrusion direction. These cylinders are about 150 Å in diameter and are separated by a similar distance. They are infinitely long in relation to their diameter. Each phase remains amorphous, but the overall structure is highly anisotropic as might be expected.

1.3.6 Summary of orientation in synthetic polymers

Sections 1.3.2–1.3.5 above show that orientation is most important in the thermoplastics technologically, and we can now summarise the differences between orientation in different types of polymers by a table.

TABLE 2
ORIENTATION BEHAVIOUR OF DIFFERENT TYPES OF POLYMERS

Class of polymer	Requirement for orientation	Reversibility of orientation
Thermoplastics a. ductile	Draw at room or elevated temperature	Some or all of orientation is lost on heating
b. brittle	Draw at elevated temperature	
Thermosets	Draw precursor polymer before cross-linking	Irreversible
Rubbers	Apply stress at room temperature and produce elongation of material	Reverses on releasing stress, i.e. allowing shrinkage
Thermoplastic elastomers	Extrude and anneal	

1.4 SOME STRUCTURAL FEATURES OF POLYMERS

Before considering in detail the molecular basis of orientation, it is helpful to have a clear picture of how the molecules are arranged in the original polymer before the molecules are oriented. We begin by considering a material which is initially isotropic, and for convenience we divide polymers into the usual two classes, amorphous and semi-crystalline. These will now be discussed separately.

1.4.1 Amorphous polymers

It is customary to make a sharp distinction between amorphous and semi-crystalline polymers. The former are typified by atactic polystyrene and polymethylmethacrylate, the latter by polyethylene, polypropylene and nylon. Amorphous polymers show no crystalline order when examined by X-ray diffraction techniques, and only a glass transition when a property such as specific volume is plotted against temperature. In contrast to this, semi-crystalline polymers give abundant signs of crystalline order under X-ray (and usually also under optical) examination, and show a crystalline melting point as well as a glass transition temperature. However, there is important evidence that this simple picture of an amorphous polymer with its disordered liquid-like structure is far from the truth. The following facts suggest that there is significant ordering in amorphous polymers.

Robertson[10] in an important paper has discussed polymer density in relation to polymer order. He has shown that the ratio of the density of the amorphous material to the density of the same polymer in the crystalline state is much higher than would be anticipated on the basis of the random (wet spaghetti) model which is predicted to give a density ratio d_a/d_c of 0·65. This density ratio varies in practice from a low of 0·855 to a high of 0·943 in passing from polyethylene to isotactic polystyrene. The results of Robertson's calculations support the suggestion of a high degree of ordering in amorphous materials, possibly involving clustering into bundles with some chain folding. These may constitute domains of approximate length 100 Å measured in the direction of the chain axis.

Kargin and his school[11] in a large number of papers, have discussed the structural features of polymers as revealed by examination under the electron microscope. It is suggested that, even in amorphous glasses, aggregation into long, thin bundles of chains is possible.

On the basis of the foregoing and other recent evidence, it seems reasonable to assume that there is a basic structural unit in the so-called amorphous polymers. This is the molecular bundle, which may or may not include some chain folding. If one postulates the existence of such bundles, it becomes a reasonable corollary to suggest that—between the bundles—there will exist truly amorphous material. This suggestion, that there will exist regions of more and less ordered material—lies in the realm of speculation, but it may form a starting point for considering the mechanism of orientation in amorphous polymers. Indeed such ideas have earlier been implicitly assumed by Ward and co-workers[12] in applying an aggregate model to the orientation of several amorphous polymers (*e.g.* polymethylmethacrylate, glassy polyethylene terephthalate) in both mechanical and structural studies.

1.4.2 Semi-crystalline polymers

In contrast to our ignorance of the details of the fine structure of amorphous polymers, we have a reasonably complete picture of the structure of semi-crystalline polymers. Crystalline polymers are properly called semi-crystalline, because they are seen to be only partly crystalline when assessed by a crystallinity-sensitive property such as density. Furthermore, they show some of the properties which are commonly associated with random chain arrangements such as long-range elasticity, and the existence of a glass transition, in addition to a crystallisation transition, is evidence of the presence of amorphous material. It should be noted that the degree of crystallisation in a given polymer is not invariant, but depends upon experimental conditions such as rate of cooling.

For crystallisation to occur in a polymer, the requirement is that the chain should be chemically sufficiently regular. In addition, there should be sufficient chain mobility at the melting point for the actual process of crystallisation to take place. For that reason, bulky side groups reduce the rate of crystallisation.

In the case of polyethylene, with its very simple structure, the crystallinity is largely controlled by the number and distribution of branches along the chain. This is shown in Fig. 5, based on the data of Faucher and Reding.[13] By varying the number of branches, the crystallinity can vary from around 35% to 90%. The lower figure is found with certain grades of high pressure polyethylene, made by a free radical process, giving from 5 to 0·1 branches per hundred backbond carbon atoms. The higher figure is found with Marlex and similar grades of low pressure polyethylene, with 0·5 to 0 branches per hundred carbon atoms.

Another aspect of the chemical composition of a polymer which affects crystallisation and hence physical properties is steric isomerism and stereo regularity. For simplicity, consider only vinyl polymers such as polypropylene or polystyrene where a substituent group (in these cases

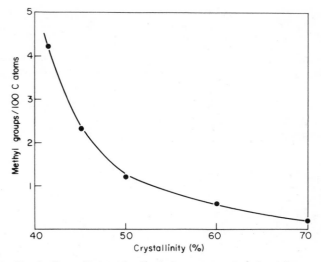

Fig. 5. Crystallinity of polyethylene versus chain branching.

CH_3 or C_6H_5) is attached to every alternate carbon atom. If the substitution is regular the polymer is said to be stereoregular. For all
substituent groups on the same side of the polymer chain the polymer
is termed isotactic, for regular alternation syndiotactic and where there
is no regularity atactic. If stereoregular sequences predominate, crystallisation becomes possible. Thus atactic polypropylene, polystyrene and
polymethylmethacrylate do not crystallise, whilst in their isotactic form
they do, although the process of crystallisation may be slow as with
isotactic polystyrene. On the other hand, stereoregularity is not abso-

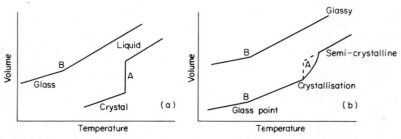

*Fig. 6. First and second order transitions. (a) Behaviour of low molecular weight
compounds. (b) Behaviour of glassy and semi-crystalline polymers.*

lutely essential if the molecules can pack easily. Polyvinyl alcohol will crystallise in its atactic form, because the –OH groups are sufficiently small not to disturb the crystal lattice.

A convenient method of discriminating between the processes of glass formation and of crystallisation is dilatometry. There is no abrupt volume change at T_g, whilst crystallisation is accompanied by a volume change. The former thus constitutes a second-order transition, whilst the latter gives a first-order transition on a volume—temperature plot. This distinction is shown diagrammatically in Figs. 6 (a) and (b).

1.4.3 Polymer crystals and spherulitic structure

In considering the structure of semi-crystalline polymers in more detail it is convenient to begin with polymer crystals. Since the original work of Fischer, Keller and Till in 1957,[14-16] polymer crystals have received a great deal of attention. In some ways they provide more definitive structures for study than the large semi-crystalline aggregates in which we encounter crystallising polymers commercially, and although polymer crystals are not wholly representative of large-scale crystallisation behaviour, they have yielded much important information on the morphology of semi-crystalline polymers.

When a polymer of sufficiently regular structure is crystallised from a dilute solution, single crystals can be obtained which are in the form of regular lamellae of uniform thickness. Furthermore, the perfection and size of these crystals can be used as a criterion of chain regularity. The surprising feature is that (a) the chains are more or less perpendicular to the basal plane of the lamellae, and (b) the lamellae thickness—of the order of 100 Å—is independent of the length of the molecule which itself will be of the order of 10^5 Å long. This immediately leads to the conclusion that the molecules must be folded backwards and forwards on themselves to produce a structure which is represented in diagrammatic form in Fig. 7. In principle, such a molecule could fold 1000 times.

The fold length is determined by the crystallisation temperature or, more exactly, by the degree of supercooling. The higher the crystallisation temperature, the longer is the fold length. The fold length thus becomes an important structural parameter in polymer science. The discovery of chain folding raises in turn a number of important problems, which have been well reviewed by Mandelkern[17] and by Keller.[18] These include the origin of chain folding, and the nature of the fold surface, *i.e.* to what extent is this disordered as the result of non-re-entrant

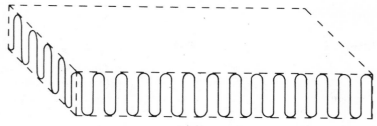

Fig. 7. Single crystal and chain folding.

adjacent folds? The importance of this fold surface problem is apparent when the structure of semi-crystalline material is considered.

Proof that chain-folded crystals exist in semi-crystalline samples which have been bulk crystallised from the melt is given by the effects of selective chemical attack on such materials. Concentrated nitric acid attacks the amorphous material preferentially,[19] to produce lamella-like crystals of low molecular weight. These so-called crystallites are therefore seen to be basic structural units in semi-crystalline polymers. Detailed studies of the rate of attack by Peterlin and others (see Ref. 26) provides evidence against the existence of a homogeneous, sharp, adjacent re-entry folded structure in these crystallites. As might be expected, the crystallites formed when the material is crystallised from the melt are smaller than the corresponding single crystals grown under optimum conditions from a dilute solution.

In the bulk crystallised material, the crystallites are contained within larger units called spherulites. The relationship between the spherulite and crystallites is quite complicated, and is shown up in Fig. 8 (after Sharples[20]). The spherulite is made up of fibrils which are arranged in a radial pattern. The fibrils themselves are made up of crystallites with the chains folded at right angles to the fibril length. They contain a proportion of non-crystalline defects, but the bulk of amorphous material which is rejected during the process of crystallisation is found between the fibrils. Other amorphous material will be found between the spherulites themselves. The size and number of spherulites depends upon the nucleation processes at work and thus can be controlled within limits. They may vary in size between an upper limit of millimetres and a lower limit of microns, *i.e.* by a factor of 1000. If the spherulites can be represented by the picture shown in Fig. 8, it remains to be seen how they are deformed during the process of orientation, when a considerable amount of molecular rearrangement must obviously take place.

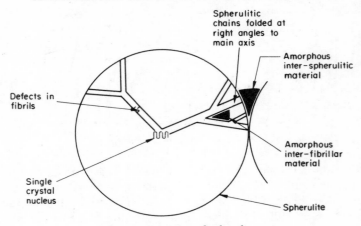

Fig. 8. Structure of spherulite.

1.4.4 Extended chain crystals

The foregoing brief survey of semi-crystalline polymers, with its emphasis on the importance of chain folding would not be complete without a brief mention of extended chain crystals, which are encountered under completely different conditions of crystallisation. Anderson[21] showed that under conditions of very slow crystallisation at temperatures close to the melting point, the polymer chains can crystallise in extended form, with no sign of folding. In this case the material was a sharp fraction of relatively low molecular weight (1000 Å chain length; 20000 m. wt.). Later work has shown that crystallisation of polyethylene melts under high pressure (c. 5000 atm) also produces extended chain crystals.[22] More recently again, it has been shown that extended chain crystals can be present in very highly oriented polyethylenes of very high modulus.[23]

It also appears that extended chain crystals can be produced in a totally different type of synthetic polymer, of which polyparabenzamide is a key example. In this case, fibres containing extended chain crystals are produced directly from a solution spinning stage. Such fibres are of extremely high modulus and are of sufficient scientific and technological importance to warrant a whole chapter (Chapter 13) later in this book.

1.5 THE DESCRIPTION OF ORIENTATION IN A POLYMER AND ITS RELATIONSHIP TO ANISOTROPY OF PROPERTIES

1.5.1 The definition of orientation

So far there has been much discussion in a general sense of orientation in polymers, how it may be achieved and some of its effects on properties. Now we must ask how orientation can be defined in precise terms, so that quantitative measures of orientation can be obtained and discussed.

For preciseness, let us begin by considering an aggregate of crystallites, *i.e.* an aggregate of small regions each defined by a crystal unit cell. We then ask how we would define the orientation of a single unit cell of lowest symmetry, *i.e.* a triclinic unit cell.

Consider a rectangular cartesian system of co-ordinate axes XYZ (Fig. 9). The orientation of the c-axis of the triclinic unit cell can be described by the two angles θ and ϕ, which define the angles between c-axis and the z-axis and between the projection of c in the XY plane and the x-axis. We must also describe the orientation of the unit cell around its c-axis, *i.e.* the orientation in a plane whose normal lies along c. This can be done, for example, by defining the angle ψ say, which the projection of the a-axis in this plane normal to the c-axis makes with a chosen direction in this plane. This procedure is equivalent to defining the orientation of a set of orthogonal reference axes $\{UVW\}$ fixed in the crystal unit cell, with respect to the reference axes (XYZ). The three angles, ϕ, θ, ψ, are called the Eulerian angles and define the three successive rotations which bring the two sets of axes into coincidence (Fig. 10).

The orientation of a single unit cell can therefore be defined in terms of three angles, θ, ϕ, ψ. The orientation distribution function for all the

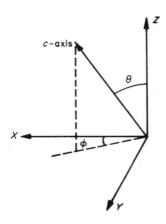

Fig. 9. The orientation of the c-axis is defined by the polar angles θ and ϕ.

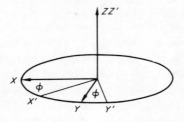

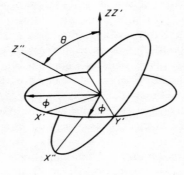

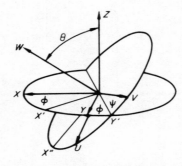

Fig. 10. The specification of the orientation of the axes of a structural unit {uvw} with respect to the reference frame {xyz}. The Euler angles (ϕ, θ, ψ) represent successive rotations about the z-axis, the y'-axis and z"(w)-axis.

unit cells in this hypothetical crystalline aggregate will take the form $\rho(\theta, \phi, \psi)$ where

$$\int_0^{2\pi} \int_0^{2\pi} \int_0^{\pi} \rho(\theta, \phi, \psi) \sin\theta \, d\theta \, d\phi \, d\psi = 1$$

We can see that the description of orientation even in a crystal aggregate is a formidable problem. It is also by no means clear that a unique description can be obtained from macroscopic measurements, which for a crystal aggregate would most appropriately be X-ray diffraction (see Chapter 3). From the X-ray data the orientation distributions for various planes in the unit cell could be obtained. These results then have to be fitted together to provide a description of the overall orientation distribution, which may not be unique.

Polymers are at best semi-crystalline and in many cases amorphous. The crystalline regions can be studied by X-ray diffraction techniques, but the non-crystalline or 'amorphous' regions are mostly studied by optical and spectroscopic methods. The details of these techniques are discussed in later chapters. A few general points are, however, worth emphasising at this juncture.

Consider, for example, infra-red spectroscopy where the measurement could, for example, be the dichroism of a C–H stretching vibration absorption in polyethylene. For a single unit cell the dichroism will be determined by the direction of the dipole moment vector for this C–H vibration, *i.e.* by the orientation of a line in the unit cell. The dichroism of the partially oriented polymer will depend on the orientation distribution of such lines throughout the specimen. Two observations are worthy of note. First, to define the orientation of the polymer by infra-red spectroscopy, measurements on several absorption lines will in general be required. Secondly the infra-red measurements will not distinguish between the crystalline and non-crystalline regions of the polymer unless absorptions are selected which are characteristic of one particular region. Similar considerations apply to other spectroscopic techniques such as laser Raman spectroscopy and nuclear magnetic resonance.

Stated in these very general terms, it is by no means obvious that the problem of adequately describing orientation in a polymer can ever be completely solved. Indeed, there are certainly situations where it can be shown to be insoluble in terms of the available information. However, in many if not most of the oriented polymer situations which have been studied in detail, simplifications are possible. The most sweeping simplification is to assume transverse isotropy, *i.e.* that there is no preferential orientation in a plane perpendicular to a single direction, often the draw direction or fibre axis if we are dealing with fibres. Again, for most semi-crystalline polymers the molecular chain axis lies along or very close to one of the unit cell axes. For example in polyethylene the chain axis coincides with the c-axis. A simplifying assumption would then be to assume that there is preferred orientation of the chain axis only, which for the crystalline regions in polyethylene means that the c-axis is preferentially oriented, with the a-axes and b-axes randomly oriented around the c-axis. Our theoretical distribution function $\rho(\theta,\phi,\psi)$ then reduces to $\rho(\theta)$, and definite measures of the orientation can be obtained from X-ray and spectroscopic methods, providing that a suitable X-ray reflection (e.g. 002 for polyethylene) can be found and that the angle between the dipole moment vector and the chain axis or some corresponding quantity is known in the spectroscopic case.

1.5.2 The measurement of orientation: orientation functions

We have already discussed to some extent the type of information which can be obtained from wide angle X-ray diffraction and infra-red spectroscopy concerning orientation in polymers. We now wish to consider in detail the knowledge which can be gained concerning the orientation distribution function $\rho(\theta,\phi,\psi)$. This involves consideration of the way in which different properties reflect orientation.

Consider, for example, infra-red spectroscopy and for simplicity it will be assumed that there is fibre symmetry as discussed above. Moreover we will assume that we can choose to measure the absorption of a molecular vibration where the change in dipole moment takes place along the molecular chain axis. For a single molecular unit or group the absorption A is proportional to $(\mu \cdot E)^2$ where E is the electric vector and μ the dipole transition moment vector. If we consider a particularly simple situation where the electric vector lies along the fibre axis (Fig. 11(a)) and the dipole moment vector makes an angle θ with the latter, the absorption will be proportional to $\cos^2 \theta$. In the gas phase approximation, where it is assumed that the absorption from the whole polymer is just the linear sum of the absorption of independent molecular groups, infra-red measurements will give $\overline{\cos^2 \theta}$, the average value of $\cos^2 \theta$ for the aggregate of molecules. It is now pertinent to consider how these ideas are modified if the dipole moment vector makes an angle (α, say) with the molecular chain axis (Fig. 11(b)). In fact, it turns out that the problem has a very neat analytical solution. If we consider the optical densities $D_\parallel$ and $D_\perp$ for the electric vector of the polarised infra-red

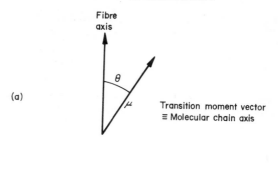

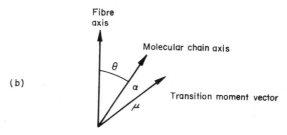

Fig. 11. Infra-red absorption. (a) Simplest situation where transition moment vector is parallel to the molecular chain axis. (b) Transition moment vector makes an angle α with the molecular chain axis.

radiation being parallel and perpendicular to the fibre axis respectively, it can be shown that

$$\frac{D_{\parallel} - D_{\perp}}{D_{\parallel} + 2D_{\perp}} = \tfrac{1}{2}(3\cos^2\alpha - 1)\tfrac{1}{2}(\overline{3\cos^2\theta} - 1)$$

The measurements therefore yield the product of a constant term $\tfrac{1}{2}(3\cos^2\alpha - 1)$, which should be known from structural considerations, and an orientation function $\tfrac{1}{2}(\overline{3\cos^2\theta} - 1)$.

Measurements of optical anisotropy, *i.e.* birefringence give the same type of orientation function. In this case, again assuming fibre symmetry, the polymer is considered to be an aggregate of polarisable units with transverse isotropy. Each unit is defined by the second rank polarisability tensor

$$\begin{bmatrix} p_1 & 0 & 0 \\ 0 & p_1 & 0 \\ 0 & 0 & p_2 \end{bmatrix}$$

and the angle θ which its unique axis (assumed to be the molecular axis) makes with the fibre axis. The components of polarisability along and perpendicular to the fibre axis will therefore depend on trigonometrical functions such as $\cos^2\theta$, but of no greater complexity. If we assume additivity of polarisabilities for the aggregate of molecular groups composing the partially oriented polymer, and use the Lorentz–Lorenz equation to convert to refractive indices, the fibre birefringence Δn is given by

$$\Delta n = \Delta n_{max} \tfrac{1}{2}(3\overline{\cos^2\theta}-1)$$

where Δn_{max} is the maximum birefringence at full orientation when $\overline{\cos^2\theta} = 1$.

This result was derived by Hermans[24] and the orientation function $\tfrac{1}{2}(3\overline{\cos^2\theta}-1)$ or $(1-3/2\overline{\sin^2\theta})$ is sometimes called the Hermans orientation function.

We see that infra-red spectroscopy and optical birefringence provide exactly the same type of information regarding the distribution of orientations. We will now consider two other techniques which yield more complex information. First, Raman spectroscopy. Here the incident electric field produces an induced dipole moment, and we are concerned with the change in polarisability due to the molecular vibration which gives rise to scattered radiation. We therefore define a *differential* polarisability tensor which describes this change in polarisability. The orientation effects now involve *two* angles, the angle between the incident electric vector and the differential polarisability axis (for simplicity we assume a polarisability tensor with cylindrical symmetry) and the angle between this axis and the direction of polarisation chosen to sample the scattered radiation. These are the angles β_1 and β_2 respectively in Fig. 12(a) and the intensity of the scattered radiation with polarisation in the direction defined by β_2 is proportional to $\cos^2\beta_1\cos^2\beta_2$. Even for a polymer system showing transverse isotropy, the Raman orientation effects are very complicated. In Fig. 12(b) we show the situation where the axis of the differential polarisability tensor makes an angle α with the molecular chain axis, which in turn makes an angle θ with the Z-axis, the fibre axis. The direction of polarisation of the incoming radiation and the scattered radiation are defined by the angles γ_1 and γ_2 respectively. It is not difficult to appreciate that the intensity of the scattered radiation now depends on products of terms involving $\cos^2\alpha$, $\cos^2\theta$, $\cos^2\gamma_1$ and $\cos^2\gamma_2$. This means that the intensity of the scattered radiation will depend on (1) the angles γ_1 and γ_2 which can be chosen experimentally, (2) the angle α, which we might hope to

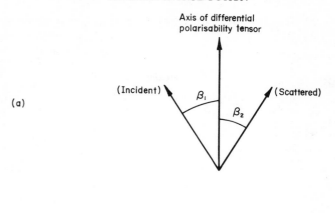

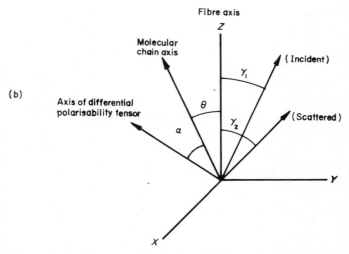

Fig. 12. (a) Raman scattering for a single unit. (b) The orientation problem for transverse isotropy.

determine from a series of Raman measurements on different samples and (3) for an aggregate, orientation functions $\overline{\cos^2 \theta}$ and $\overline{\cos^4 \theta}$.

The important point is that $\overline{\cos^4 \theta}$ is now determined, and that this enables the orientation distribution function to be more closely defined.

Another technique which is of considerable importance in determining orientation in polymers is broad line nuclear magnetic resonance. In a solid polymer at low temperatures when molecular motions are quenched,

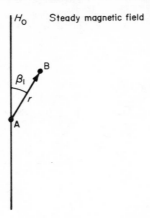

*Fig. 13. Nuclear magnetic resonance: the internuclear vector **r** makes an angle β_1 with the steady magnetic field direction.*

~~the fourth moment, but this is more complicated) of the NMR absorp-~~
the NMR absorption line shape depends on the spatial arrangement of the static resonant nuclei (usually protons). The broadening of the line arises from magnetic dipole interactions. For an isolated pair of identical magnetic nuclei of magnetic moment μ (A and B in Fig. 13), the local field H_{loc} at one nucleus due to the other is $\mu/r^3(3\cos^2\beta_1 - 1)$, where β_1 is the angle which the internuclear vector makes with the direction of the steady magnetic field H_0. The local field determines the line width, because each magnetic nucleus sees the sum of the external magnetic field H_0 and H_{loc} the field due to its neighbours. Hence it follows that in an oriented polymer, the line width will depend on the orientation of the polymer with respect to H_0. It turns out that the second moment (and the fourth moment, but this is more complicated) of the NMR absorption curve can be described quantitatively in terms of the dipolar interactions. The second moment is essentially the moment of inertia of the absorption curve and thus relates to

$$H_{loc}^2 = \left\{\frac{\mu}{r^3}(3\cos^2\beta_1 - 1)\right\}^2$$

We can see that the NMR second moment depends on quantities such as $\cos^4\beta_1$ and $\cos^2\beta_1$. It is therefore easy to appreciate, in the spirit of the discussion of Raman spectroscopy above, that the second moments will define orientation functions $\overline{\cos^4\theta}$ as well as $\overline{\cos^2\theta}$.

In Table 3 the orientation information which can be obtained from these various structural techniques is summarised. This table also shows the part of the molecular structure which is being characterised, and some of the theoretical and experimental limitations of each method. A further technique, that of polarised fluorescence has been added. This technique is exactly analogous in its orientation aspects to Raman spectroscopy. The distinction between the two techniques lies in the fact that in the Raman effect, the lifetime of the process is of the order of the vibrational period ($\sim 10^{-12}$s) whereas fluorescence occurs after much longer occupancy of the transition state ($\sim 10^{-7}$s).

A final point should be made regarding the determination of orientation functions. For simplicity, the above discussion has been couched in terms of $\overline{\cos^2 \theta}$ and $\overline{\cos^4 \theta}$. If we wish to describe $\rho(\theta)$ say, in terms of such functions, it is desirable to choose an orthogonal set of functions. It is therefore customary to carry out these analyses in terms of spherical harmonic functions, in this case

$$P_2(\cos\theta) = \frac{1}{2}(3\cos^2\theta - 1) \quad \text{and} \quad P_4(\cos\theta) = \frac{35}{8}\cos^4\theta - \frac{15}{4}\cos\theta + \frac{3}{8}$$

The distribution function is then given by

$$\rho(\theta) = \sum_{n=0}^{\infty} (n+\tfrac{1}{2})\overline{P_n(\cos\theta)}P_n\cos\theta$$

and the spherical harmonic orientation functions by

$$\overline{P_n(\cos\theta)} = \int_0^{\pi} \rho(\theta)P_n\cos(\theta)\sin\theta\,d\theta$$

Other advantages of working in terms of spherical harmonic functions are that for cases with fibre symmetry, the Legendre addition theorem can be used, and affords considerable algebraic simplifications (see for example Ref. 25), and that for lower symmetries, the treatment can readily be generalised. It should be mentioned that the exact definitions of $P_2(\cos\theta)$, etc., and $\rho(\theta)$, can differ in different treatments due to the adoption of different normalisation procedures (see, for example, Chapter 5, Section 5.2).)

1.5.3 Mechanical anisotropy
At this juncture we will give a very brief discussion of low strain mechanical anisotropy, first to indicate its complexity and secondly to hint at its relationship to the discussion of the measurements of orientation in the previous section.

Table 3

INFORMATION ABOUT ORIENTATION DERIVED FROM PHYSICAL MEASUREMENTS

Method	Part of structure being characterised	Orientation functions	Theoretical limitations	Experimental limitations
Birefringence	Aggregate property, result weighted towards high polarisability groups, e.g. benzene rings, carbon–oxygen bonds	$\overline{\cos^2\theta}$, $\overline{\cos^2\theta\cos^2\phi}$	Assumes additivity scheme for polarisabilities; particularly uncertain for carbon–carbon bonds. Form birefringence also may occur	Transparent specimens required; also may require sectioning technique
X-ray diffraction	Crystalline regions: very specific in that plane normal directions determined	Complete characterisation —any orientation function can be calculated		Background scattering can be difficult to estimate
Infra-red dichroism	Molecular bond directions: can be specific to crystalline or non-crystalline regions, perhaps even fold bands	$\overline{\cos^2\theta}$, $\overline{\cos^2\theta\cos^2\phi}$	Usually assumes dipole moment change corresponds to bond direction—may not always be correct; results only as good as band assignments	Thin specimens required. Difficulties in overlapping bands; base-line determination
Laser–Raman spectroscopy	Molecular bond directions: can be specific to crystalline or non-crystalline regions, parts of structure	$\overline{\cos^2\theta}$, $\overline{\cos^4\theta}$, $\overline{\cos^2\theta\cos^2\phi}$, etc. (hyper-Raman might give higher moments)	To relate polarisability changes to directions in molecules may not be as straightforward as assuming a bond direction correspondence	Transparent specimens, with little or no fluorescence required

Broad-line nuclear magnetic resonance	Aggregate property, may be able to distinguish crystalline and non-crystalline regions in favourable cases, possibly parts of molecule: determines internuclear vectors	Second moment gives $\overline{\cos^2\theta}$, $\overline{\cos^4\theta}$, $\overline{\cos^2\theta}$, $\overline{\cos^2\phi}$; etc. Fourth moment gives $\overline{\cos^6\theta}$. $\overline{\cos^8\theta}$.	Requires knowledge of structure to give model for positions of magnetic nuclei	Aggregate property means that magnetic anisotropy is not always large enough
Polarised fluorescence	Aggregate property, if use dye molecules weighted to non-crystalline regions	$\overline{\cos^2\theta}$, $\overline{\cos^4\theta}$, etc.	Difficulty of relating fluorescence axes to direction polymer chain	Either polymer must fluoresce or must dye with fluorescent molecule

The mechanical properties of an anisotropic elastic solid for small strains are defined by the generalised Hooke's law

$$\varepsilon_{ij} = s_{ijkl}\sigma_{kl}$$
$$\sigma_{ij} = c_{ijkl}\varepsilon_{kl}$$

relating the tensor strains ε_{ij} and ε_{kl} to the stresses σ_{kl} and σ_{ij} through the compliance constants s_{ijkl} or the stiffness constants c_{ikjl} (note that the strains ε_{ij} in this formulation are not engineering strains).

In practice an abbreviated notation is often used in which

$$e_p = S_{pq}\sigma_q$$
$$\sigma_p = C_{pq}e_q$$

where S_{pq} and C_{pq} are the compliance and stiffness matrices respectively, and p, q take the values $1, 2, \ldots, 6$. e_1, e_2, e_3 are the extensional strains; e_4, e_5, e_6 the engineering shear strains, σ_1, σ_2, σ_3, the normal stresses and σ_4, σ_5, σ_6 the shear stresses (the conversion rules from the S_{ijkl} and C_{ijkl} notation to the abbreviated notation are given in standard texts.)[1]

For simplicity, the elastic properties are usually referred to a set of cartesian co-ordinate axes which coincide with symmetry axes in the material. Thus for a material with fibre symmetry we have

$$
\begin{array}{cccccc}
S_{11} & S_{12} & S_{13} & 0 & 0 & 0 \\
S_{12} & S_{11} & S_{13} & 0 & 0 & 0 \\
S_{13} & S_{13} & S_{33} & 0 & 0 & 0 \\
0 & 0 & 0 & S_{44} & 0 & 0 \\
0 & 0 & 0 & 0 & S_{44} & 0 \\
0 & 0 & 0 & 0 & 0 & 2(S_{11}-S_{12})
\end{array}
$$

If we now wish to obtain the extensional compliance at an angle θ to the symmetry axis (3-axis) we recall from the definition of a fourth rank tensor that this will depend on terms which involve the product of four direction cosines
i.e.
$$S_{pqmn} = a_{pi}a_{qj}a_{mk}a_{nl}S_{ijkl}$$
and we have

$$S_\theta = \sin^4\theta S_{1111} + \cos^4\theta S_{3333} + 2\sin^2\theta\cos^2\theta S_{1133} + 4\sin^2\theta\cos^2\theta S_{1313}$$

In the abbreviated notation this equation becomes

$$S_\theta = \sin^4\theta S_{11} + \cos^4\theta S_{33} + (2S_{13} + S_{44})\sin^2\theta\cos^2\theta$$

The Young's modulus in a direction making an angle θ with the draw direction is $E_\theta = 1/S_\theta$. Polar diagrams of Young's modulus for poly-

methylmethacrylate[26] at various draw ratios and for low density polyethylene[27] are shown in Fig. 14(a) and (b). Two features of these results
are of note. First, low density polyethylene shows much greater
mechanical anisotropy than polymethylmethacrylate. This is typical of
the difference between a crystalline and an amorphous polymer, the
latter rarely achieving high degrees of orientation. Secondly the Young's
modulus of low density polyethylene shows a curious star-like pattern
with very low modulus in the 45° direction. The equation for S_θ shows
that this is due to the large value of $(2S_{13} + S_{44})$ compared with S_{11} and
S_{33}. Properties such as birefringence which relate to simpler second-rank
tensors cannot show such unusual features unless the material is
inhomogeneous.

It is also worth commenting on the relationship to the discussion of
orientation functions. Even on a simple model where the polymer is
regarded as an aggregate of anisotropic units whose compliance or stiffness constants can be averaged (see Chapter 8) the mechanical anisotropy
will relate to quantities such as $\overline{\cos^4 \theta}$ and $\overline{\sin^4 \theta}$, i.e. to both $\overline{\cos^2 \theta}$ and

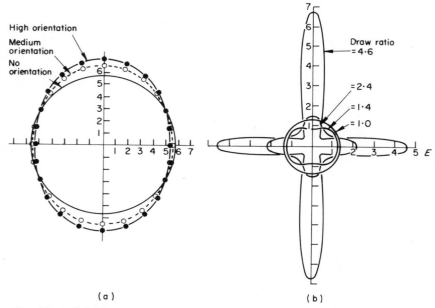

(a)   (b)

Fig. 14. Polar diagrams of Young's modulus for (a) polymethylmethacrylate;
(b) polyethylene (not to same scale). After Treloar[26], and Raumann and Saunders[27].

$\overline{\cos^4\theta}$ or $\overline{P_2(\cos\theta)}$ and $\overline{P_4(\cos\theta)}$. This point again brings out the complexity of mechanical anisotropy compared with for example optical anisotropy. It also emphasises the desirability of developing methods to determine $\overline{\cos^4\theta}$ as well as $\overline{\cos^2\theta}$ if we wish to achieve any quantitative understanding of mechanical anisotropy in oriented polymers.

1.5.4 Theoretical schemes for orientation in polymers

The discussion in Section 1.5.1 above outlines the complexity of determining orientation in a polymer in the most general situation. There is therefore considerable merit in examining theoretical schemes for predicting the orientation from the macroscopic deformation which occurs during the drawing or forming process.

One of the best-known of such schemes is the AFFINE deformation model for rubbers.[28] The rubber is considered to be a network of flexible chains, and the macroscopic strain is imagined to be transmitted to the network such that lines joining the network junction points rotate and translate exactly as lines joining corresponding points marked on the bulk material. If we assume that the flexible chains consist of rotatable segments called 'random links', and that some statistical model can describe the configurational situation, it is then possible to obtain explicit expressions which relate the segmental orientation to the macroscopic deformation.

Such ideas formed the basis of the Kuhn and Grün model for the stress-optical behaviour of rubbers.[29] The birefringence of a uniaxially stretched rubber is given by

$$\Delta n = \frac{2\pi}{45}\frac{(n^2+2)^2}{n}N(p_1-p_2)\left(\lambda^2-\frac{1}{\lambda}\right)$$

where n is the mean refractive index;

N is the number of chains per unit volume;

p_1 and p_2 are the polarisabilities of the random link along and perpendicular its length respectively; and

λ is the extension ratio or draw ratio.

This expression has been shown to provide a quantitative understanding of the behaviour of rubbers. Because of the many configurational possibilities for the chains between the network junction points, the segmental orientation for strains $\sim 100\text{--}200\%$ is much less than the orientation of the lines joining the junction points. The development of birefringence which reflects this segmental orientation is therefore slow, particularly at low extensions where the chains are still close to the

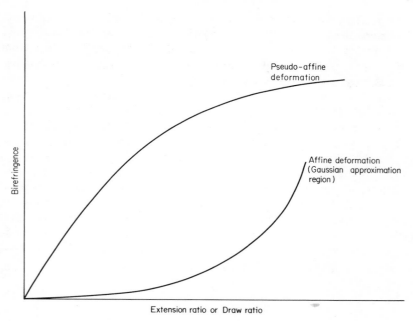

Fig. 15. The development of birefringence with increasing extension ratio for two theoretical deformation schemes.

crumpled configurations which they adopt in the isotropic state. This is shown in Fig. 15 (lower curve).

The deformation of polymers in the glassy or tough state, *i.e.* in cold drawing, is very different from that in the rubbery state. This is also usually true for commercial hot drawing processes. In particular, there are elements of the structure, such as crystallites, which partly retain their identity during deformation. This, of course, implies that the deformation process is no longer homogeneous. In one sense, therefore, it is unreasonable to expect a continuum model to describe the orientation process with any precision. However, it has been found that in many glassy and semi-crystalline polymers a rather simple deformation scheme holds to a first approximation. In this scheme it is assumed that the polymer consists of transversely isotropic anisotropic units whose symmetry axes rotate on stretching in the same manner as lines joining pairs of points in the bulk material, which again deforms at constant volume. This assumption is similar to the affine deformation scheme but it ignores the required change in length of the units on deformation. It is equivalent to describing orientation in terms of the lines joining the network junction points in the model for the deformation of a rubber. It can therefore be appreciated that this new deformation scheme, originally proposed by Kratky[30] to describe crystallite orientation, does give a more rapid initial orientation. A typical birefringence/draw ratio plot is shown in Fig. 15 (upper curve). We have called this scheme the 'pseudo-affine' deformation scheme (Ref. 5, p. 257).

It is quite straightforward to calculate $\overline{\cos^2 \theta}$ and $\overline{\cos^4 \theta}$ as a function of draw ratio, using these two deformation schemes. The results are shown in Fig. 16. It is to be noted that the clearest distinction can be observed between the development of orientation as a function of draw ratio; the difference between the nature of the distribution orientation functions is not so great. In spite of their limitations, calculations of this type have proved extremely valuable in understanding orientation

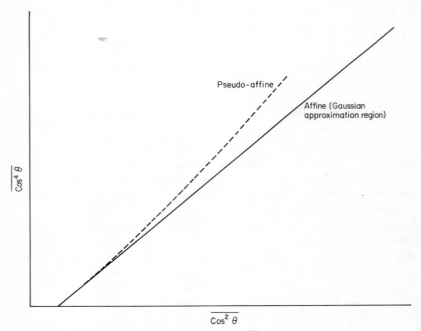

Fig. 16. The relationship between $\overline{\cos^2 \theta}$ and $\overline{\cos^4 \theta}$ for two theoretical deformation schemes.

effects in polymers. They form a basis for first order calculations of the relationships between optical and mechanical anisotropy and for the interpretation of complex structural techniques such as broad line nuclear magnetic resonance and polarised Raman spectroscopy.

REFERENCES

1. Nye, J. F. (1957). *Physical Properties of Crystals,* Clarendon Press, Oxford.
2. Unckel, H. A. (1953). *Iron Age,* **170.**
3. ICI Corporate Laboratory/Runcorn (private communication).
4. Anderton, G. E. and Treloar, L. R. G. (1971). *J. Mater. Sci.,* **6,** 562.
5. Ward, I. M. (1971). *Mechanical Properties of Solid Polymers,* Wiley, London.
6. Andrews, E. H. (1968). *Fracture in Polymers,* Oliver & Boyd, Edinburgh, p. 53.
7. Holliday, L., Mann, J., Pogany, G. A., Pugh, H. L. D. and Gunn, D. A. (1964). *Nature,* **202,** 381.
8. Carter, G. B. (1970). *Rep. Prog. Appl. Chem.,* **55,** 96.
9. Keller, A., Dlugosz, J., Folkes, M. J., Pedemonte, E., Scalisi, F. P. and Willmouth, F. M. (1971). *J. Phys.,* **32,** C5a, 295.
10. Robertson, R. E. (1965). *J. Phys. Chem.,* **69,** 1575.
11. Kargin, V. A. (1966). Special Lectures presented at the International Symposium on Macromolecular Chem. in Prague 1965, Butterworth, London, p. 35.
12. Ward, I. M. (1962). *Proc. Phys. Soc.,* **80,** 1176.
13. Faucher, J. A. and Reding, F. P. (1965). In *Crystalline Olefin Polymers,* (Ed. R. A. V. Raff and K. W. Doak), Interscience, New York, p. 681.
14. Fischer, E. W. (1957). *Naturforsch.,* **12a,** 753.
15. Keller, A. (1957). *Phil. Mag.,* **2,** 1171.
16. Till, P. H. (1957). *J. Polymer Sci.,* **24,** 301.
17. Mandelkern, L. (1970). In *Progress in Polymer Science,* (Ed. A. D. Jenkins), Vol. 2, pp. 165 *et. seq.,* Pergamon Press, Oxford.
18. Keller, A. (1968). *Prog. Rep. Phys.,* **31,** 623.
19. Palmer, R. P. and Cobbold, A. J. (1964). *Makromol. Chem.,* **74,** 174.
20. Sharples, A. (1966). *Introduction to Polymer Crystallisation,* Arnold, London, p. 25.
21. Anderson, F. R. (1964). *J. Appl. Phys.,* **35,** 64.
22. Wunderlich, B. and Arakawa, T. (1964). *J. Polymer Sci.,* **A2,** 3697.
23. Capaccio, G. and Ward, I. M. (1973). *Nature Physical Science,* **243,** 143.
24. Hermans, P. H. (1946). Contribution to the *Physics of Cellulose Fibres,* Elsevier, Amsterdam, p. 195.
25. McBrierty, V. J. and Ward, I. M. (1968). *Brit. J. Appl. Phys.,* **21,** 1529.
26. Treloar, L. R. G. (1970). *Plastics and Polymers,* **39,** 29.
27. Raumann, G. and Saunders, D. W. (1961). *Proc. Phys. Soc.,* **77,** 1028.
28. Treloar, L. R. G. (1958). *The Physics of Rubber Elasticity,* Clarendon Press, Oxford.
29. Kuhn, W. and Grün, F. (1942). *Kolloid-Z.,* **101,** 248.
30. Kratky, O. (1933). *Kolloid-Z.,* **64,** 213.

CHAPTER 2

MOLECULAR ASPECTS OF ORIENTED POLYMERS

A. PETERLIN

2.1 INTRODUCTION

The fundamental difference between polymeric and any other type of material is the consequence of the geometry of molecular structure. Linear macromolecules are extremely long flexible chains with the monomers covalently bound in the chain direction. The chain diameter is a few angstroms and the extended chain length can be 10^4 times larger. The strength of the covalent bond in the chain direction is that of the C–C bond in a diamond. The elastic forces resulting from bond stretching and valency angle deformation yield an elastic modulus $E_{c\parallel} = 2 \cdot 4 \times 10^{11}$ N/m^2 in the chain direction of crystalline polyethylene.[1] With one chain per 18 Å^2 that corresponds to a force constant $k = F/x = 34 \cdot 6$ N/m^2 per chain. Perpendicular to the chain, however, the interaction with adjacent molecules is of the usual intermolecular type, *i.e.* mainly by relatively weak van der Waals forces diminishing with the 6th power of intermolecular distance. Although in special cases they are enhanced by hydrogen bridges (polyamides), electric dipole (polyacrylonitrile) and polar (ionomers) forces they are always some orders of magnitude smaller than the forces of covalent bonding in the chain. In the case of polyethylene crystals the elastic modulus perpendicular to the chain direction[1] $E_{c\perp}$ is about 4×10^9 N/m^2. In the amorphous state the lateral forces are substantially smaller as a consequence of the larger intermolecular distances yielding an elastic modulus between 10^6 and 10^9 N/m^2.

This extreme anisometry of the linear macromolecules and the anisotropy of cohesive forces are unique for polymeric material and have no counterpart in low molecular weight material, as for instance metals,

36

ionic and molecular solids. They dominate not only the structure of the liquid and the morphology of the solid state but also the mechanical properties of both phases.

One of the most important consequences of the anisometry of linear polymers is the prevalence of amorphicity in the solid state. Upon cooling the melt the polymer increases rapidly in viscosity and usually solidifies as an amorphous glass. Crystallisation occurs only with very regular polymers, *e.g.* polyolefins and some sterically regular polyvinyls, but is never complete. A final fraction of the material remains amorphous.

In the amorphous state the linear macromolecule is more or less randomly coiled with the root mean square end-to-end distance proportional to the square root or a slightly higher power of molecular weight. Any deviation from this average value yields a mechanical force by which the sample tends to restore the average value of chain conformation. This entropic force is the basis of rubber elasticity.

On a short range scale the situation is more complicated because some thermodynamically permitted sequences of conformations are simply not possible as a consequence of the presence of other macromolecules. The requirement of an almost as perfect space filling as in the crystals— the density of amorphous (ρ_a) and crystalline (ρ_c) materials differ by about 10%—one finds parallel alignment of short segments of adjacent chains more often than a completely random arrangement. Such 'nodulae' of higher order and density detectable by electron microscopy[2,3] may have dimensions of about 30 Å, *i.e.* about 20 chain atoms in the chain direction and about 5 chains in the lateral direction. The molecules of one nodula diverge and enter more or less at random adjacent nodulae with different orientation. Such a fringed micelle type model of the amorphous state with the nodulae as micelles seems to emanate rather convincingly from recent electron microscopy studies of thin films of polystyrene, polyethyleneterephthalate, polycarbonates, polyisobutylene, and polymethylmethacrylate. It applies not only to completely amorphous, uncrystallisable polymers in the glassy and rubbery phase but also to crystallisable polymers where the nodulae may be considered as precursors of true crystals.

The amorphous supercooled melt obtains shape stability by chemical or mechanical cross-links and then behaves like an isotropic rubber with low elastic modulus and high recoverable or even reversible strain. The cross-links transform the melt into a network of N individual chain links per volume which yield a restoring force coefficient

$3NkT \sim 10^6$ N/m^2 for $N = 10^{26}$ m^{-3} which corresponds to about 400 chain atoms per link. With between four and ten chain atoms per link as found in the loops at the surface of polymer crystals the elastic modulus increases about 100 times to 10^8 N/m^2 in good agreement with experimental data.

If the cross-linking is performed in a strained state the rubbery polymer becomes anisotropic with a finite chain orientation in the direction of the principal strain. The amorphous chain orientation characterised by

$$f_a = (3\langle \cos^2 \theta \rangle - 1)/2$$

where θ is the angle between the chain segment and the direction of preferred orientation of the sample, can be measured by birefringence and dichroism, sonic modulus and to some extent also by X-ray scattering. In the rubber with the high chain mobility the anisotropy of mechanical properties is the consequence of a larger fraction of chain links connecting the sample in the preferred direction than perpendicular to it.

In glasses the main effect of orientation on mechanical properties comes from the anisotropy of mechanical forces in the direction of the chain segment and perpendicular to it, usually characterised by a longitudinal, $E_{a\parallel}$, and transverse, $E_{a\perp}$, amorphous modulus.[4] Their difference is less than in the crystals because the segments are never in full *trans*-conformation. Each gauche conformation reduces drastically the force transmission in the chain direction and hence the longitudinal modulus below the extreme value $E_{c\parallel}$. In polypropylene at room temperature $E_{a\perp}$ is about 10^9 N/m^2.

2.2 CRYSTALLINE POLYMERS

As already mentioned the polymer solid is never completely crystalline. From the density ρ one can deduce the volume crystallinity α

$$\alpha = (\rho - \rho_a)/(\rho_c - \rho_a) < 1$$

That applies even to single crystals grown from dilute solution. The amorphous component is partially distributed all over the crystal in form of crystal defects, *i.e.* vacancies and kinks, and to a larger extent on the surfaces of the crystal lamellae and mosaic blocks.

The terms 'crystallinity' and 'amorphous' apply to an ideal two-component system, ideal crystal and ideal supercooled melt. In the actual sample the crystal lattice is not ideal but contains all types of crystal defects and the non-crystalline component is not ideally amorphous. Hence the density of the former is less than the ideal crystallographic density ρ_c and of the latter larger than ρ_a, the density of ideal supercooled melt. The opposite applies to the heat content. Hence crystallinity, crystalline and amorphous component are just abbreviated terms for the first characterisation of the sample and not an exact description of its physical state.

The normal polymer single crystal is a very thin lamella or any other crystallographically permitted shape, as for instance hollow pyramids in the case of polyethylene. In all cases the extension L in the chain direction is very small, about 100 Å, with the chains folding back at the surface. The folds with adjacent or random re-entry do not fit the crystal lattice of the straight sections. Together with the free chain ends not incorporated in the interior of the crystal they constitute the amorphous layers on both faces of the single crystal lamellae. From the radial width of the equatorial wide-angle X-ray reflections one has concluded that the crystal lattice has a finite lateral coherence length L_1 proportional to L which can be interpreted as the lateral width of mosaic blocks of which the lamella is composed. The boundary between adjacent blocks must be to some extent disordered for interruption of scattering coher-

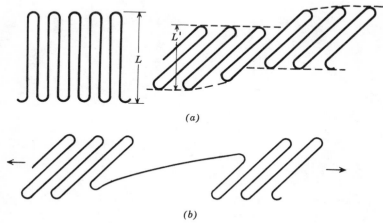

Fig. 1. Chain tilt and slip (a) in a crystal lamella leading to the formation of a crack, (b) bridged by the partially unfolding macromolecules (Peterlin[6]).

ence of the crystal lattice. It makes an additional contribution to the amorphous component of the single crystal.

In the lamellar or pyramidal, single crystals having the very small extension, about 100 Å, in the chain direction and no limitation for lateral extension, the lateral van der Waals forces in a polymer crystal are about

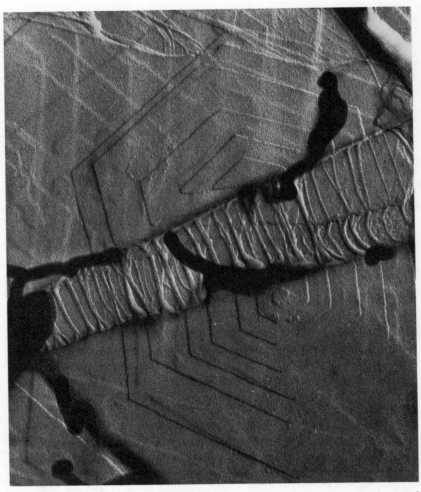

Fig. 2. Electron micrograph of a multilayer crystal of polyoxymethylene with a crack bridged by a great many microfibrils (Ingram[7]).

100 times smaller than the covalent forces in the chain direction. One can imagine that with such a geometry and force field the isolated lamella is no mechanical giant. Phase change and/or twinning involving chain rearrangement in the plane perpendicular to the chain direction easily occur because they are opposed merely by the weak van der Waals but not by the strong covalent forces. A particularly important easy motion is translation of the chain in its own direction which is indeed, together with a 180° flip-flop rotation, the basic mechanism of mechanical relaxation of polyethylene close to the melting point (α absorption peak).[5] Under tension in the lamella plane with the slightest shear component, the crystal is easily deformed by chain tilt and slip (Fig. 1) up to the formation of a crack which under favourable conditions is bridged by a great many microfibrils (Fig. 2).

2.3. POLYCRYSTALLINE SOLID

In contrast with low molecular weight material the crystalline polymer solid is never an agglomeration of randomly oriented single crystals. This is mainly the consequence of the fact that polymer crystals as a rule are very thin lamellae. Such a shape favours parallel packing into single crystal stacks which in turn may be and usually are randomly oriented in space. Under favourable crystallisation conditions with few primary nuclei the stacks of parallel lamellae grow freely without interfering with each other. By non-crystallographic branching they develop into spherulites (Fig. 3). At sufficient distance from the primary nuclei the stacks of lamellae are radially oriented with the chains perpendicular to the radius.

With a high number of primary nuclei so many growth centres exist that the growing stacks impinge on each other before the spherulitic structure can develop. The sample contains randomly oriented stacks of parallel lamellae which can be considered as frustrated spherulite embryos. Such a microspherulitic structure is obtained from a highly seeded (heterogeneous nuclei) or quenched melt. The macrospherulitic structure results from slow cooling or isothermal crystallisation (small number of homogeneous nuclei) of a very pure melt with as small content as possible of heterogeneous nuclei.

The spherulitic structure has no macroscopic orientation. But the stacking of parallel lamellae (Fig. 4) produces a high local order and orientation in a range of a few lamella thicknesses ($\sim 0.1 \div 1\mu$) which

Fig. 3. Model of spherulite structure showing the stacks of parallel lamellae, the embryo of the spherulite (inside the broken line quadrangle), the non-crystallographic branching, the low strength radial boundaries between stacks of lamellae and the still less strong boundaries between adjacent spherulites (Peterlin[8]).

shows up in the small-angle X-ray scattering (SAXS). The scattering maxima yield the lamella thickness (the long period L). If by any means the stacked lamellae are oriented a macroscopically oriented polymer is obtained with anisotropic physical properties.

Wide-angle X-ray scattering (WAXS), SAXS, optical birefringence, IR dichroism and light scattering can be used for the detection and measurement of local and macroscopic orientation. One has mainly the following four cases of oriented crystalline polymer solid:

(1) oriented single crystal mat,
(2) transcrystalline structure, obtained by crystallisation in a temperature gradient or by annealing of rolled films,
(3) row nucleated material from strained melt or solution (shish-kebab structure),
(4) fibrous structure obtained by plastic deformation (drawing and rolling).

The chain orientation and hence the anisotropy is maximum in cases (1) and (4). But they differ enormously in mechanical properties. They are indeed the best example demonstrating the fallacy of connecting mechanical properties with crystal orientation without paying attention to more intricate features of sample morphology and to the most important amorphous matrix the crystals are imbedded in.

The amorphous component of the crystalline polymer solid contains beyond the amorphous component of single crystals, *i.e.* crystal defects (linear vacancies, kinks, and interstitials), chain folds and free chain ends, as new elements the rejected non-crystallisable impurities and tie molecules. The former concentrate on the outer boundaries of lamella stacks and spherulites, the latter in the amorphous layers separating the lamellae of the same stack. With the exception of impurities all other components of the amorphous phase are intimately connected with the crystals and cannot be physically separated from them or moved independently of them.

The tie molecules in samples crystallised from melt or solution are sections of long chain molecules which cannot crystallise because the molecules have started independently to crystallise in two different crystals so that the intervening section cannot be included in any one of

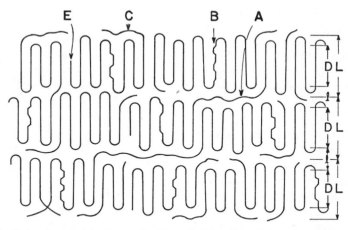

Fig. 4. Molecular model of a stack of parallel lamellae of the spherulitic structure; A, interlamellar tie molecule; B, boundary layer between two mosaic blocks; C, chain end in the 'amorphous' surface layer (cilium); D, thickness of the crystalline core of the lamella; E, linear vacancy caused by the chain end in the crystal lattice; L, long period; l, thickness of the 'amorphous' layer (Peterlin[9]).

them. Occasionally, the strained tie molecules may give rise to filament crystallisation which provides very strong fibrillar links between the two crystals connected.[10] In row nucleated samples the tie molecules are occasionally so much bundled that the lamellae are firmly stapled on such places. In drawn material the extremely large number of taut tie molecules originates from partially unfolded chains during micronecking which tears folded chain blocks from the lamella of the spherulitic material and incorporates them into the microfibrils. They firmly connect subsequent crystal blocks and thus make the long and narrow microfibril the strong basic element of the fibrous structure.[9]

The two-component system—crystal lamellae or blocks alternating with amorphous layers which are reinforced by tie molecules—results in a mechanism of mechanical properties which is drastically different from that of low molecular weight solids. In the latter case it is based on crystal defects and grain boundaries. In the former case it depends primarily on the properties and defects of the supercrystalline lattice of lamellae alternating with amorphous surface layers (in spherulitic, transcrystalline or cylindritic structure) or of microfibrils in fibrous structure, and on the presence, number, conformation and spatial distribution of tie molecules. It matters how taut they are, how well they are fixed in the crystal core of the lamellae or in the crystalline blocks of the microfibrils and how easily they can be pulled out of them. In oriented material the orientation of the amorphous component (f_a) is a good indicator of the amount of taut tie molecules present[10] and hence an excellent parameter for the description of mechanical properties. In fibrous structure it directly measures the fraction and strength of microfibrils present and therefore turns out to be almost proportional to elastic modulus and strength in the fibre direction.

2.3.1 Oriented single crystal mats

Under special conditions of precipitating single crystals grown from dilute solution one obtains well oriented single crystal mats (Fig. 5(a)). According to wide and small-angle X-ray scattering the orientation is almost as good as in drawn material of moderate draw ratio. But the mechanical properties are completely different. As a consequence of the preparation technique there is no material connection between superimposed crystals. No tie molecules going from one crystal to the next one are bridging the folds containing 'amorphous' surface layers sandwiched between the crystal cores. Also the lateral fit of lamellae is poor. Hence such mats have very poor mechanical properties. They are brittle with

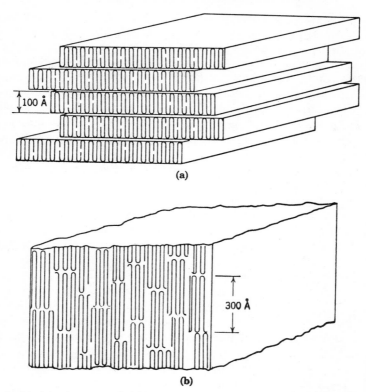

Fig. 5. Model of a mat of folded-chain single-crystals (a) as grown, (b) after annealing (Stratton[12]).

low strength. Annealing, however, enormously enhances their mechanical properties by improving the lateral fit and producing by the long period growth, an interpenetration of lamellae in the chain direction (Fig. 5(b)). Such mats can be rolled and drawn yielding an extremely strong fibrous structure.

2.3.2 Transcrystalline structure

The crystallisation from melt in a temperature gradient[13] or in contact with a surface acting as a nucleating substrate[14,15] may result in a transcrystalline structure[16] with the growth axis of the crystals, the *b*-axis in the case of polyethylene,[17] the hydrogen bonds in the case of nylon[18] parallel to the gradient or perpendicular to the substrate, respectively. The orientation of the other two axes is at random in the plane perpendicular to the growth direction. A basic requirement for the formation of such a structure is the absence of heterogeneous nuclei which would start a random nucleation in the supercooled melt and thus interfere with the well aligned crystal growth initiated by homogeneous nucleation in the cooler section of the melt or on the nucleating substrate. The high concentration of primary nuclei in the growth plane very soon limits the lateral growth of spherulites so that the subsequent growth is confined to the direction perpendicular to the plane of primary nuclei.

A very similar orientation is obtained by annealing of rolled or drawn low density polyethylene sheets.[19] The chain (c) axis is perpendicular to the sheet, the a-axis is oriented in the original draw direction and the b-axis is perpendicular to it in the plane of the sheet. Very little is known about the orientation of the amorphous component and about the number of tautness of tie molecules.

2.3.3 Row nucleated structure (cylindrites)

The crystallisation from strained melt as for instance in a blown film or in the jet during fibre spinning produces a row nucleated structure.[20-22] Linear nuclei are formed parallel to the strain direction. They contain more or less extended polymer chains. Secondary epitaxial nucleation on the surface of such linear row nuclei produces folded chain lamellae which are oriented perpendicular to the strain (Fig. 6). In such a case the sample exhibits a high uniaxial orientation of chain axes in the strain direction with random orientation of the a- and b-axes perpendicular to it. If the growing lamellae exhibit a helical twist the chain orientation in the strain direction is very soon replaced by the orientation of the axis of maximum growth rate (b-axis in the case of polyethylene) perpendicular to the strain direction and a more random orientation of the remaining two axes (a- and c-axes in the case of polyethylene) with a maximum in the strain direction. Such a row nucleated structure has parallel cylindrical spherulites (cylindrites) as its basic supercrystalline element.

The row nucleated structure contains two types of crystals: a small fraction of fibrillar crystals (row nuclei) with partially or even fully extended chains and the normal type folded chain lamellae. The existence of two types of crystals is detectable by calorimetry and the resistance to fumic nitric acid attack, high in the row nuclei and low in the surface layers of lamellae.[22] The number of tie molecules between consecutive

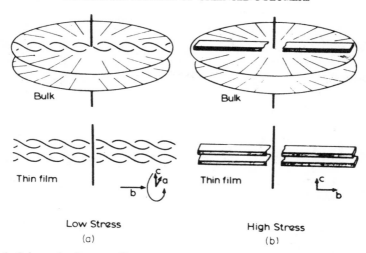

Fig. 6. Schematic diagram illustrating row nucleated crystallisation under (a) low stress and (b) high stress for polyethylene (Keller and Machin[21]).

lamellae is probably of the same order of magnitude or a little higher than in the stacked parallel lamellae of the conventional spherulitic structure. The row nuclei with partially extended chains are by far the strongest element of the structure.

Since the row nuclei in the crystallising strained melt act as reinforcing frame structure carrying most of the load in the strained liquid, the not yet crystallised melt inside the frame is able to relax almost completely. It hence either crystallises in conventional manner on the surface of the row nuclei yielding the cylindrites or forms new conventional primary nuclei yielding spherulites as shown by electron microscopy of thick nylon 6 fibres as spun.[23] The coexistence of both structural elements in the fibre as spun with a very small fraction of material in row nuclei explains the poor mechanical properties of such a material and the similarity of fibrous structure obtained after drawing with that obtained from purely spherulitic films.

Under special thermal and strain conditions, the radial growth of lamellae of a cylindrite favours the formation of highly localised tie molecule bridges (Reneker web) between two adjacent lamellae.[24,25] With sufficiently strong bridges the lamellae are as stapled (Fig. 7). Under tensile stress perpendicular to the lamellae the whole structure may deform like a honeycomb with the free sections of lamellae acting

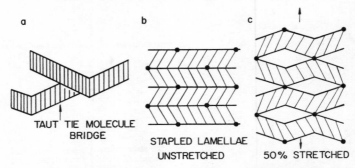

Fig. 7. *Model of hard elastomer: (a) taut tie molecule bridge deposited during growth; (b) Reneker web type stapling of parallel lamellae; (c) deformation of lamellae between the bridges producing large holes (Clark*[26]*).*

as elastic springs.[26] Completely reversible deformations up to a few hundred per cent are attainable in such hard elastomers of polyoxymethylene, polypropylene, and polypivalolactone. The high elasticity is based on the reversible bending or shearing of the crystal lamellae. Since it is of energetic origin it decreases with increasing temperature in contrast with the entropic elasticity of rubber. The opening of the gaps between adjacent lamellae yields an extremely high porosity enhancing up to 10^7 times the transport of gases over that of compact material.

2.3.4 Fibrous structure

Highly drawn fibres and films exhibit an almost complete orientation of the crystal lattice, a high orientation of the amorphous component and a fairly good orientation of lamellae either perpendicular or at a finite angle to the fibre axis. The same anisotropy applies to mechanical properties with a high elastic modulus in the fibre axis and a smaller one perpendicular to it.[27-29] Technically the most important feature is the much higher values of the axial modulus, yield stress, and tensile strength, by one decade or even more, than observed in unoriented spherulitic material and a proportional reduction of draw range and elongation at break. As a rule, the effects increase linearly with the draw ratio.[27-35] One is tempted to assume that this significant modification of mechanical properties is caused by chain orientation because both increase with increasing draw ratio. A closer analysis shows that such a parallelity obtains for the amorphous but not for the crystalline component. It is the consequence of the fact that in a polymer sample with increasing draw ratio the fraction of fibrous structure is steadily increasing and both the amorphous orientation and mechanical properties are proportional to this ratio.

The morphology of fibrous material reveals as the basic structural element the microfibril and not the lamella.[9] Electron micrographs of surface replicas of slightly etched fibres and films with fibrous structure show exclusively highly aligned very long microfibrils with lateral dimensions between 100 and 200 Å. The length is in the tens of micrometres. The microfibrils can be rather easily separated from each other as seen in Fig. 8. A higher organisational unit is the fibril (Fig. 9), *i.e.* a bundle of parallel microfibrils formed during drawing from a stack of parallel lamella of the starting material and differing in draw ratio from the adjacent fibril originating from a stack with different lamella orientation. That makes the boundary between adjacent fibrils weaker and favours the formation of longitudinal voids.

During micronecking folded chain blocks are broken off from the lamella and incorporated in the microfibril which therefore consists of alternating crystalline and amorphous regions. A great many taut tie molecules, up to 30% of chains of the crystal lattice in highly drawn polyethylene,[38,39] originating from partial chain unfolding during separation of the blocks are bridging the amorphous layer and connecting the blocks thus imparting to the microfibril a great axial tensile strength and elastic modulus. Hence, the microfibril acts much more as a

Fig. 8. Electron micrograph of a surface replica of isotactic polypropylene drawn at 100°C to a draw ratio $\lambda = 5$. Note the highly aligned microfibrils of lateral dimensions of 200 Å all over the sample and the disorientation of the microfibrils bridging the longitudinal voids (Peterlin[8]).

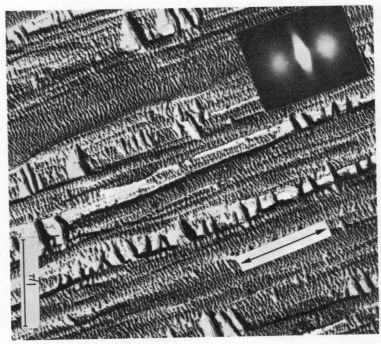

Fig. 9. Surface replica electron micrograph and small-angle electron diffraction pattern of polyethylene film drawn at 60°C to a draw ratio 16 and subsequently annealed for 2 h at 120°C. Etched for 6 h in fuming nitric acid at 80°C. The transformed areas belong to fibrils of lower draw ratio than the rest (Peterlin and Sakaoku[37]).

crystal with locally weakened areas than as an alternation of crystalline and amorphous layers.

The lamellar structure of the fibrous material is actually an optical artifact produced by the fact that the crystal blocks of adjacent microfibrils tend to be rather well aligned laterally so that more or less wavy lamellae are formed (Fig. 10). The waviness explains the two point and the line diagram of small-angle X-ray scattering. A four point diagram indicates a discrete orientation at a finite angle to the fibre axis. But there is practically no material connection between adjacent blocks of the same lamella in contrast with the great many tie molecules connecting blocks of the same microfibril. By prolonged etching one can remove the amorphous component in the layers between the blocks and eventually the not completely crystalline tie molecule bridges connecting the blocks in axial direction. That which remains is indeed the stacking of parallel lamellae without visible remains of microfibrillar structure.[37,41] The persistence of lamellae also demonstrates that there is less amorphous material on the lateral than on the axial surfaces of the folded chain blocks.

The orientation of the crystal lattice f_c increases very fast during drawing and soon reaches a value close to 1. But it is not a good parameter for the description of mechanical properties. It fails completely during annealing at higher temperatures, which only marginally reduces the crystal orientation, particularly if the sample is clamped so that it cannot shrink, but drastically reduces the mechanical properties down to the values of the unoriented material.[42] Concurrently IR dichroism[11] shows a nearly complete loss of amorphous orientation which seems to indicate that the taut tie molecules either have almost completely relaxed and assumed a nearly equilibrium random conformation or have been reduced in number. Electron micrographs of surface replicas indeed show a rotation of lamellae around an axis perpendicular to the fibre axis and a lateral growth beyond the boundary of the fibril.[36] One has the impression that the lamellae have attained a physical individuality

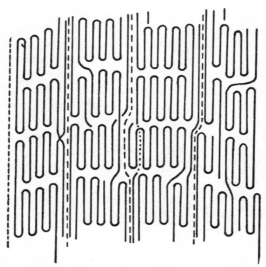

Fig. 10. Microfibrillar model of fibrous structure (Peterlin[40]).

not existing in the material as drawn and a substantial independence from each other which is only possible if the axial connection by tie molecules has been lost and the microfibrils have ceased to exist.

In a stress field the microfibril reacts almost as a rigid structure much stronger than any other element of the polymer solid. This is a consequence of the large fraction of taut tie molecules connecting consecutive crystal blocks like a quasicrystalline bridge and of the special dense packing of aligned microfibrils but not a consequence of some higher crystal perfection of the blocks. Indeed one knows that the crystal defect concentration is increasing with draw ratio yielding a lower than ideal crystal density and consequently also a lower strength and elastic modulus of the crystals.[43] The removal of crystal defects and restoration of crystal perfection during annealing makes the crystals stronger but cannot prevent the drastic reduction of elastic modulus and strength of the fibrous material caused by the disappearance of tie molecules.

The dense packing of microfibrils with the enormous surface ($4l_{mf}d$) to cross-section (d^2) ratio ($4l_{nf}/d \sim 4000$) seems to overcompensate the role of crystal defects. The longitudinal displacement of a single microfibril is opposed by the sum of van der Waals forces along the whole microfibril which adds to so much that such a displacement is less easy than in the case of stacked lamellae. Moreover, it blocks the individual crystal deformation because such a deformation always demands some longitudinal displacement of the whole or of a part of the microfibril. Consequently the deformation of the fibrillar structure is much more a co-operative than a highly localised effect of a single crystal block. That not only makes the microfibril more resistant to deformation but also homogenises the stress and strain field. The combination of all these effects results in the higher tensile strength and elastic modulus of fibrous material. As long as one can avoid longitudinal void formation the elastic modulus in the transverse direction too is larger than in a spherulitic sample[27] because in spite of the fact that the crystal lattices of adjacent microfibrils are not in crystallographic register the van der Waals forces between them are stronger than in the thicker amorphous layer between two stacked lamellae.

In the special case of branched polyethylene one observes an exceptionally small elastic modulus at 45° to the fibre axis which means a very high shear compliance along the fibre axis.[27-29] Such a deformation involves shearing displacement of adjacent microfibrils. Since a similarly high shear compliance does not occur with linear polyethylene the difference may reside with the substantial difference in draw ratio,

4.5 in branched, 20 in linear polyethylene, yielding proportionately shorter microfibrils in the former than in the latter case. The shorter the microfibril the smaller the surface to cross-section ratio and hence the resistance to shear displacement. Nylon 6 with a maximum draw ratio between 4 and 5 does not show the effect in spite of the small draw ratio. Very likely one can expect that the hydrogen bridges so efficiently increase the forces between adjacent microfibrils that in spite of the much shorter microfibrils their autoadhesion is similar to that in linear polyethylene and polypropylene with at least twice as long microfibrils.

From the morphology of fibrous structure one can conclude that also the plastic deformation will first involve the sliding motion of microfibrils which is opposed by the friction resistance in the boundary. In spite of the weakness of the van der Waals forces the friction resistance over the 1000 times larger length than the thickness of crystal lamella will be much larger than the forces opposing plastic deformation in spherulitic material. Hence the strain hardening effect in the fibrous material is a straightforward consequence of the large surface to volume ratio of the strong microfibrils which efficiently hampers shearing displacement of microfibrils and explains the increase of elastic modulus and stress to break by a factor nearly equal to the draw ratio.

The bundling of microfibrils into fibrils yields two interdependent modes of plastic deformation of fibrous structure under tensile load: sliding motion of fibrils upon each other (pure dilatation as described by the affine transformation) and shear deformation of fibrils by sliding motion of microfibrils. The former mode is responsible for most of macroscopically observed deformation and also for the final failure by microcrack formation along the outer boundaries of the fibrils. The coalescence of such microcracks leads to physical separation of individual fibrils which can be indeed observed on fracture surface of high strength fibrous material.[44] The sliding motion of the fibrils also produces a shear stress on more or less skewed fibrils. The shear stress is proportional to the skewedness and the resistance to deformation proportional to the length of the fibril. Hence one expects larger shear deformation in nylon with shorter and wider fibrils than in polyethylene with longer and thinner fibrils.

The shear displacement of adjacent microfibrils extends enormously the interfibrillar tie molecules. If the fibril is 20 microfibrils, i.e. between 2000 and 4000 Å wide and 5 μm = 50 000 Å long then a shear displacement by 1000 Å increases the length by 20 000 Å, i.e. by 40%, but increases the number of interfibrillar tie molecules per amorphous layer by a factor 10, *i.e.* 1000%. This estimate is based on a long period of 100 Å. The tie molecule originally passing through one amorphous layer passes now through 10 of them if it is unfolded over 1000 Å. This example is supposed to be valid for nylon with a small draw ratio (3) at completed transformation from spherulitic to fibrous structure. If modified for polyethylene with a 3 times larger draw ratio (9) at completed transformation and hence about 15 μm long fibrils and a long period of 200 Å one finds an extension of 13% and an increase of interfibrillar tie molecules by a factor of 5. The difference becomes still larger if one considers that at equal stress the strain is 3 times smaller in polyethylene than in nylon because the length of microfibrils and hence the resistance to shear displacement is 3 times larger in the former than in the latter

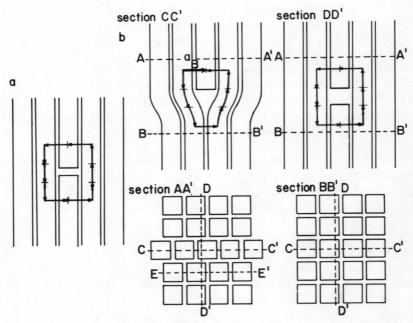

Fig. 11. Microfibrillar superlattice with a vacancy defect: (a) and a point dislocation; (b) viewed in two perpendicular planes CC' and DD' through the point dislocation and two planes AA' and BB' perpendicular to the fibre axis. For simplicity of scale the microfibrils are assumed to have a quadratic cross-section. The Burgers vector a_B is one microfibril thickness long in the CC' plane and is zero in the DD' plane or in any plane EE' parallel to CC' and not going through the inserted microfibril.

case. That reduces the extension of the fibril to 4·4% and the increase of tie molecule to a factor of 2.

As a consequence of shear deformation of fibrils during plastic deformation of fibrous structure one obtains a large fraction of interfibrillar tie molecules connecting adjacent microfibrils. The effect is larger the smaller the draw ratio $\lambda*$ at which the transformation of spherulitic into fibrous structure is completed because the length of the fibrils is proportional to $\lambda*$ and the lateral dimensions to $\lambda*^{-1/2}$. The almost fully extended taut tie molecules are not in crystallographic register and hence form a gradually thickening non-crystalline boundary layer between the crystal blocks of adjacent microfibrils.

The ends of the microfibrils create about 10^{21} m^{-3} point vacancies in the microfibrillar superlattice (Fig. 11). Under applied tensile load they may fail first, eventually by microcrack formation so that the adjacent microfibrils have to carry a heavier load than the rest of the sample.[45] Hence they are first candidates for rupture detectable by the radicals formed at the rupture of tie molecules in at least one amorphous layer of the microfibril affected. Depending on the ratio of axial strength to lateral adhesion of the microfibrils the microcracks will grow parallel (high ratio) or perpendicular (low ratio) to the fibre axis yielding a large number of broken chains and radicals in the former and a small one in the latter case. Nylon is an example of the former and linear polyethylene of the latter type.

REFERENCES

1. Sakurada, I., Ito, T. and Nakamae, K. (1966). *J. Polymer Sci., C.*, **15**, 72.
2. Yeh, G. S. Y. (1972). *Crit. Rev. Macromol. Sci.*, **1**, 173.
3. Geil, P. H. *J. Macromol. Sci., Phys.* (in press).
4. Samuels, R. J. (1965). *J. Polymer Sci.*, **A3**, 1761.
5. Olf, H. G. and Peterlin, A. (1970). *J. Polymer Sci. A-2*, **8**, 753, 771, 791.
6. Peterlin, A. (1969). *Kolloid-Z&Z Polymere*, **233**, 857.
7. Ingram, P., unpublished work.
8. Peterlin, A. (1971). *J. Polymer Sci., C* **32**, 297.
9. Peterlin, A. (1971). *J. Mater. Sci.*, **6**, 490; (1965). *J. Polymer Sci. C*, **9**, 61.
10. Keith, H. D., Padden, F. J., Jr. and Vadimsky, R. G. (1971). *J. Appl. Phys.*, **42**, 4585.
11. Glenz, W. and Peterlin, A. (1970). *J. Macromol. Sci.*, **B4**, 473; (1971). *J. Polymer Sci. A-2*, **9**, 1191; (1971). *Makromol. Chem.*, **150**, 163; (1971). *Kolloid Z&Z Polymere*, **247**, 786.
12. Statton, W. O. (1967). *J. Appl. Phys.*, **38**, 4149.
13. Eby, R. K. (1964). *J. Appl. Phys.*, **35**, 2720.
14. Schonhorn, H. and Ryan, F. W. (1968). *J. Polymer Sci. A-2*, **6**, 231.
15. Fitchmun, D. R. and Newman, S. (1970). *J. Polymer Sci. A-2*, **8**, 1545.
16. Jenckel, E., Teege, E. and Hinrichs, W. (1952). *Kolloid-Z.*, **129**, 19.
17. Hoffman, J. D. (1964). *SPE Trans.*, **4**, 315.

18. Barriault, R. J. and Gronholz, L. F. (1955). *J. Polymer Sci.*, **18**, 393.
19. Hay, I. and Keller, A. (1966). *J. Mater. Sci.*, **1**, 44.
20. Kobayashi, K., as quoted in P. H. Geil (1963). In *Polymer Single Crystals*, Wiley, New York, pp. 465–475.
21. Keller, A. and Machin, M. J. (1967). *J. Macromol. Sci.*, **B1**, 41.
22. Keller, A. and Machin, M. J. (1967). In *Polymer Systems: Deformation and Flow*, Macmillan, London, p. 97.
23. Sakaoku, K., Morosoff, N. and Peterlin, A. (1973). *J. Polymer Sci. A-2*, **11**, 25.
24. Garber, C. H. and Clark, E. S. (1970). *J. Macromol. Sci.*, **B4**, 499.
25. Quinn, R. G. and Brody, H. (1971). *J. Macromol. Sci.*, **B5**, 721.
26. Clark, E. S. (1972). ASC Meeting, Boston, April.
27. Raumann, G. and Saunders, D. W. (1961). *Proc. Phys. Soc., London*, **77**, 1028.
28. Gupta, V. B. and Ward, I. M. (1967). *J. Macromol. Sci.*, **B1**, 373; (1968). **B2**, 89.
29. Stachurski, Z. H. and Ward, I. M. (1968). *J. Polymer Sci. A-2*, **6**, 1083; (1969). *J. Macromol. Sci.*, **B3**, 445.
30. Sheehan, W. C. and Cole, T. B. (1964). *J. Appl. Polymer. Sci.*, **8**, 2359.
31. Haward, R. N. and Mann, J. (1964). *Proc. Roy. Soc., London*, **A282**, 120.
32. Meinel, G. and Peterlin, A. (1968). *J. Polymer Sci. A-2*, **6**, 587.
33. Haward, R. N. and Thackray, G. (1968). *Proc. Roy. Soc., London*, **A302**, 453.
34. Daniels, B. K. (1971). *J. Appl. Polymer Sci.*, **15**, 3109.
35. Peterlin, A. (1972). *Tex. Res. J.*, **42**, 20.
36. Peterlin, A. and Sakaoku, K. (1967). *J. Appl. Phys.*, **38**, 4152.
37. Peterlin, A. and Sakaoku, K. (1970). In *Clean Surfaces* (Ed. G. Goldfinger), M. Dekker, New York, p. 1.
38. Meinel, G. and Peterlin, A. (1967). *J. Polymer Sci.*, **B5**, 197; (1968). *A-2*, **6**, 587.
39. Meinel, G., Peterlin, A. and Sakaoku, K. (1968). *Analytical Calorimetry* (Ed. R. S. Porter and J. F. Johnson), Plenum Press, New York, p. 15.
40. Peterlin, A. (1969). *J. Polymer Sci. A-2*, **7**, 1151.
41. Hay, I. and Keller, A. (1964). *Nature*, **204**, 802.
42. Meinel, G. and Peterlin, A. (1967). *J. Polymer Sci.*, **B5**, 613.
43. Fischer, E. W., Goddar, H. and Schmidt, G. F. (1968). *Makromol. Chem.*, **118**, 144.
44. Olf, H. G. and Peterlin, A. *J. Polymer Sci.*, in press.
45. Peterlin, A. (1971). *Inter. J. Fract. Mech.*, **7**, 496; (1972). *J. Macromol. Sci.*, **B6**, 583.

CHAPTER 3

PHYSICO-CHEMICAL APPROACHES
TO THE MEASUREMENT OF ANISOTROPY

R. S. STEIN and G. L. WILKES

3.1 OPTICAL ANISOTROPY AND BIREFRINGENCE

3.1.1 Introduction

The measurement of optical anisotropy is one of the simplest and most exploited methods of studying orientation in polymers. It depends on the fact that a given material has associated with it unique optical properties determined by its polarisability.

Since polarisability is a tensor quantity, the resulting optical properties may also be directionally dependent unless this tensor is isotropic. A simple example is illustrated by Fig. 1 where one can envisage that the interaction of the bond electrons will be greater for the imposed field that is oscillating in a plane parallel to the direction of the bond than for a field oscillating in a plane perpendicular to the bond, *i.e.* where the polarisability is highest. This interaction leads to a decrease in the velocity of the incident wave by an amount defined by the refractive index, n. For a non-absorbing system, the polarisability is related to the refractive index by the Lorenz–Lorentz equation:

$$\frac{n^2 - 1}{n^2 + 2} = \frac{4}{3} \pi P \tag{1}$$

where P is defined as $N\alpha$, N being the number of molecules per unit volume and α is the polarisability of the molecular unit and has the dimensions of volume.

Recalling Fig. 1 it is now clear that the refractive indices (or their difference) along two orthogonal directions can be used as a measure of orientation. A convenient index of orientation is the birefringence

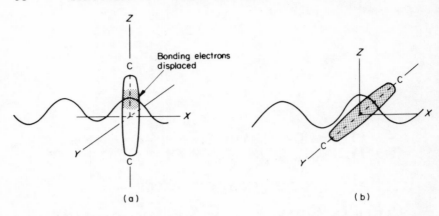

Fig. 1. *Schematic illustrating the interaction of electrons of a carbon–carbon bond with an electromagnetic wave—note there is a directional dependence to this interaction.*

Δn_{ij} where i and j refer to the two orthogonal axes in question. For a cartesian co-ordinate system three birefringence indices exist, two of which are independent.

$$\Delta n_Y = \Delta n_{ZX} = n_Z - n_X$$

$$\Delta n_X = \Delta n_{ZY} = n_Z - n_Y \tag{2}$$

$$\Delta n_Z = \Delta n_{XY} = n_X - n_Y = \Delta n_{ZY} - \Delta n_{ZX}$$

For uniaxial symmetry as might occur about a stretching axis, Z, $\Delta n_X = \Delta n_Y$ and $\Delta n_Z = 0$ so, only a single birefringence value is necessary to describe the orientation.

We have indicated that a finite value of Δn may provide a means of detecting orientation but its magnitude and sign may vary considerably from system to system and will depend on the chemical nature and material composition (amorphous v. crystalline). The chemical structure determines whether the refractive index (polarisability) is higher along the chain (polyethylene) or perpendicular to the chain (polystyrene-rubbery state). The composition in turn is of importance because the intrinsic optical character of a crystal may differ from that of an amorphous chain. Furthermore, polarisability interactions known as internal field and form effects may also influence the observations and hence must be considered in many instances.

3.1.2 Measurement of birefringence

In solids there are two general methods used for the determination of Δn. One is the transmission method while the other is a compensator technique. Each has advantages and disadvantages and thus the method chosen depends highly on the experimental circumstances as well as on the material itself. It is worth pointing out that the above discussion applies to both uniaxially and biaxially oriented systems. Two other methods not to be discussed here are interference microscopy[1] and refractometry.[2]

Most of the methods depend upon the optical retardation, R, defined by

$$R = \frac{d}{\lambda}\Delta n \qquad (3)$$

where d is the thickness of the sample, λ is the incident wavelength and Δn is as defined earlier. Using an optical set-up as illustrated in Fig. 2, and, where the sample deformation axis is at 45° to the crossed polarisers, it can be shown[3] that R is related to the transmitted intensity, T, by

$$T \sim \sin^2 (\pi R) \qquad (4)$$

By monitoring T using a photomultiplier, R can be determined. Note that the sign of R is unknown since T varies with the sine squared of R; hence, one must determine the sign by another method. Also T is a multivalued function of R so that the 'order' of the retardation must be established via another route.

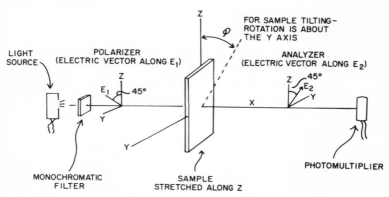

Fig. 2. Schematic of a typical experimental apparatus for measuring birefringence.

Although the transmission method suffers somewhat from its lack of sensitivity to very small changes in R, one great advantage of this technique is its application where high speed deformation is involved.[4,5] If one can monitor the thickness change with extension ratio then Δn can be calculated as a function of elongation.

The number of compensator methods are numerous and will not be discussed in detail here. The interested reader is referred to Refs. 6–8. Basically what is involved is that a known retardation is used to nullify or compensate the retardation induced by the sample. This amounts to putting some birefringent (anisotropic) material into the light path, e.g. a wedge or plate of quartz or calcite. By changing the thickness of such a material the degree of optical retardation can be controlled—recall eqn. (3).

Some of the advantages of the compensator methods are: they can be highly sensitive, they can determine the sign of the birefringence, and are conveniently used in conjunction with microscopy for determining localised birefringence (e.g. spherulitic). They generally are not suitable, however, for following instantaneous birefringence changes during a deformation process. In cases where biaxial orientation exists (blow-moulded films) or where a radial distribution of orientation may exist (fibres), the compensator methods are generally utilised.[9] The treatment of uniaxial systems is simplified in most cases since the anisotropy can be assumed to be cylindrically symmetric about the deformation axis thereby facilitating the measurements. In the case of biaxial orientation where one may be required to obtain two of the three birefringence parameters (eqn. (2)) use of more sophisticated methods may be required.[10] This also may be the case where skin-core effects exist as can occur in fibres.[9]

3.1.3 Types of birefringence
When the literature is surveyed a number of different terms arise to describe the type of birefringence observed. We will try to establish at the onset what some of these categories are and to comment on their origin and relative importance in polymeric materials.

(A) Orientation birefringence
This arises when there is a physical ordering of optically anisotropic elements (e.g. chemical bonds) along some preferential direction—see Fig. 3. This can occur in polymers by aligning amorphous or crystalline chains as by an extension or drawing deformation. Orientation birefringence is the quantity that is most generally measured (or desired)

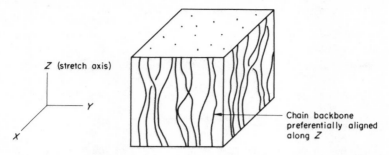

Fig. 3. Simple illustration depicting molecular orientation which will result in orientation birefringence.

when characterising orientation but the investigator must be aware that other origins of birefringence exist and may influence the overall value considerably.

Intrinsic birefringence, Δn°, is a special case of orientation birefringence and is, by definition, the maximum in orientation birefringence. It follows that the intrinsic birefringence will depend on whether one has an amorphous chain or a crystal. In fact, for a crystal, three intrinsic birefringence values exist due to the three dimensionality of a unit cell.[11] However, only two of these are independent.

(B) *Deformation birefringence*
This quantity may at times be confused in the literature with orientation birefringence. Deformation birefringence, however, can occur in a collection of optically isotropic particles as well as in anisotropic systems. One can picture, for example, that an axial dilatation or compression (*e.g.* hydrostatic force) could change the 'lattice' spacing as shown in Fig. 4 thereby resulting in a refractive index difference between axes Z and Y. Distortion of bond angles and/or bond lengths from equilibrium may also result in a finite birefringence without orientation (*e.g.* within glasses).

(C) *Form birefringence*
Until recently, this origin of apparent optical anisotropy had been generally neglected for birefringence measurements on polymeric solids. The phenomenon arises when the medium contains at least two phases each having a different refractive index and at least one dimension

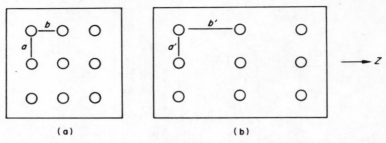

(a)   (b)

Fig. 4. Illustration showing how deformation birefringence may arise as a result of a change in packing caused by an external deformation.

comparable with the wavelength of light. To give an idea of the magnitude of this effect let us consider one of the classical cases treated by Wiener.[12] He showed that if one had an isotropic two component system as shown in Fig. 5, then a finite birefringence (of form origin) could result. Specifically, if the isotropic cylindrical rods (phase 1 having refractive index, n_1) are parallel and separated by the isotropic matrix material (phase 2 having refractive index, n_2) such that the intercylinder distance between surfaces is small relative to the wavelength of light, then the resulting *apparent sample* refractive indices parallel, n_a, and perpendicular, n_o, to the rod axes will be

$$n_a = \phi_1 n_1{}^2 + \phi_2 n_2{}^2 \tag{5}$$

$$n_o = \frac{n_2{}^2(\phi_1 + 1)n_1{}^2 + \phi_2 n_2{}^2}{(\phi_1 + 1)n_1{}^2 + \phi_2 n_1{}^2} \tag{6}$$

where the ϕ refers to the volume fraction of phases 1 and 2. Using eqns. (5) and (6) Folkes and Keller[13] have shown that the resulting form birefringence from such a system is given by

$$\Delta n = n_a - n_o = \frac{\phi_1 \phi_2 (n_1{}^2 - n_1{}^2)^2}{2n_a\{(\phi_1 + 1)n_2{}^2 + \phi_2 \phi_1{}^2\}} \tag{7}$$

where it is clear that Δn becomes zero when $n_1 = n_2$. It is also apparent that the sign of the form birefringence is positive whether n_1 is greater or less than n_2. Other geometries can also lead to form effects of equal importance.[12,14]

Recently, Folkes and Keller[13] showed that all the observed birefringence in an extruded cylinder of an ABA block copolymer was due to

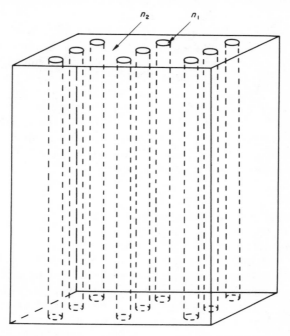

Fig. 5. Illustration showing one possible geometrical arrangement of isotropic (or anisotropic) elements that can lead to form *birefringence.*

form effects and not molecular orientation. Form effects have also been shown to contribute as much as ten per cent to birefringence in simple semi-crystalline polymers such as polyethylene.[15] The reader is therefore cautioned to be aware of this effect, particularly if one is dealing with composite or multicomponent systems where one of the components is highly dispersed and, where there is an appreciable refractive index difference between phases. By swelling the material with inert fluids having differing refractive index and which do not alter morphology, one can evaluate the degree of form birefringence.[14,15] At the point where the refractive index of the imbibing fluid matches that of the phases which do not swell (n_1 or n_2) there will be a minimum in the birefringence.† It is at this point that form birefringence is assumed to be zero.

† If the sample axes shown in Fig. 5 are rotated 90° within the plane of the paper then the form birefringence will be negative and hence in a swelling experiment the birefringence will pass through a maximum rather than a minimum.

The correlation of molecular orientation behaviour with birefringence data requires an understanding of the origin of inherent molecular anisotropy; in general, two approaches can be taken. The first of these is to approximate the polymer chain by a statistical segment model.[16,17] Assuming additivity of segment polarisabilities and using polymer chain statistics, one can calculate chain polarisabilities and their differences thereby giving the birefringence. The second approach is to try to treat the real chain by making use of additivity of chemical bond polarisabilities and chain statistics to determine the average spatial configuration of a single chain or groups of chains.[18,19] We will attempt to briefly review the characteristics of each of these treatments.

3.1.4 Statistical segment model

Although a hypothetical model, the statistical segment approach originated by Kuhn and Grün[16] has been useful for understanding the orientation and optical behaviour of a real chain. The initial approach is to consider that a linear polymer chain can be subdivided into cylindrical statistical segments, each possessing only two principal axes of polarisability—these being parallel, b_1, and perpendicular, b_2, to the segment axis. With this model and use of statistical procedures, one can show that for an undeformed chain there will be an inherent birefringence of $3/5(b_1 - b_2)$ for the undeformed chain as measured parallel to the displacement vector separating the chain ends. Constructing a network of such linear statistical chains can then be used as an analogue for a rubberlike polymer. With certain simplifying yet, many times reasonable assumptions, it has been shown that upon orientation of such a three-dimensional system, the individual chains tend to orient along the stretching direction. If the system has N_c network chains per unit volume one can show that the *orientation* birefringence is to a good approximation

$$\Delta n = \frac{2}{45} \pi \frac{(\bar{n}^2 + 2)^2}{\bar{n}} N_c (\lambda^2 - 1/\lambda)(b_1 - b_2) \qquad (8)$$

where $\bar{n}$ is the average refractive index of the system and λ is the extension ratio.

The above equation has found high applicability by combining it with the relationship for the true stress, σ_a, as derived from the Gaussian theory of rubber elasticity. Specifically this theory[20] leads to

$$\sigma_a = N_c kT(\lambda^2 - 1/\lambda) \qquad (9)$$

where k is the Boltzmann constant and T is the absolute temperature. Forming the ratio of eqns. (8) and (9) gives the well-known stress optical law

$$\text{SOC} \equiv \frac{\Delta n}{\sigma_a} = \frac{2}{45} \frac{\pi}{kT} \frac{(\bar{n}^2 + 2)^2}{\bar{n}} (b_1 - b_2) \tag{10}$$

which states that this ratio, commonly known as the stress optical coefficient (SOC), is independent of extension ratio and cross-link density and depends only on $(b_1 - b_2)$ and $\bar{n}$—both characteristics of the material. This equation has been applied as a means of characterising the degree of Gaussian rubber elasticity behaviour and, to obtain the optical anisotropy $(b_1 - b_2)$. This latter term, in conjunction with the theoretically calculated optical anisotropies of the repeat unit, has been used as a measure of the stiffness of the chain.[20,21]

In conjunction with the stress optical law and use of eqn. (10), it is noteworthy to point out that this relationship is based on the Gaussian theory of rubber elasticity and assumes equilibrium thermodynamics. In practice the approach to an equilibrium state following deformation or a temperature change is hastened by a swelling medium and in fact Gaussian behaviour is best illustrated when the volume fraction of rubber is 0·2 or less. However, the value of the SOC may vary although it remains constant for a given swelling medium. Recent studies by Gent,[22] Fukuda et al.[23] and Ishikawa and Nagai[24,25] have shown that this apparent anomalous effect is related to the molecular optical anisotropy of the swelling medium and that the SOC reaches its correct value when the swelling agent is optically isotropic, e.g. carbon tetrachloride— see Fig. 6.

In conjunction with the above equations it can be shown that if N_c, $\bar{n}$ and $(b_1 - b_2)$ are known, one can use birefringence to determine the average angle, ϕ, that a segment makes with the stretch axis. Specifically, to a good first approximation

$$\Delta n = \frac{2\pi}{9} \frac{(\bar{n}^2 + 2)^2}{\bar{n}} N_c (b_1 - b_2) \left(\frac{3\langle \cos^2 \phi \rangle - 1}{2} \right) \tag{11}$$

where, as in the X-ray section, the last bracketed term is identically equal to the Herman's orientation function. It therefore follows that birefringence provides only the second moment of the orientation distribution as do the dichroism or sonic techniques.

The above discussion has been only concerned with homopolymer

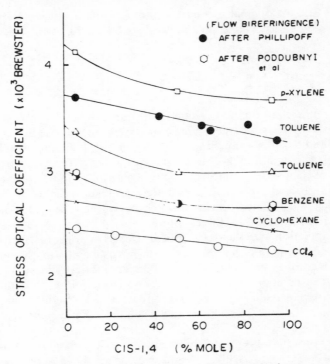

Fig. 6. The variation of the stress optical coefficient with cis *content for 1,4-polybutadiene rubbers swollen with a number of solvents. The data include streaming birefringence data of Poddubnyi et al.[26] and of Phillipoff.[27] (From Ref. 23.)*

systems. In light of the growing number of copolymeric materials (block, graft, etc.) as well as different rotational isomer states within a given homopolymer, it is reasonable to suspect that more than one statistical segment may be required in order to interpret the optical behaviour. Shindo and Stein have extended the Kuhn and Grün treatment to multicomponent systems and have shown that the multisegment consideration is indeed useful.[17] One important result obtained is that the degree of orientation of a statistical segment depends on the square of its length. Since the length is related to stiffness, one can understand the ramification of this result with respect to the orientational behaviour of multicomponent systems.

3.1.5 Anisotropy and the real chain

The utilisation of the parameters of the real chain and its repeat units in trying to analyse the inherent anisotropy of a chain or, in fact, the anisotropy of a deformed network of chains, is clearly non-trivial. To approach this problem one requires some knowledge of the repeat unit structure and the bond polarisabilities of its constituents. One also needs to know the distributions and energy considerations of different rotational isomers (if different ones exist) since these will affect the overall chain configuration and anisotropy. One *generally* assumes that bond polarisabilities are additive and no interactions (internal field effects) occur between units of an amorphous chain. As can be easily realised, to account for all parameters completely and accurately is difficult, particularly in complex systems where polar interactions (*e.g.* hydrogen bonding) affect conformation and neighbour–neighbour packing arrangements. The result is that the mathematics for treating the real chain become considerably complex and require the use of digital computer calculations. There has been, however, a good degree of success in this approach and the interested reader should consult Refs. 18 and 19 for further details.

3.1.6 Birefringence of multicomponent systems

If we assume additivity of polarisabilities it seems also reasonable to think of additivity of birefringences arising from difference phases, *i.e.*

$$\Delta n_\mathrm{T} = \Delta n_\mathrm{f} + \sum_i \phi_i \Delta n_i \tag{12}$$

where i refers to the ith component, ϕ_i its volume fraction and Δn_f is that part of the total birefringence, Δn_T, that arises from form or deformation effects rather than orientation. As an example, for a semi-crystalline polymer we could write eqn. (12) as

$$\Delta n_\mathrm{T} = \Delta n_\mathrm{f} + \phi_\mathrm{am} f_\mathrm{am} + \phi_\mathrm{cry} f_\mathrm{cry} \tag{13}$$

where the subscripts 'am' and 'cry' refer to the amorphous and crystalline components. This equation can also be written (as can eqn. (12)) in the form

$$\Delta n_\mathrm{T} = \Delta n_\mathrm{f} + \phi_\mathrm{am} f_\mathrm{am} \Delta^\circ n_\mathrm{am} + \phi_\mathrm{cry} f_\mathrm{cry} \Delta^\circ n_\mathrm{cry} \tag{14}$$

where $\Delta^\circ n$ terms refer to the intrinsic birefringence and f the orientation function of the individual components. Since a crystal has three axes all of which have an intrinsic birefringence, generally $\Delta°n_{\text{cry}}$ refers to the axis of the crystal being most parallel to the chain axis since it is this axis which generally aligns along the stretch direction during orientation. If, however, biaxial orientation exists, particular caution must be taken due to planar orientation effects which may influence the value of $\Delta°n_{\text{cry}}$ considerably.

From eqn. (14), if one can specify the values of $\Delta°n, \Delta n_f$ and one of the orientation functions, f, then the other may be determined by measuring Δn_T. This procedure has been used for semi-crystalline systems where X-ray diffraction is used to determine f_{cry} of all crystalline axes. Thus one can determine f_{am}, a quantity which, in general, is not easy to obtain from two-component systems. The results plotted in Fig. 7 are an example of the data obtained by combining both wide-angle X-ray diffraction and birefringence to uniaxially drawn low density polyethylene.[28] It is clear

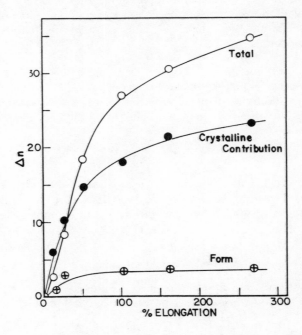

Fig. 7. Birefringence as a function of percentage elongation for DuPont Alathon-2 Polyethylene. The amorphous and crystalline contributions to the total birefringence have been determined by using both birefringence and X-ray methods. (From Ref. 28.)

from this figure that the degree of amorphous orientation is not as large as that of the c-axis of the crystal. Another application of eqn. (13) could be for amorphous copolymer systems, *e.g.* block or graft copolymers. Let us represent the birefringence from a two-block system as

$$\Delta n_T = \Delta n_f + \phi_A f_A \Delta°n_A + \phi_B f_B \Delta°n_B \tag{15}$$

where we recall that the Δn_f term may be of considerable importance if the refractive indices of the individual components (A and B) are considerably different.[13] If Δn_f is known along with ϕ_A, $\Delta°n_B$, $\Delta°n_B$ then f_A or f_B can be obtained if one of the two can be determined by another method. Since we stated that the system was amorphous, a method other than X-ray diffraction would be required to obtain one of the f values, *e.g.* dichroism measurements—see section on dichroism. In fact, if all values of f could be obtained by the dichroism method, the values would likely be more precise than those calculated by the above procedure since this approach requires accurate determination of Δn_f, $\Delta°n_A$ and $\Delta°n_B$.

In view of the commercial growth of multi-phase systems based on additives, fillers, etc., it should be pointed out that birefringence techniques have been utilised to interpret the points of stress concentration about filler particles. Specifically, Kotani and Sternstein[29] have developed a comprehensive theory to explain inhomogeneous swelling in filled elastomers. Their theory predicted that asymmetry in the stress tensor existed for points in the matrix media near the boundaries of a filler particle and found the factors on which this asymmetry depended. Correlated with the asymmetric stress tensor is an asymmetric polarisability tensor. Kotani and Sternstein also verified the initial theory by making local birefringence measurements via microscopy near the interface region.

3.1.7 Concluding remarks

We have attempted to present the basic principles of the origin and measurement of birefringence. It was not our intention to go into great depth on any of the presented material and therefore the reader pursuing further information should consult the references given in the text. It should be pointed out that two areas where the use of birefringence techniques are important, yet which were not discussed here, are: (1) dynamic birefringence and (2) relaxation birefringence. The former method amounts to monitoring both the birefringence and the stress (instantaneously) while a sample is oscillated at a known small strain amplitude and at a fixed frequency.[30,31] These measurements lead to information regarding the time response of orientation by the components of the system, e.g. crystal and amorphous material. This technique has been combined with dynamic X-ray diffraction and dynamic light scattering to aid in the separation of component orientation. Furthermore, phenomenological theories correlating birefringence measurements with molecular parameters have been developed and applied.

Relaxation birefringence is meant to imply the measurement of birefringence during stress relaxation of a deformed material. The typical application has concerned rubbers and rubberlike materials where the objective is to test the degree of ideal Gaussian rubber elasticity by utilising the stress optical law given earlier in eqn. (10).

3.2 WIDE ANGLE X-RAY DIFFRACTION

3.2.1 Introduction

Wide angle X-ray diffraction is another technique which has been of vital importance in the understanding of oriented polymers. In this section we shall illustrate the usefulness of the X-ray method for characterising the crystalline orientation in polymers. It is intended that the discussion will be mostly descriptive, and the reader who wishes a review of the principles of measuring the intensity of diffracted X-ray radiation is referred to Refs. 32–34.

As with low molecular weight compounds, crystalline polymers may be characterised by specifying the unit cell parameters for the crystal lattice. Two examples of unit cell configurations are shown in Fig. 8 for polyethylene and Nylon 6,6. In each, the c-axis lies parallel to the chain axes; this is not true for all polymer crystals. Planes may be drawn through the crystal cell at various angles with respect to the axes and these planes are specified by the Miller indices (hkl) which are derived by forming the reciprocal of the intercepts of a given plane with the a, b and c-axes. These reciprocals are then reduced to the smallest integer values resulting in the Miller indices for that set of planes—see Fig. 9. When considering the orientation of the polymer chain, it is imperative to know the chain direction in the unit cell in order to obtain meaning from the diffraction data. However, if only the orientation of certain planes is desired, e.g. shear planes, one needs to know only the characteristics of the unit cell.

Rather than emphasising the determination of crystal structure we shall

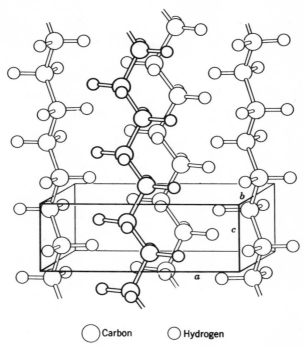

○ Carbon   ○ Hydrogen

Fig. 8(a). Unit cell structure of polyethylene (after Bunn[35,36]).

be concerned with the use of diffraction for the characterisation of the anisotropy of semi-crystalline polymeric materials or more specifically to characterise the orientation of the various crystallographic axes *a*, *b* and *c* with respect to some specified set of sample axes *X*, *Y* and *Z* as shown in Fig. 10. At the onset it should be made clear that any induced anisotropy (by drawing, rolling, etc.) in polymeric solids is never perfect, that is to say there is really a distribution of orientations that would be found if one could look at all unit cells in the solid. This is particularly apparent when one considers the vector form of the Bragg law[32]

$$H_{hkl} = \frac{s' - s_o}{\lambda} = \frac{s}{\lambda} \tag{16}$$

where H_{hkl} is a reciprocal lattice vector that can be shown to be normal to a given set of (*hkl*) planes and has a magnitude equal to the reciprocal

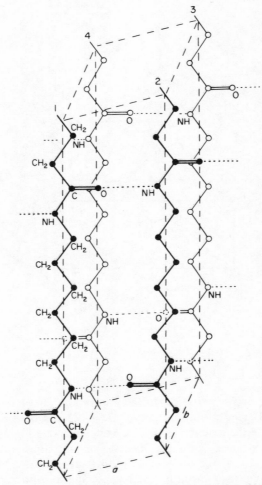

Fig. 8(b). Unit cell structure of Nylon 6,6 (after Bunn[35,36]).

of the interplanar spacing d. The vectors s' and s_0 are unit vectors along the incident and diffracted beams respectively while λ is the wavelength of the beam. From this equation it follows that for diffraction from a given set of (hkl) planes, the normal to those planes must have the direction and magnitude of s and is therefore parallel to ρ_{hkl}. This strict requirement implies that for a distribution of unit cell orientations

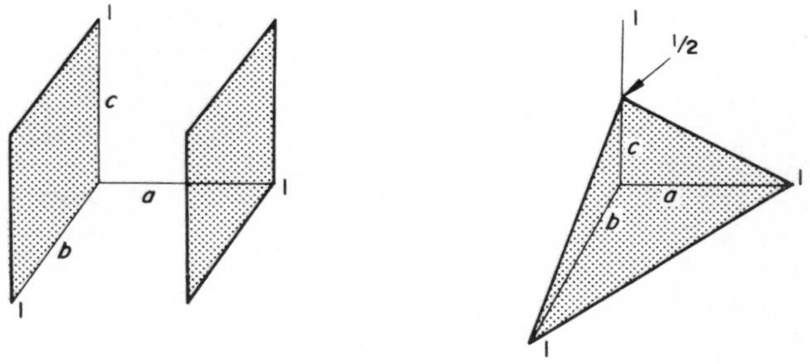

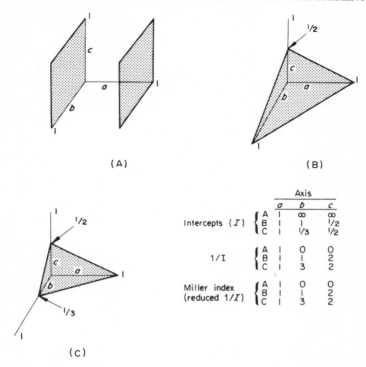

(A) (B)

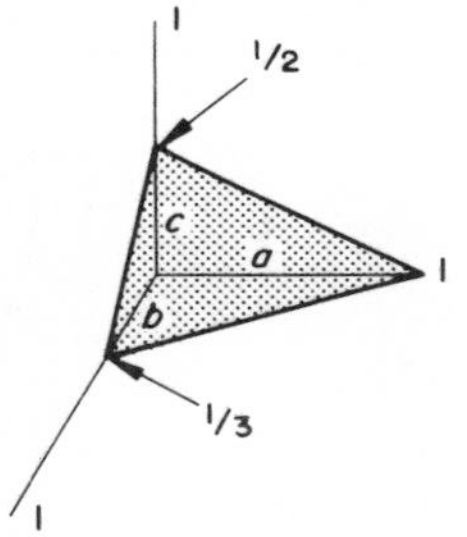

(C)

		Axis		
		a	*b*	*c*
Intercepts (*I*)	A	1	∞	∞
	B	1	1	1/2
	C	1	1/3	1/2
1/I	A	1	0	0
	B	1	1	2
	C	1	3	2
Miller index (reduced 1/*I*)	A	1	0	0
	B	1	1	2
	C	1	3	2

Fig. 9. Cartesian co-ordinate system illustrating how the Miller indices are determined for particular reflecting planes.

very few will diffract at any specified $\boldsymbol{s}$ and serves as the basis for allowing a critical evaluation of crystalline orientation. This is seen in Fig. 11 which shows that for the chosen (*hkl*) planes, diffraction will occur only when the (*hkl*) normal ($\boldsymbol{\rho}_{hkl}$) is oriented as shown. Therefore to obtain the diffraction (and orientation distribution) from a given set of (*hkl*) planes, at a fixed Bragg angle, θ, one must either rotate the sample about the angles χ and ω while holding the beam and detector fixed or vice versa—clearly the former is easier and is the method practised. Depending on the symmetry of the orientation, simplifications are possible, *e.g.* for uniaxial symmetry, as in fibres, only the angle χ need be scanned, *i.e.* only an azimuthal dependence of the diffraction is required. In cases of biaxial stretching (*e.g.* blow moulding or rolling) both χ and ω scans are required.

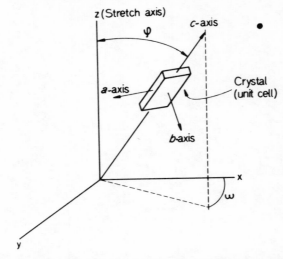

Fig. 10. Schematic illustrating the spatial orientation of a unit cell relative to the sample axes x, y and z. Other unit cells in the sample would generally have other orientations.

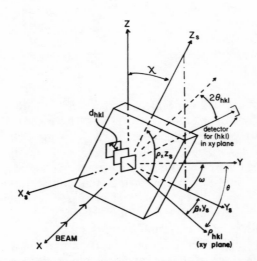

Fig. 11. Schematic illustrating the principles of a diffraction experiment. Sample co-ordinate system is labelled X_s, Y_s and Z_s while the fixed laboratory co-ordinates are X, Y and Z. ρ_{hkl} is the normal to the hkl planes in question.

3.2.2 Experimental techniques

In practice, several very elaborate experimental devices exist for studying orientation. There are also some very simple procedures that can be applied but usually with far less quantitative accuracy. Because of their simplicity we will consider the latter methods first.

When a crystalline polymer is oriented, the random circular film pattern (random orientation) transforms to a collection of defined reflection arcs that are correlated with particular (hkl) planes that can be identified based on the crystal structure and Bragg relationship (see Fig. 12a–b). It follows that the magnitude of the azimuthal spread ($\chi/2$) of these reflections is indicative of the degree of orientation. (The breadth, κ, of the reflection is related to crystal size and imperfection—see Ref. 32.) Also, the location of the reflection with respect to the sample axes indicates the orientation of the crystallographic planes. For example Fig. 5(a) and (b) show two X-ray photographs of polyethylene that had been cold rolled. From the (200) reflection in sample (a) one sees that the a-axis is aligned preferentially normal to Z whereas in (b) there are two distinct orientations of the a-axis—one along Z and one normal to this.

The single diffraction pattern method is in many cases suitable for uniaxially deformed materials. Where biaxial deformation occurs, however, more than a single film pattern may be necessary to define the orientation to any good approximation.[37] X-ray patterns taken with the beam along the respective sample axes (X, Y, Z) may be required. (In many cases only two patterns are necessary.) One does this by simply cutting and stacking pieces of the material so that a sufficient quantity is available to cover the X-ray beam. Figure 13 gives an example of X, Y, and Z-axis patterns taken from polyethylene that had been cold rolled along Z. Biaxial orientation is apparent since the Z-axis (roll axis) pattern shows an azimuthal dependence of the diffraction from the various (hkl) planes.

While photographic techniques may allow one to obtain some average orientation values for a deformed crystalline polymer[38] there is need for a quantitative measure of this distribution. In general, the distribution in orientation is determined for a single (hkl) plane—usually a $(h00)$, $(0k0)$ or $(00l)$ plane if sufficient diffraction exists. The data are then presented either in a pole figure or may be used to determine the Herman's orientation function defined as

$$f_{hkl,\,i} = \frac{3\langle\cos^2\phi\rangle_{hkl,\,i} - 1}{2} \tag{17}$$

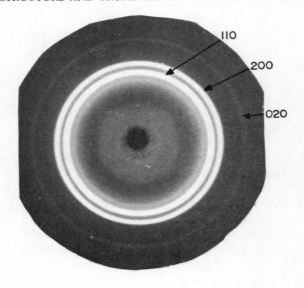

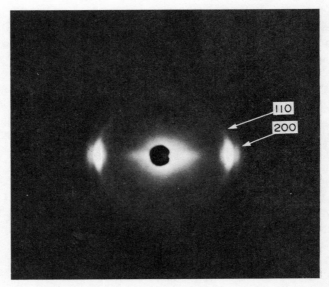

Fig. 12. Wide angle X-ray patterns of (a) undeformed high density polyethylene, (b) cold rolled high density polyethylene ($\lambda = 5$)—beam directed along the X-axis of Fig. 13.

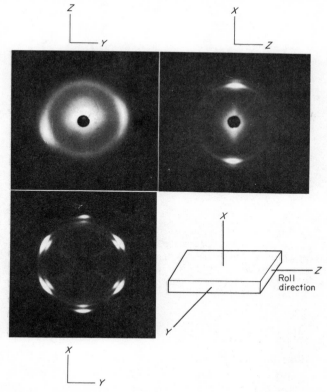

Fig. 13. Wide angle diffraction patterns taken on a sample of low density polyethylene that had been cold rolled along Z to an extension ratio of 4 and then annealed without restraint. X-ray beam was directed along the respective axis indicated with an arrow.

where $\langle \cos^2 \phi \rangle_{hkl, i}$ is the average angle that the normal to the set of (hkl) planes makes with some specified ith axis—usually the principal deformation direction of the sample. For perfect alignment ($\phi = 0°$), $f_{hkl, i}$ is $+1$; for perpendicular alignment ($\phi = 90°$), $f_{hkl, i}$ is $-1/2$. For random orientation $\langle \cos^2 \phi \rangle_{hkl, i}$ is $1/3$ and hence $f_{hkl, i}$ is zero.

Although f is a convenient index to characterise orientation, one must note that it is only an average value—specifically the second moment of the whole orientation distribution for the specific set of (hkl) planes utilised. Hence, two different orientation distributions could exist having

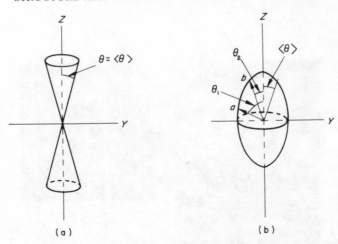

Fig. 14. Two different hypothetical uniaxial orientation distributions: (a) all chains lie at an angle θ with respect to the Z-axis; (b) a more realistic ellipsoidal distribution. Here the length of any line a, b, . . . is proportional to the number of chains lying at $\theta_1, \theta_2 \ldots$ to the Z-axis. The average angle over this distribution could lead to the same average angle as in distribution (a), yet, the overall distribution is very different.

the same value of $\langle \cos^2 \phi \rangle$ and therefore identical f's. For an example of this inspect Fig. 14. To separate these distributions and more fully characterise, the orientation envelope would require higher moments of the total distribution and, in fact, all moments would be necessary for a complete description. What is necessary for *any* moment to be calculated? In general such a calculation requires the evaluation of the following

$$\langle \cos^n \phi \rangle = \frac{\int \cos^n \phi F(\phi) \sin \phi \, d\phi}{\int F(\phi) \sin \phi \, d\phi} \tag{18}$$

where we have expressed the equation for the nth moment in terms of $\cos \phi$. $F(\phi)$ is the probability distribution function for the normals of the (hkl) plane in question and expressed also in terms of ϕ. Just how can we obtain this function $F(\phi)$? Recall that the diffracted intensity, as measured by a detector at a fixed Bragg angle (θ), depends on the concentration of unit cells being in a *suitable position* for diffraction for a given set of the angles (χ, ω) where χ and ω are the azimuthal and rotation angles of the *sample* relative to some fixed axis and plane—see Fig. 11. From this figure the value of $F(\phi)$ will be proportional to $I(90-\chi,\beta)$ where I refers to the diffracted intensity and in the context of Fig. 11, angle ϕ is $(90-\chi)$.[†] Utilising Fig. 11, eqn. (19) becomes

$$\langle \cos^2 \rangle \phi \rangle = \frac{\displaystyle\int_0^{2\pi} \int_{-\pi/2}^{\pi/2} I(\pi/2-\chi,\beta)\cos^2(\pi/2-\chi)\sin(\pi/2-\chi)\,d\chi\,d\beta}{\displaystyle\int_0^{2\pi} \int_{-\pi/2}^{\pi/2} I(\pi/2-\chi,\beta)\sin(\pi/2-\chi)\,d\chi\,d\beta} \quad (19)$$

For uniaxial deformation one can generally assume symmetry about the deformation axis and therefore neglect the integration over β. It now is obvious that the complete distribution of normals to some (hkl) plane can be viewed simply as a polar plot (or stereographic plot) of $I(\chi,\omega)$;[‡] such a plot is called a pole figure and was first introduced by Decker and Harker[39] for studying orientation in cold rolled metals. The reader can easily picture the physical meaning of a pole figure by recalling Fig. 11 which gives the geometrical considerations of diffraction.

Examples of pole figures for oriented polyethylene are given in Fig. 15(a) and (b) where the contours are isointensity lines; higher values imply a higher concentration of (hkl) normals. Figure 15(a) shows that the (200) or a-axis is preferentially oriented along Y (transverse direction) while the (020) b-axis (Fig. 15(b) is oriented along X (thickness direction). Recalling that the polyethylene crystal is of orthorhombic symmetry requires that the c-axis (normal to an 001 plane) must therefore preferentially lie along Z (stretch axis).

The general method used today for collecting intensity data for orientation determination is to use a pole figure device such as that shown in Fig. 16. Using both Figs. 11 and 16 the reader can picture how the $I(\chi,\omega)$ data are collected. Due to the development of reliable computer controlled interface systems, manual control systems are rapidly becoming obsolete.

3.2.3 Inverse pole figures
Although the use of pole figures allows one to obtain considerable information regarding the orientation of specific crystalline *axes* or *planes*,

---

[†] ω is the tilt angle between the transverse direction (Y_s) of the sample and the fixed Y-axis. The angle between the (hkl) normal and the transverse direction (Y_s) is given by Fig. 11 to be β, i.e. θ-ω. This shows again that the (hkl) planes must be tilted at the Bragg angle, θ, for diffraction to occur.

[‡] In the context of Fig. 11 one would plot $I(90-\chi,\beta)$ to obtain the polar plot of the orientation of the (hkl) normals.

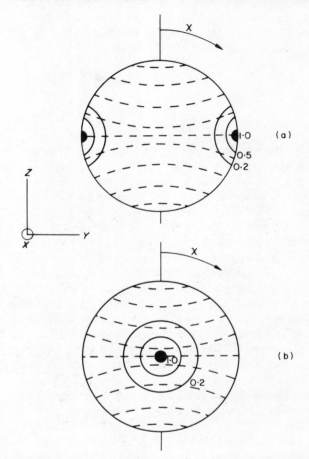

Fig. 15. Hypothetical wide angle X-ray pole figure—expressed for diffraction from a (h00) and (0k0) plane. The numbers indicate the relative density of hkl normals and therefore illustrate that (a) the a-axis is aligned along the Y-axis and (b) the b-axis along X.

there is some degree of inconvenience as well as inaccuracy in deciphering the *unit cell* orientation with respect to a chosen reference axis. Generally the degree of inconvenience is relatively minor and the degree of inaccuracy is small so that pole figure data suffice unless considerable emphasis is directed at obtaining very quantitative orientation data.

Fig. 16. One example of X-ray instrumentation for measuring X-ray orientation. (Courtesy of Philips Corp.)

When higher accuracy is desired one can resort to using inverse pole figures.[40] The term 'inverse' is used since in constructing such a figure, one converts the orientation distribution of the poles of specific planes (determined with respect to some reference axis) to an orientation distribution of the reference axis with respect to a fixed position of the crystallographic axes of the unit cell. To do this accurately, however, requires the determination of pole figures from *several* (*hkl*) planes, the number depending on the degree of crystallographic symmetry. This method has been discussed in detail for oriented metals[41] and has also been applied to oriented polymers.[42]

3.2.4 Orientation function diagrams

So far we have found that the orientation of a given crystal axis, i, with respect to some reference axis, j, may be specified by use of the Herman's orientation function, f_{ij}, or, more completely by a pole figure. In many cases one not only desires to note how a given axis changes with some variable (*e.g.* temperature, extension ratio) but also how all other axes change. Certainly, pole figures for three axes as a function of some variable would contain all the data but might prove cumbersome in presentation. For this reason the use of the values of f_{ij} may be utilised and plotted in an orientation diagram.[43]

Consider Fig. 17 which shows an equilateral triangle having a point, P, within its perimeter. If we allow the line that bisects each apex angle and which extends to the opposing side to have a length of unity, then the distance along or parallel to each of these bisectors can be related to

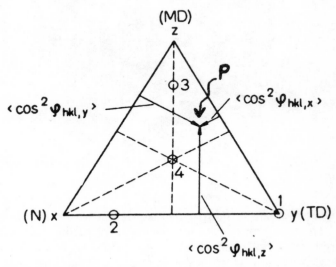

Fig. 17. Hypothetical equilateral triangle orientation plot—discussed in text.

the value of $\langle \cos^2 \phi \rangle$—the parameter used in the orientation function f. The length of δ_z would be equal to $\langle \cos^2 \phi \rangle_{i,z}$ where i refers to the ith crystal axis and Z refers to the reference axis. Similar definitions follow for δ_x and δ_y. By the law of cosines

$$\delta_x + \delta_y + \delta_z = \langle \cos^2 \phi \rangle_{i,x} + \langle \cos^2 \phi \rangle_{i,y} + \langle \cos^2 \phi \rangle_{i,z} = 1 \qquad (20)$$

Therefore, once two values are determined the third is specified. Hence, a single point on such a triangular plot defines the second moment orientation parameter of a given crystal axis with respect to all three sample axes (or some other reference cartesian co-ordinate system). To further establish the meaning of this method, consider points 1–4 shown in Fig. 17, each of which represents a particular orientation of some (hkl) normal with respect to axes X, Y, and Z. Point 1 refers to the case where there is perfect alignment along the Y-axis of the normal to a chosen (hkl) plane. Point 2 represents preferential orientation about the X-axis relative to the Y-axis and no orientation along Z. Point 3 represents preferred orientation along Z and with uniform (uniaxial) distribution about this same axis. We note, for any axis about which the orientation is uniaxial, that the location of the point describing the orientation will fall on a line which bisects the apex angle associated with the preferred axis. Point 4 falls at the centre of the triangle (intersection of the three lines bisecting the apexes) and therefore refers to random orientation, *i.e.*

$$\langle \cos^2 \phi \rangle_{i,x} = \langle \cos^2 \phi \rangle_{i,y} = \langle \cos^2 \phi \rangle_{i,z} = 1/3 \qquad (21)$$

In some crystalline systems where there are no reflecting planes normal to the desired crystallographic direction, one can still obtain the orientation for this direction providing the crystalline structure is defined. This method has been developed by Wilchinsky,[44] where he showed that by measuring the orientation functions associated with various known (hkl) planes, one can calculate the value of the desired orientation function. This method should not be confused with the application of the direction cosines relationship as discussed earlier. A specific example is given for the case of polypropylene in eqn. (22) where the value of $\langle \cos^2 \phi \rangle_{001}$, can be found (and therefore the orientation of the chain axis) by measuring the diffraction from the (110) and (0k0) planes.[44]

$$\langle \cos^2 \phi \rangle_{001,z} = 1 - 0 \cdot 91 \langle \cos^2 \phi \rangle_{0k0,z} - 1 \cdot 099 \langle \cos^2 \phi \rangle_{110,z} \qquad (22)$$

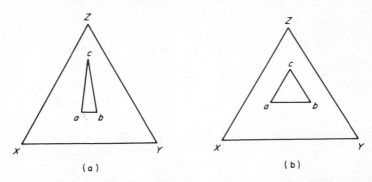

Fig. 18. Equilateral triangle orientation plots of sample data—discussed in text.

One can apply the above procedure for constructing orientation diagrams to all three crystallographic axes if data are available. The result leads to triangular plots as shown in Fig. 18(a) and (b) where Fig. 18(a) shows preferential orientation of c along Z while the a and b-axes are equally oriented along X and Y implying a uniaxial orientation of c as might occur upon drawing a fibre of polyethylene, Fig. 18(b) on the other hand shows c to lie preferentially along Z while a is along X and b is along Y. This behaviour might occur during biaxial orientation. In cases where orthogonality exists for the crystallographic axes, only two axial orientation values are independent, the third value for any reference axis, i, following from geometrical consideration that

$$\langle \cos^2 \phi \rangle_{a,j} + \langle \cos^2 \phi \rangle_{b,j} + \langle \cos^2 \phi \rangle_{c,j} = 1 \qquad (23)$$

or expressed in terms of orientation functions

$$f_{a,j} + f_{b,j} + f_{c,j} = 0 \qquad (24)$$

3.2.5 Concluding remarks

We have attempted to present a basic introduction to the use of wide angle X-ray diffraction in studying the orientation in polymer materials. We have not, however, discussed some of the newer or more novel approaches utilising this same method. Specifically, the use of dynamic X-ray diffraction[45] and crystalline orientation relaxation as studied by high speed X-ray detection[46] has not been included. These two methods are still in the development stages but they have already been shown to contribute to our understanding of the time response of crystal orientation during deformation or during oscillatory strain. For complete details, the interested reader should consult the references given above.

3.3 SCATTERING TECHNIQUES: LOW ANGLE X-RAY SCATTERING AND LIGHT SCATTERING

3.3.1 Géneral introduction

The application of scattering techniques to studies of orientation in polymers forms an important subject in its own right. The use of low angle X-ray scattering is a very well-established technique, and is often used as a complementary tool to wide angle X-ray diffraction to define the structure and morphology of oriented crystalline polymers. Light scattering from solid polymers is of more recent origin, but promises to be extremely valuable also. Because of the close theoretical relationship between these two techniques, it is desirable to provide a general theoretical introduction, and then present a more descriptive account of the two methods in subsequent separate sections. The reader who is only interested in low angle X-ray diffraction will, however, find the section dealing with this topic wholly self-contained, and it may therefore be read very largely without reference to the theoretical introduction.

3.3.2 General theory of scattering

(A) *Physics of scattering*[47]

An electromagnetic wave consists of an electrical and a magnetic field which vary periodically with position and time, and which travel through space at a velocity $c = 3.0 \times 10^{10}$ cm/s (in vacuum). The electrical field strength may be given by:

$$E = E_0 \exp\left[i(\omega t - kd)\right] \tag{25}$$

where $\omega = 2\pi v$, v is the frequency, t the time and k the wave number given by $2\pi/\lambda$, where λ is the wavelength in the medium. The distance, d, is that which the wave travels from some arbitrary reference plane as shown in Fig. 19. When such a wave passes through matter, the atoms become polarised leading to an induced dipole moment:

$$\mathbf{m} = |\alpha|\mathbf{E} \tag{26}$$

where $|\alpha|$ is the polarisability tensor and is a measure of the electronic mobility in the matter. For isotropic matter $|\alpha|$ is a scalar and $\mathbf{m}$ and $\mathbf{E}$ are collinear.

Consequently, $\mathbf{m}$ varies with time and changes essentially in phase with $\mathbf{E}$ if the frequency of the light is sufficiently below the resonant frequency for

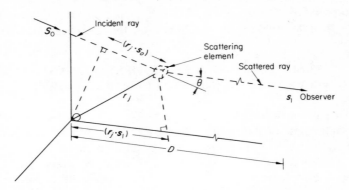

Fig. 19. The optical path length for a ray scattered from the jth volume element.

electronic motion. Such an oscillatory dipole serves as a source of electromagnetic radiation having a scattered amplitude given by:[47]

$$E_s = \frac{-\ddot{m}}{c^2 r} \cos \gamma = \frac{\alpha \omega^2 \cos \gamma}{c^2 r} E_0 \exp\left[i(\omega t - kd)\right] \qquad (27)$$

where r is the distance from the scattering system to the observer and γ is the angle between the dipole and the plane normal to the scattered ray. The scattered amplitude decreases with distance from the scattering centre and increases with ω^2 for this case of Rayleigh scattering. This leads to the blueness of the scattered light from the sky since the higher frequency blue light is scattered more than the lower frequency red components. For higher frequency radiation such as X-rays, inertial contributions of the electrons must be considered and the resulting phase lag leads ultimately to an independence of scattered amplitude upon frequency for the case of Thompson scattering.

For isotropic systems, the plane of polarisation of the scattered light is parallel to that of the incident light and its amplitude depends upon the cos γ value which describes the variation of the component of the incident amplitude along the direction of polarisation of the scattered ray. For unpolarised incident light, the relative scattering of the polarised components depends upon the scattering angle θ. At $\theta = 0°$ both components are scattered equally while at $\theta = 90°$, only one component is scattered leading to plane polarisation for isotropic media. For anisotropic media, where the induced dipole moment m is not necessarily collinear with E, the 90° light will not be completely polarised and the depolarisation serves as a measure of the anisotropy. Thus, the light from the blue sky originating primarily from the scattering by anisotropic nitrogen and oxygen molecules, is partially polarised.

For a system consisting of a collection of particles, each of which is small as compared with the wavelength of light, the total scattered amplitude is the sum of the amplitudes from the particles and is:

$$E_s = \sum_j (E_s)_j = \frac{\omega^2 \cos \gamma}{c^2 r} E_0 \sum_j \alpha_j \exp\left[i(\omega t - kd_j)\right] = K_1 \sum_j \alpha_j \exp\left[-kd_j\right] \quad (28)$$

where α_j is the polarisability of the jth particle and d_j is the optical path length for rays scattered from this particle. It is assumed that the differences in path length are small compared with r so that a common r can be used for all of the particles. It is also assumed that the electric field acting on each particle is the external field E_0. For concentrated or condensed systems, the field must be modified to account for the local fields of neighbouring particles and must be multiplied by an internal field factor K_j to correct for this effect.

The factor K_1 of eqn. (28) is given by:

$$K_1 = \frac{\omega^2 \cos \gamma}{c^2 r} E_0 \, e^{i\omega t} \quad (29)$$

The optical path length seen in Fig. 19 may be given by:

$$d_j = D + (r_j \cdot s) \quad (30)$$

where r_j is a vector from an arbitrary origin to the scattering particle, and s is a vector defined by:

$$s = s_o - s_i \quad (31)$$

where s_o and s_i are unit vectors along the incident and scattered rays. D is an arbitrary phase factor dependent upon the location of the origin.

The scattered intensity is related to the amplitude by:

$$I_s = \frac{c}{4\pi}(E_s \cdot E_s{}^*)$$
$$= K_2(F \cdot F^*) \quad (32)$$

where F is the structure factor given by:

$$F = \sum_j d_j \exp\left[ik(r_j \cdot s)\right] \quad (33)$$

F^* is the complex conjugate of F and

$$K_2 = \frac{\omega^4 \cos^2 \gamma}{4\pi c^3 r^2} E_0{}^2 \quad (34)$$

The scattered intensity is often described in terms of the Rayleigh ratio given by:

$$\mathscr{R} = \frac{I_s r^2}{I_0 V} \tag{35}$$

where I_0 is the incident intensity $(c/4\pi)E_0^2$ and V is the scattering volume of the system.
Then:

$$\mathscr{R} = K_3 F F^* \tag{36}$$

where

$$K_3 = \frac{\omega^4 \cos^2 \theta}{c^4} \tag{37}$$

The summation for F is over a unit volume of scattering material.

The structure factor F relates the structure of the scattering system to the angular dependence of scattering and depends upon the co-ordinates of the scattering atoms or centres, d_j. Two approaches to the evaluation of the scattered intensity are (a) the model approach and (b) the statistical approach. The model approach involves the evaluation of F by assuming that the scattering atoms are arranged in space in some definite way so as to constitute a scattering object having a defined structure; whereas, the statistical approach evaluates the product FF^* directly in terms of statistical parameters characterising the distribution of matter in the system.

(B) *Particulate scattering*

For scattering elements which are arranged so that the entire collection takes the form of a larger particle of definite shape, it is convenient to replace the sum over these units by an integral over the volume of the particle. Thus, eqn. (33) becomes:

$$F = \int \alpha(r) \exp\left[ik(r \cdot s)\right] dr \tag{38}$$

where $\alpha(r)$ is the polarisability of the three-dimensional volume element dr. The integral represents a three-dimensional integral over dr and requires a knowledge of how $\alpha(r)$ varies with r for its evaluation. This integral over dr represents a triple integral over the magnitude and orientation angles of r.

For spherically symmetrical systems, $\alpha(r)$ depends only upon r, the scalar distance of the volume element from the origin and eqn. (38) reduces to:

$$F = 4\pi \int_{r=0}^{\infty} \alpha(r) \frac{\sin hr}{hr} r^2 dr \tag{39}$$

where $h = ks = (4\pi/\lambda)\sin(\theta/2)$

(C) *Scattering from spheres*[48]

For a uniform sphere, $\alpha(r) = \alpha_0$ for $r \leq R$, the radius of the sphere and is zero for larger r. Thus:

$$F = 4\pi\alpha_0 \int_{r=0}^{R} \frac{\sin hr}{hr} r^2 dr \tag{40}$$

$$= V_s \Phi(U)$$

where $V_s = (4/3)\pi R^3$ is the volume of the sphere and $\Phi(U)$ is the sphere scattering function:

$$\Phi(U) = [3/U^3][\sin U - U \cos U] \tag{41}$$

and $U = hR = 4\pi(R/\lambda)\sin(\theta/2)$

This gives rise to a variation of scattered intensity with angle shown in Fig. 20. The intensity is a maximum at $\theta = U = 0$ and at first decreases with increasing U or θ. The reason for this variation is apparent from Fig. 21. For scattering at $0°$, all rays travel the same path length and constructively interfere, leading to an intensity maximum. With an increase in θ, the path lengths become different at a rate which depends upon (R/λ). The bigger the particle, the more rapidly will the intensity fall off with increasing scattering angle. Hence, this decrease may serve as a measure of particle size, provided that the dimensions of the particle are comparable with the wavelength of the radiation. Thus, X-ray scattering is useful for learning about structure of molecular dimensions while light scattering, for which the wavelength is about a thousand times that of X-rays is useful for the study of much larger structures.

The inverse relationship between the size of structure and the scattering angle is evident. For large structures, scattered intensity falls off to small values at small angles. Consequently, the technique of small-angle X-ray scattering is used for the study of structures of size 100–10 000 Å with X-rays of wavelength of ~ 1 Å and low-angle light scattering is used for the study of structures of size of several micrometres with visible light of wavelength of $0·5$ μm.

At sufficiently large angles when path differences become of the order of half a wavelength, destructive interference occurs leading to the intensity minimum seen in Fig. 20. At higher angles when path differences reach one wavelength, constructive interference again occurs and a maximum is seen. With increasing wavelength, additional maxima and minima are seen. With white light, these maxima and minima occur at different angles for different wavelengths so that the colours of the Tyndall spectra are seen.

These maxima and minima or Tyndall colours can only be observed if all of the scattering spheres are of closely the same size. If there is heterogeniety in size the maxima and minima arising from different size spheres occur at different angles so that only a monotonic decrease of intensity with angle is observed.

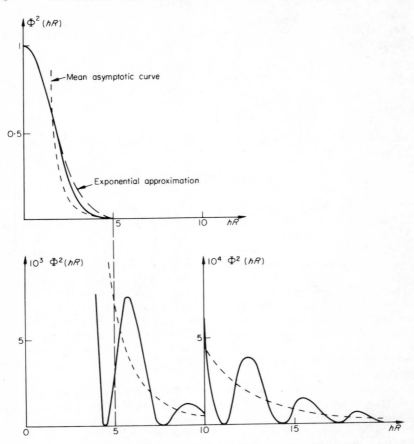

Fig. 20. The variation of scattered intensity with angle for an isotropic sphere. The angular variation is plotted in terms of the reduced variable $U = 4\pi(R/\lambda)\ \sin(\theta/2)$. *(From Gunier, A. and Fournet, G. (1955).* Small Angle Scattering of X-Rays, *Wiley, New York, p. 20.)*

In the above equation, internal field effects are neglected and it is assumed that the electric field acting upon a volume element is the same as the external field. This will be a good approximation for X-rays where the refractive indices for most materials are close to unity, but it is only accurate for visible light scattering if the spheres are suspended in a medium of closely matching refractive index. Otherwise, the more complicated Mie theory must be used.[48,49]

If the scattering power of a sphere is not uniform but depends upon the

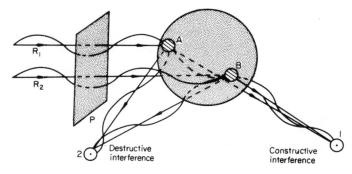

*Fig. 21. The interference of scattered waves from an isotropic sphere. (From Zimm, B. H., Stein, R. S. and Doty, P. (1945). Polymer Bull., **1**, 90.)*

distance from its centre, then the variation of $\alpha(r)$ with r must be considered in the integration of eqn. (38). For example, if:

$$\alpha(r) = \alpha_0 \exp{(-r^2/a^2)} \tag{42}$$

then:

$$F = \pi\alpha_0 a^3 \exp{(-a^2 h^2/4)} \tag{43}$$

This gives rise to a uniformly decreasing intensity with θ which depends upon the characteristic dimension a. This approach may be used, for example, to describe the X-ray scattering from an atom where the electron density varies with radius by an amount described by quantum mechanics. Thus the scattering depends upon the radial distribution function of electronic density. The agreement between the predicted variation of scattered intensity with that which is measured serves as a test of the correctness of the quantum mechanical description.

The calculated form factor for a single atom obtained using eqn. (38) for summing over the contribution of the volume elements of the atom is often designated by f and is called the 'atomic form factor'.

(D) Ellipsoids
The techniques used for the calculation of the scattered intensity for spheres can be extended to other shapes of particles. One case of particular interest is that of ellipsoids for which the principal axis is oriented in a particular direction. This may arise for example from a system originally containing spherical particles which become deformed in an affine fashion. An example is the extension of a high-impact polystyrene containing initially spherical rubber particles.

The integration of eqn. (38) may be readily accomplished in ellipsoidal co-ordinates leading to the identical equation to $(41)^{50}$ for spheres except that the variable U is redefined as:

$$U = (4\pi R/\lambda) \sin(\theta/2)[1 + (v^2 - 1)\cos^2(\theta/2)\cos^2\Omega] \tag{44}$$

where v is the axial ratio of the ellipsoidal particle and Ω is the azimuthal angle of scattering (Fig. 22). In this case the scattered intensity is no longer cylindrically symmetrical about the incident beam but is more intense in a direction perpendicular to the direction of extension of the particle (again resulting from the reciprocity between particle dimensions and scattering angle).[51] This anisotropy in the shape of the scattering pattern serves as a measure of the deformation of the particle.

(E) *Rods*
The scattering integral may also be readily evaluated for rod-shaped particles in which case, the intensity is dependent upon the angle of orientation of the rod

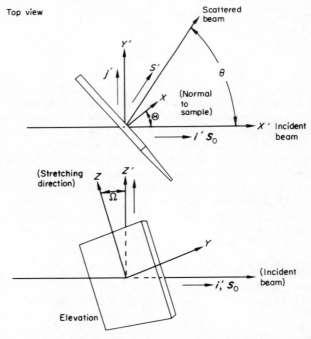

Fig. 22. The azimuthal scattering angle Ω and the film tilt angle ϕ. (From Stein, R. S. and Hotta, T. (1964). J. Appl. Phys., 35, 2237.)

axis. In the simple case of thin rods of length L lying in a plane perpendicular to the incident beam, the scattered intensity is given by:[52]

$$I(\theta, r, \alpha) = K_4 L^2 \{\sin(kaL/2)/(kaL/2)\}^2 \tag{45}$$

where $a = -\sin(\alpha+\Omega)\sin\theta$ and α is the angle between the axis of the rod and the normal to the plane of measurement of the scattering angle, θ and K_4 are a combination of physical constants.

A plot of this equation reveals that the scattering pattern is extended in a direction perpendicular to the rod axis. Real systems often contain rodlike particles which are distributed in orientation. If the particles scatter independently of each other (see next section) the total intensity may be found by adding intensities from the individual particles.

$$I(\theta, \Omega) = \int_{\alpha=0}^{2\pi} N(\alpha)I(\theta, r, \alpha)\,d\alpha \tag{46}$$

where $N(\alpha)\,d\alpha$ is the number of rods oriented at angles between α and $\alpha+d\alpha$. If the rod orientation is random, the scattering pattern becomes independent of Ω and is cylindrically symmetrical. If there is preferred orientation in some direction, the scattering intensity becomes greater in a direction perpendicular to this.

(F) *Scattering from an assembly of particles*

In the preceding theory only the scattering from individual particles has been considered. For an assembly of particles, the structure factor of eqn. (38) should be calculated by integrating over all particles in a unit volume as well as the surroundings of the particles.

Suppose r_{jk} is the vector which locates the jth volume element of the kth particle with respect to an origin within the particle and R_k is the vector to this origin from the overall origin of the co-ordinate system (Fig. 23). Then, in general, a vector from an arbitrary origin to any volume element can be written in the form:

$$r_j = R_k + r_{jk} \tag{47}$$

Thus the structure factor for the entire system becomes:

$$F = \sum_k \left[\int_k \alpha(r_{jk})\exp[ik(r_{jk}\cdot s)]\,dr_{jk} \right] \exp[ik(R_k\cdot s)] + \int_s \alpha(r_s)\exp[ik(r_s\cdot s)]dr_s \tag{48}$$

where the integral over k is taken within the bounds of the kth particle and that over s is over the surroundings of all of the particles. Now $\alpha(r_s)$ is the polarisability of the surroundings which may be taken as constant and designated α_s.

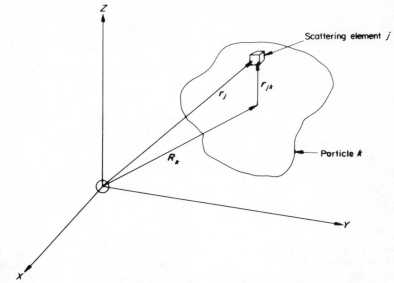

Fig. 23. The co-ordinate system for describing the scattering from a collection of particles.

Then:

$$\int_s \alpha(\mathbf{r}_s) \exp\left[ik(\mathbf{r}_s \cdot \mathbf{s})\right] d\mathbf{r}_s = \alpha_s \int_\infty \exp\left[ik(\mathbf{r} \cdot \mathbf{s})\right] d\mathbf{r}$$
$$-\sum_k \left[\int_k \alpha_s \exp\left[ik(\mathbf{r}_{jk} \cdot \mathbf{s})\right] d\mathbf{r}_{jk}\right] \exp\left[ik(\mathbf{R}_k \cdot \mathbf{s})\right] \quad (49)$$

where the subscript, ∞, designates an integral over all space. This integral of a medium of constant polarisability over all space is zero (a uniform medium does not scatter). Thus, upon substituting eqn. (49) into eqn. (48) one obtains:

$$F = \sum_k \left\{ \int_k \left[\alpha(\mathbf{r}_{jk}) - \alpha_s\right] \exp\left[ik(\mathbf{r}_{jk} \cdot \mathbf{s})\right] d\mathbf{r}_{jk} \right\} \exp\left[ik(\mathbf{R}_k \cdot \mathbf{s})\right] = \sum_k F_k \exp\left[ik(\mathbf{R}_k \cdot \mathbf{s})\right] \quad (50)$$

where F_k is the structure factor of the kth particle relative to its surroundings. This equation permits one to calculate the scattering from an assembly of particles if that from the individual particles is known and the particle locations are specified.

The intensity of scattering given by eqn. (36) is then for centrosymmetrical particles:

$$\mathcal{R} = K_3 FF^* = K_3 \left[\sum_k F_k \exp\left[ik(R_k \cdot s)\right]\right]\left[\sum_l F_l^* \exp\left[-ik(R_l \cdot s)\right]\right] \quad (51)$$

$$= K_3 \left\{ \sum_k F_k F_k^* + \sum_k \sum_{k \neq l} F_k F_l^* \cos\left[k(R_k - R_l)\cdot s\right] \right\}$$

For a system in which particles are arranged with spherical symmetry, the second term may be averaged over angle to give:

$$\mathcal{R} = K\left[\sum_k F_k F_k^* + \sum_k \sum_{k \neq l} F_k F_l^* \frac{\sin hR_{kl}}{hR_{kl}} \right] \quad (52)$$

where $R_{kl} = |R_k - R_l|$

If particles are randomly arranged so that all values of R_{kl} are equally likely, the second term becomes zero and:

$$\mathcal{R} = K_3 \sum_k F_k F_k^* = \sum I_k \quad (53)$$

where $I_k = K_3 F_k F_k^*$ is the intensity scattered by the kth particle alone. Thus for such independent scatterers where there is no phase relationship among scattering from different particles, the total scattered intensity is the sum of the intensities scattered from the individual particles. If the particles are atoms, $F_k = F_k^* = f_k$ (the form factor for centrosymmetrical particles may be shown to be real). Then one obtains:

$$\mathcal{R} = K_3 \left\{ \sum_k f_k^2 + 4\pi \sum_k \sum_{k \neq l} f_k f_l \frac{\sin hR_{kl}}{hR_{kl}} \right\} \quad (54)$$

which is the basic equation for the X-ray scattering for an assembly of atoms.

In general, for an isotropic distribution of identical centrosymmetrical particles:

$$\mathcal{R} = 4\pi K_3 F^2 \int_0^\infty P(R) \frac{\sin hR}{hR} R^2 dR \quad (55)$$

where $P(R)R^2 dR$ is the probability that the centres of two particles will be a distance R apart.

In such a case this radial distribution function for particle centres may be obtained by Fourier inversion of this equation for the Rayleigh ratio.[51]

(G) *The effect of anisotropy*

The preceding theory applies to the scattering from isotropic systems where the polarisability is a scalar. Most cases of X-ray scattering fulfil this requirement since the anisotropy of polarisability for X-rays is usually negligibly small. (This is so because X-ray scattering involves the inner electrons of atoms which are not perturbed by molecular bonding. Chemical bonding involves the outer or valence electrons and leads to anisotropy for visible light.) However, for visible light, the tensor nature of the polarisability must be considered and the dipole moment induced in a volume element may be given by:[53]

$$m = \delta(E \cdot a)a + \alpha_2 E \tag{56}$$

It is assumed that the volume element has uniaxial anisotropy with a polarisability α_1 along its principal axis direction (along the unit vector a) and a polarisability α_2 perpendicular to this direction. The anisotropy, δ, is equal to $\alpha_1 - \alpha_2$. For an isotropic volume element $\alpha_1 = \alpha_2 = \alpha$ and $\delta = 0$. Thus, eqn. (56) reduces to:

$$m = \alpha E \tag{57}$$

Thus m is in the direction of E. For an anisotropic element, m lies in the plane containing m and E.

The amplitude of scattering depends upon the projection of m in a plane perpendicular to the scattered ray and in the plane passed by an analyser in the scattered ray. This is approximately given by:[53,54]

$$E_s = K_5(m \cdot O) \exp\left[i(\omega t - kd)\right] \tag{58}$$

where O is a unit vector perpendicular to the scattered ray and in the direction of polarisation passed by an analyser.

The effect of anisotropy may be best seen by examining the polarised components of scattering. In this context, vertical polarisation (designated by V) refers to polarisation of the electric vector in a plane perpendicular to that containing the incident and the scattered ray, whereas horizontal polarisation refers to polarisation perpendicular to this plane.

For both vertical polarisation of the incident and scattered beam (designated V_v) for an isotropic system, m and O are parallel and scattering occurs, whereas for vertical incident polarisation but horizontal polarisation of the scattered ray (H_v) or the converse (V_h), m and O are perpendicular and the scattered amplitude is zero. Thus the finding of a non-zero H_v and V_h component of scattering is an indication of anisotropy.

It is evident that the theory of the previous sections may be extended to anisotropic systems by replacing the polarisability with a term proportional to $(m \cdot O)$. This approach will be demonstrated in the later section on light scattering.

(H) The statistical approach[55]

In the case of condensed systems which do not consist of discrete particles, the preceding approach is not useful. It is more convenient to describe such systems in terms of statistical functions rather than in terms of sizes, shapes and locations of particles.

For this approach, we may combine eqns. (28), (30), (33) and (36) to obtain for the Rayleigh ratio:

$$\mathscr{R} = K_3 \left[\sum_j \alpha_j \exp\left[ik(\mathbf{r}_j \cdot \mathbf{s})\right] \right]\left[\sum_l \alpha_l \exp\left[-ik(\mathbf{r}_l \cdot \mathbf{s})\right] \right] \tag{59}$$

$$= K_3 \sum_j \sum_l \alpha_j \alpha_l \exp\left[ik(\mathbf{r}_{jl} \cdot \mathbf{s})\right]$$

where $\mathbf{r}_{jl} = \mathbf{r}_j - \mathbf{r}_l$.

For a heterogeneous medium, the polarisabilities of the jth and lth volume elements will be different and will deviate from the average polarisability, $\bar{\alpha}$ by fluctuations η_j and η_l where, for example:

$$\alpha_j = \bar{\alpha} + \eta_j \tag{60}$$

Thus:

$$\mathscr{R} = K_3 \sum_j \sum_l (\bar{\alpha} + \eta_j)(\bar{\alpha} + \eta_l) \exp\left[ik(\mathbf{r}_{jl} \cdot \mathbf{s})\right]$$

$$= K_3 \sum_j \sum_l \left[\bar{\alpha}^2 + \bar{\alpha}\eta_j + \bar{\alpha}\eta_l + \eta_j \eta_l\right] \exp\left[ik(\mathbf{r}_{jl} \cdot \mathbf{s})\right] \tag{61}$$

The first term represents the scattering from a hypothetical homogeneous medium of polarisability $\bar{\alpha}$ and is therefore zero. The second (and third) terms are sums over a product of η_j with a term dependent upon $\mathbf{r}_{jl}$. Since η_j may be positive or negative in a way which is not correlated with $\mathbf{r}_{jl}$, the sum represents random products of positive and negative terms and averages to zero. Consequently, only the last term remains.

One may substitute for $\eta_j \eta_l$, its average value at constant $\mathbf{r}_{jl}$ designated as $\langle \eta_j \eta_l \rangle_{\mathbf{r}_{jl}}$ to give:

$$\mathscr{R} = K_3 \sum_j \sum_l \langle \eta_j \eta_l \rangle_{\mathbf{r}_{jl}} \exp\left[ik(\mathbf{r}_{jl} \cdot \mathbf{s})\right]$$

$$= K_3 \int_{r_j} \int_{r_l} \langle \eta_j \eta_l \rangle_{\mathbf{r}_{jl}} \exp\left[ik(\mathbf{r}_{jl} \cdot \mathbf{s})\right] \mathrm{d}\mathbf{r}_j \, \mathrm{d}\mathbf{r}_l \tag{62}$$

where the double sum has been replaced by a double integral over the co-ordinates of $\mathbf{r}_j$ and $\mathbf{r}_l$. Since the integral depends only upon $\mathbf{r}_{jl}$, one may integrate over this variable rather than over $\mathbf{r}_j$ and $\mathbf{r}_l$ separately to give:

$$\mathscr{R} = K_3 V \int_{r_j} \langle \eta_j \eta_l \rangle_{r_{jl}} \exp\left[ik(\mathbf{r}_{jl} \cdot \mathbf{s})\right] \mathrm{d}\mathbf{r}_{jl} \tag{63}$$

where in the following equations, the subscripts jl or r may be neglected for brevity.

One may now define a *correlation function* as:

$$\gamma(r) = \frac{\langle \eta_j \eta_l \rangle_r}{\langle \eta^2 \rangle_{av}} \tag{64}$$

where $\langle \eta^2 \rangle_{av}$ denotes the mean square η. When $r = 0$, $\gamma(r)$ obviously becomes unity. When r is large and the volume elements are far apart there is little correlation and $\gamma(r)$ approaches zero. The way in which $\gamma(r)$ changes with r describes the structure of the system.

Upon using this correlation function, eqn. (63):

$$\mathscr{R} = K_3 V \langle \eta^2 \rangle_{av} \int_r \gamma(r) \exp\left[ik(r \cdot s)\right] dr \tag{65}$$

For spherically symmetrical systems, $\gamma(r)$ depends only upon the length of r and integration over angle leads to:[55]

$$\mathscr{R} = 4\pi K_3 V \langle \eta^2 \rangle_{av} \int_{r=0}^{\infty} \gamma(r) \frac{\sin hr}{hr} r^2 dr \tag{66}$$

For many systems, the correlation function may be fitted by an exponential function:

$$\gamma(r) = e^{-r/a} \tag{67}$$

in which case the Rayleigh ratio becomes:

$$\mathscr{R} = 8\pi K_3 V \langle \eta^2 \rangle_{av} a^3 / [1 + h^2 a^2]^2 \tag{68}$$

The angular dependence of scattering depends upon the correlation distance a, while the intensity depends as well upon the mean-square density fluctuation, $\langle \eta^2 \rangle_{av}$. A plot of $R^{-1/2}$ against h^2 leads to a straight line having a slope proportional to a^2.

Alternatively, a gaussian correlation function:

$$\gamma(r) = \exp(-r^2/a^2) \tag{69}$$

leads to a Rayleigh ratio:

$$\mathscr{R} = 4\pi K_3 V \langle \eta^2 \rangle_{av} a^3 \exp(-h^2 a^2/4) \tag{70}$$

for which a plot of $\ln \mathscr{R}$ against h^2 is linear. While it is convenient to test these simple forms for the correlations function, the actual form of the correlation function can be determined from a Fourier inversion of the variation of the Rayleigh ratio with scattering angle.

For oriented systems, the vector equation (65) must be used, in which case the equivalent to the gaussian correlation function is:[57]

$$\gamma(r) = \exp\left\{-\left[\frac{x^2}{a^2} + \frac{y^2}{b^2} + \frac{z^2}{c^2}\right]\right\} \tag{71}$$

where x, y, and z are the co-ordinates of the vector r and a, b, and c are correlation distances in these three directions.

In this case, the scattering becomes a function of the azimuthal angle as well as the scattering angle θ (Fig. 22), and the vector s becomes:

$$s = (1 - \cos\theta)i - \sin\theta\cos\Omega j - \sin\theta\sin\Omega k \tag{72}$$

The integration of eqn. (65) leads to:

$$\mathscr{R}(\theta, \Omega) = K_3\, V\langle\eta^2\rangle_{\mathrm{av}}\, abc\exp\left\{-\pi^2\left[\left(\frac{a}{\lambda}\right)^2(1 - \cos\theta)^2\right.\right.$$
$$\left.\left. + \left(\frac{b}{\lambda}\right)^2\sin^2\theta\cos^2\Omega + \left(\frac{c}{\lambda}\right)^2\sin^2\theta\sin^2\Omega\right]\right\} \tag{73}$$

where the z direction is taken to be along the orientation direction of the sample and the sample is assumed to be uniaxially oriented so that $a = b$.

The correlation distance, a, can be obtained from the variation of scattered intensity with θ at $\Omega = 0°$ in which case:

$$\mathscr{R}(\theta, 0°) = K_3 V\langle\eta^2\rangle_{av}\, abc\exp\left\{-\pi^2\left(\frac{a}{\lambda}\right)^2\sin^2(\theta/2)/4\right\} \tag{74}$$

The variation of $\mathscr{R}(\theta, 90°)$ depends upon both a and c so that c may be obtained from its change and the value of a obtained from $\mathscr{R}(\theta, 0°)$.

(I) Statistical theory for anisotropic systems[53,54]

The statistical theory for the scattering for anisotropic systems may be developed in a manner similar to that described in eqns. (59)–(63) of the preceding section except that one starts from eqn. (58) rather than eqn. (28). The derivation was carried out in very general form by Goldstein and Michalik,[54] but the special case of 'random orientation fluctuations' was considered by Stein and Wilson[53] which involves the assumption that the correlation in orientation of two optic axes depends only upon their separation vector r and not upon the angle β that the optic axes make with the vector. This gives rise to the approximate equations for the Rayleigh ratios for polarised light for spherically symmetrical systems:

$$\mathscr{R}_{V_v} = 4K_3 V\left\{\langle\eta^2\rangle_{av}\int_{r=0}^{\infty}\gamma(r)\frac{\sin hr}{hr}r^2\,dr + \frac{4}{45}\langle\delta^2\rangle_{av}\int_{r=0}^{\infty}f(r)\frac{\sin hr}{hr}r^2\,dr\right\} \tag{75}$$

$$\mathcal{R}_{H_v} = 4\pi K_3 V \left\{ \frac{1}{15} \langle \delta^2 \rangle_{av} \int_{r=0}^{\infty} f(r) \frac{\sin hr}{hr} r^2 \, dr \right\} \tag{76}$$

where $\langle \delta^2 \rangle_{av}$ is the mean-squared anisotropy for a scattering element and $f(r)$ is a correlation function for orientation defined as:

$$f(r) = [3\langle \cos^2 \theta_{jl} \rangle_r - 1]/2 \tag{77}$$

where θ_{jl} is the angle between the optic axes of the jth and lth scattering elements averaged at a constant distance of separation equal to r. Obviously, when $r = 0$, $\theta_{jl} = 0°$ and $f(r) = 1$. When $r = \infty$, there will be no correlation in orientation of optic axes and $f(r)$ will approach zero.

As with $\gamma(r)$, one often finds that $f(r)$ can be approximated as a gaussian or exponential function of r/a, where a is a correlation distance for orientation correlation which, in general, will differ from that for $\gamma(r)$. It is a measure of the size of the region in which orientation is correlated.

It follows from eqns. (75) and (76) that:

$$\mathcal{R}_{V_v} - \frac{4}{3}\mathcal{R}_{H_v} = 4\pi R_3 V \langle \eta^2 \rangle_{av} \int_{r=0}^{\infty} \gamma(r) \frac{\sin hr}{hr} r^2 \, dr \tag{78}$$

Thus, from a study of $\mathcal{R}_{H_v}$, the contribution to scattering from orientation correlations may be determined, and $f(r)$ can be obtained from its Fourier inversion, and $\langle \delta^2 \rangle_{av}$ can be obtained from insertion of the theoretical value of K and a measure of the absolute value of $\mathcal{R}_{H_v}$. If there are no orientation correlations, $\mathcal{R}_{H_v} = 0$.

Correspondingly, the density correlation contribution may be obtained from a study of $[\mathcal{R}_{V_v} - 4/3\mathcal{R}_{H_v}]$ leading to values of $\gamma(r)$ and $\langle \eta^2 \rangle_{av}$. When $\langle \delta^2 \rangle_{av} = 0$ only density fluctuations contribute and eqn. (75) reduces to the Debye–Bueche eqn. (66).

The case of random orientation fluctuations for oriented systems has been treated by Stein and Hotta[56] whose equations reduce, for example, to[57]

$$\mathcal{R}_{V_v} = \frac{2}{45} 4\pi K_3 V \langle \delta^2 \rangle_{av} (13g + 2) \cdot \frac{1}{4\pi} \int f(r) \cos k(\mathbf{r} \cdot \mathbf{s}) \, dr \tag{79}$$

at $\Omega = 0°$ and:

$$\mathcal{R}_{V_v} = \frac{4}{45} 4\pi K_3 V \langle \delta^2 \rangle_{av} (1 - g) \cdot \frac{1}{4\pi} \int f(r) \cos k(\mathbf{r} \cdot \mathbf{s}) \, dr \tag{80}$$

for $\Omega = 90°$ for the orientation correlations for uniaxial orientation. The quantity

$f(r)$ is a vector orientation correlation function which may be expressed in gaussian form as in eqn. (71).

The parameter g is an orientation factor defined by:

$$g = \frac{1}{2}[3\langle\cos^2\alpha_i\rangle_{av} - 1]$$ (81)

where α_i is the angle between the ith optic axis and the orientation direction. For unoriented systems, $g = 0$ and eqns. (79) and (80) become identical and reduce to the orientation part of eqn. (75). With increasing orientation, as g approaches unity, $\mathcal{R}_{V_v}$ appreciably increases at $\Omega = 0$ and decreases toward zero at $\Omega = 90°$. This is a consequence of the scattering being most intense when the optic axis direction lies along the polarisation direction.

Thus for unoriented systems, the scattered intensity is independent of Ω but becomes Ω dependent upon stretching as is illustrated with the experimental data for stretched polyethylene shown in Fig. 24.[58]

The preceding theory is for the case of random orientation fluctuations, where in the unoriented state, the oriented domains have spherical symmetry (Fig. 25). In many systems, this is not so and the domains may be rod or disclike in shape. In fact, in the case of more complicated morphology such as that of spherulites,

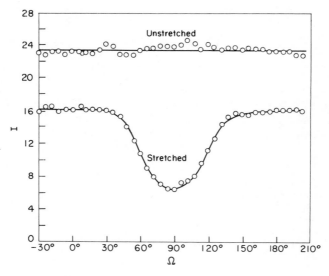

Fig. 24. The variation of the scattered intensity with Ω for stretched polyethylene. (From Keane, J. J., Norris, F. H. and Stein, R. S. (1956). J. Polymer Sci., **20**, 209.)

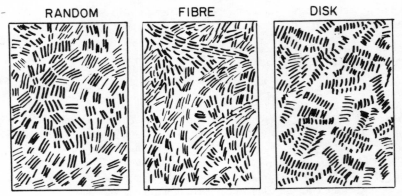

RANDOM   FIBRE   DISK

Fig. 25. The arrangement of optic axes directions for random rodlike and disclike orientation fluctuations. (From Stein, R. S. and Wilson, P. (1962). J. Appl. Phys., **33**, *1914.)*

the domains are complex in shape. In the formulation of the statistical theory, this means that the correlation in orientation between two optic axes depends upon the angle β that an axis makes with the interconnecting vector r in a manner that may vary with r. This problem has been treated in two dimensions by Stein et al.[59] and in three dimensions by van Aartsen.[60] In the simpler two-dimensional treatment, quantities like $\langle \cos 2\theta_{il} \rangle_{r,\beta}$ depend on both r and β and may be expanded in a Fourier series in β:

$$\langle \cos 2\theta_{ij} \rangle_{r,\beta} = T_0(r) + \sum_n T_n(r) \cos(n\beta) \tag{82}$$

where the $T_n(r)$ function are the Fourier coefficients which depend upon r. For random orientation correlations, all of these coefficients are zero except $T_0(r)$ which reduces to the two-dimensional analogue of the orientation correlation function:

$$T_0(r) = \langle \cos 2\theta_{ij} \rangle_r = 2\langle \cos \theta_{ij} \rangle_r - 1 \tag{83}$$

In terms of this theory, a simple case which arises for crossed-polaroid scattering at small angles is that:

$$I_+ = K_6 \langle \delta^2 \rangle_{av} \int \left\{ T_0 J_0(w) - \frac{1}{2}(T_4 - S_4) J_4(w) \cos 4\psi \right\} r \, dr \tag{84}$$

where $w = kr \sin \theta$ and $J_0(w)$ and $J_4(w)$ are Bessel functions of w. The angle ψ is that between the polarisation direction of the incident light and the normal to the

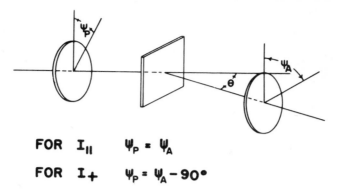

FOR $I_{\|}$ $\psi_P = \psi_A$

FOR I_+ $\psi_P = \psi_A - 90°$

Fig. 26. The angles ψ_P and ψ_A characterising the orientation of the polarisation direction of the polariser and analyser. For $I_{\|}$, approximately $\psi = \psi_A = \psi_P$, whereas for I_+, $\psi = \psi_A = \psi_P + 90°$. (From Stein, R. S., Erhardt, P., van Aartsen, J. J., Clough, S. and Rhodes, M. (1966). J. Polymer Sci., C, No. 13.)

scattering plane (in which θ is measured) (Fig. 26). (It is only approximately true, as pointed out by Gouda and Prins,[61] that I_+ designates the situation when the polariser and analyser are crossed. The exact relationships between ψ_A and ψ_P are given by Prins.) S_4 is the Fourier coefficient in the expansion of $\langle \sin 2\theta_{jl} \rangle_{r,\beta}$.

For random correlations $(T_4 - S_4) = 0$ and the scattering is independent of ψ. This corresponds to a scattering pattern which is cylindrically symmetrical about the incident beam. For non-random correlations $(T_4 - S_4)$ is finite which leads to a pattern which has four-fold symmetry in ψ as is often experimentally observed (Fig. 27). Thus in addition to the usual correlation function $T_0(r)$, the coefficient $(T_4 - S_4)$ plays the role of another correlation function which characterises the shape of the correlated region. The evaluation of these correlation functions for perfect two-dimensional spherulites has been carried out and their substitution into eqn. (84) has been shown to lead to a scattering pattern which is identical with that which is directly calculated by the amplitude summation method. Thus, this formulation enables one in principle to describe scattering from systems ranging from random to highly ordered.

It has been shown that with parallel polarisers, the expression for $I_{\|}$ involves a third correlation function $(T_2 - S_2)$ which multiplies $\cos 2\psi$ and imparts two-fold symmetry to the scattering pattern.

The non-random theory has been generalised to describe oriented systems.[62] However, a large number of parameters are required and it is difficult to associate them with physical concepts.

(J) *Scattering from systems with periodic density variation—relationship to diffraction*
Diffraction is a special case of scattering which results when the scattering objects

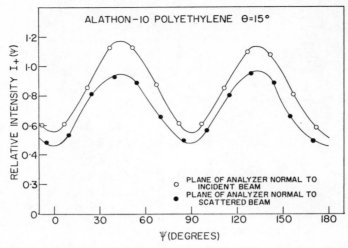

Fig. 27. The variation of I_+ with ψ for an unoriented polyethylene film. (From Stein, R. S., Erhardt, P., van Aartsen, J. J., Clough, S. and Rhodes, M. (1966).J. Polymer Sci., C, No. 13.)

are arranged regularly in space. A simple case of eqn. (51) is for two spherically symmetrical identical particles separated by distance a to give:

$$\mathscr{R} = 2KF_k^2\{1 + \cos[k(a \cdot s)]\} \tag{85}$$

It is apparent that $\mathscr{R}$ will pass through maxima whenever the argument of the cosine is an even multiple of π and will be zero when it is an odd multiple. Thus as seen in Fig. 28, there will be evenly spaced intensity maxima occurring when:

$$k(a \cdot s) = 2n\pi \tag{86}$$

In terms of the incident and scattering angle (Fig. 29) this equation becomes:

$$a(\cos\theta_1 - \cos\theta_2) = n\lambda \tag{87}$$

The equation implies that the path difference between waves scattered from the two objects is an integral number of wavelengths.

For an array of three equally spaced identical objects in a line, eqn. (51)

$$\mathscr{R} = F_k^2\{3 + 4\cos[k(a \cdot s)] + 2\cos[k(2a \cdot s)]\} \tag{88}$$

It is apparent that if $k(a \cdot s)$ is an even multiple of π, then $k(2a \cdot s)$ will also be. Thus for arrays of three atoms, the strong maxima will also be defined by eqn. (87) in which case $\mathscr{R} = 9K_3F_k^2$. When $k(a \cdot s)$ is an odd multiple of π, $\cos[k(a \cdot s)]$ will be -1 and $\cos[k(2a \cdot s)] = +1$ so $\mathscr{R} = K_3F_k^2$. When $k(2a \cdot s)$ is an odd multiple of π

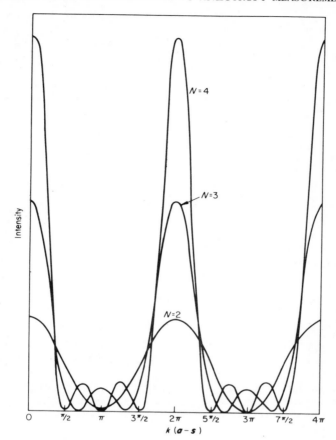

Fig. 28. The variation of Rayleigh ratio with $k(\mathbf{a} \cdot \mathbf{s})$ *for an array of scattered points.* N *designates the number of points in the linear array.*

the third term is -1 and the second is zero so $\mathscr{R} = K_3 F_k^2$. A plot of $\mathscr{R}$ as a function of $k(\mathbf{a} \cdot \mathbf{s})$ is included in Fig. 28. It is seen that in going from two particles to three, the intensity of the principal maxima are increased and smaller subsidiary maxima occur in between. The effect is to narrow the principal maxima.

With increasing numbers of particles, the principal maxima becomes still narrower and higher. Thus a large one-dimensional lattice leads to discrete diffraction maxima whose width are related to the size of the lattice. The

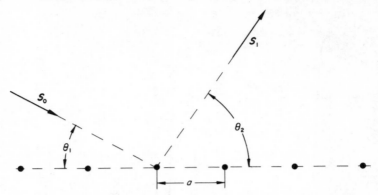

Fig. 29. The angles θ_1 and θ_2 defining the scattering from an array of atoms.

relationship is that the half-width of the diffraction peak, β, is related to length of the lattice, d, by:

$$\beta = \frac{K_7 \lambda}{d \cos \theta_\beta} \tag{89}$$

where K_7 is a constant of order unity.

The three-dimensional equivalent of this equation is the Scherrer equation[63] which may be used to determine crystal size from the width of the diffraction peaks.

The locus of diffraction for a given value of θ_1 in eqn. (87) is a cone of apex angle θ_2 (Fig. 30). Thus a one-dimensional lattice does not lead to diffraction

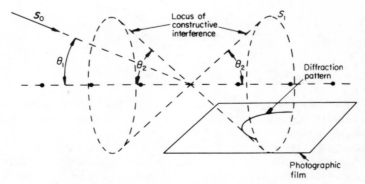

Fig. 30. The locus of diffraction for a one-dimensional lattice.

spots but rather to parabolic streaks resulting from the intersection of these cones with a photographic film.

Equation (87) may be written as:

$$\frac{a_1 \cdot s}{\lambda} = h_1 \tag{90}$$

where a_1 is the spacing of the lattice in a particular direction and h_1 is an integer. For a three-dimensional lattice, constructive interference must also occur among atoms in the other two directions giving rise to (Fig. 31):

$$\frac{a_2 \cdot s}{\lambda} = h_2 \tag{91}$$

$$\frac{a_3 \cdot s}{\lambda} = h_3 \tag{92}$$

where a_2 and a_3 are the lattice spacings in the other two directions and h_2 and h_3 are the other integers. The simultaneous solution of these three equations leads to the vector Bragg equation. A reciprocal lattice vector H is defined as:

$$H = h_1 b_1 + h_2 b_2 + h_3 b_3 \tag{93}$$

where:

$$b_1 = \frac{a_2 \times a_3}{[a_1 a_2 a_3]} \tag{94}$$

$$b_2 = \frac{a_3 \times a_1}{[a_1 a_2 a_3]} \tag{95}$$

$$b_3 = \frac{a_1 \times a_2}{[a_1 a_2 a_3]} \tag{96}$$

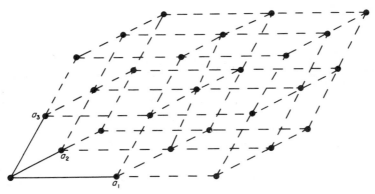

Fig. 31. A three-dimensional lattice corresponding to the unit cell vectors $\mathbf{a}_1$, $\mathbf{a}_2$ and $\mathbf{a}_3$.

where b_1, b_2 and b_3 are the unit reciprocal lattice vectors and $[a_1 a_2 a_3]$ is the triple product of the three unit cell vectors $a_1 \times a_2 \cdot a_3 = a_1 \cdot a_2 \times a_3$ and is the volume of the unit cell. Then it may be shown that the simultaneous solution of eqns. (90)–(92) leads to:

$$H = \frac{s}{\lambda} \tag{97}$$

The vector H is normal to the crystal plane having the Miller indices h_1, h_2 and h_3 and has a length equal to $1/d$ where d is the interplane distance. The vector s as previously defined (eqn. (31)) depends upon the orientation of the incident ray in the direction of the unit vector s_0 and the diffracted beam in direction s_1. The vector s bisects the angle between these two unit vectors (Fig. 32) and according to eqn. (97) must be parallel to H. This leads to the Bragg condition that both the incident and diffracted beam must lie at the same angle, θ_B, to the crystal plane.

It is evident from Fig. 32 that the length of s is $2 \sin \theta_B$ so that if one equals the magnitudes of the two sides of eqn. (97), one obtains:

$$1/d = 2 \sin \theta_B / \lambda$$

or

$$\lambda = 2d \sin \theta_B \tag{98}$$

which is the scalar form of Bragg's equation which relates the Bragg angle to the interplane spacing.

Equation (97) relates the diffraction direction to the orientation of the crystal plane. One of its consequences is that for a given crystal plane normal, the angle α

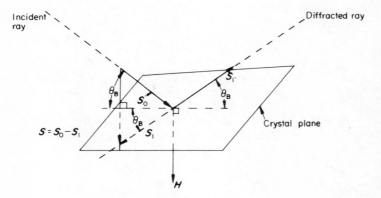

Fig. 32. The relationship between the vectors H, s, s_i, and s_0.

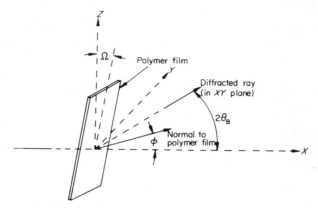

Fig. 33. The diffraction of X-rays from a polymer film.

between the crystal plane normal and a reference direction, say the stretching direction of an oriented polymer film, is given by:

$$\cos \alpha = \cos \Omega \cos(\theta_B - \phi) \tag{99}$$

where Ω is the azimuthal angle of diffraction and ϕ is the tilt angle between the sample normal and the incident beam (Fig. 33). If the sample is tilted at θ_B to the incident beam, then $\alpha = \Omega$ so that the azimuthal angle of diffraction equals the azimuthal angle of orientation of the crystal plane. This is the basis for the determination of orientation by diffraction as discussed in Section 3.2.

It is of interest to point out that the substitution of eqns. (90–92) and (93) into eqn. (50) along with the relationship that the location of the jth atom in the unit cell:

$$r_j = x_j a_1 + y_j a_2 + z_j a_3 \tag{100}$$

leads to the result that the structure factor for a crystal is:

$$F = \sum_j f_j \exp\left[-2\pi i(h_1 x_j + h_2 y_j + h_3 z_j)\right] \tag{101}$$

where f_j is the form factor for the jth atom. In obtaining this equation, one uses the reciprocity relationships that $(a_j \cdot b_k)$ is one if $j = k$ and zero if $j \neq k$. The co-ordinates x_j, y_j and z_j locate the jth atom in the unit cell in terms of fractional unit cell distances. The summation in eqn. (101) is over all of the atoms in the unit cell.

This equation is the basis of the calculation of the intensities of X-ray diffraction from a knowledge of the locations of the atoms in the cell and provides a means for experimentally testing a postulated crystal structure.[64] If F can be determined experimentally for all crystal reflections, then the Fourier inversion of eqn. (101) permits the establishment of the location of all atoms. The details of such procedures may be found in books on the X-ray determination of crystal structure.

3.3.3 Low angle X-ray scattering

(A) *Introduction*

As discussed in Section 3.3.2, there is an inverse relationship between scattering angle θ and the ratio of the size of the scattering structure to the wavelength of radiation which is used. Most scattering equations may be expressed in terms of a reduced variable of the form:

$$U^* = C^*(a/\lambda)\sin(\theta/2) \tag{102}$$

where a is some dimensional parameter of the scattering structure. C^* is a constant characteristic of the experiment which is close to one. Measurements are usually in the range of convenient observation when U^* is of the order of unity. For example, for the diffraction from crystal planes, maxima occur when $U^* = n$ if $C^* = 2$, $a = d$, the interplanar distance and $\theta/2 = \theta_B$, the Bragg angle. For light scattering from spherulitic polymers (Section 3.3.4) maxima occur when $U^* = 4\cdot1$ if $C^* = 4\pi$.

For X-rays, λ is in the range of 1–5 Å whereas for visible light it is in the range of 3000–8000 Å. While infra-red scattering could be used with longer wavelengths, it is usually not, at least for the study of solids, since structures having sizes in the range where it would be useful are in the range of direct microscopic observation. Unfortunately, the range of wavelengths of 10–3000 Å is not readily usable because of the high absorption of such radiation by most materials. Consequently, to study sizes in this range by scattering, one usually uses X-rays, which necessitates making measurements at quite low angles in the range of 0·001–0·1 radians. Special techniques and means of interpretation have been developed for this region of low angle X-ray scattering (LAXRS) so that their separate discussion is warranted.

(B) *Experimental*

As with wide angle X-ray studies, experimental methods may be classified into photographic and photometric ones. The photographic techniques have the advantage of exhibiting the entire scattering pattern at once, so that its general form and positions of scattering maxima can be seen. This is advantageous for qualitatively establishing the existence and nature of orientation and the sizes of discrete spacings. The photometric method is essential for determination of the actual angular variation of intensity which is necessary for the interpretation of diffuse scattering. This is particularly desirable for low angle scattering studies, since sharp discrete scattering is less common, and correction for slit smearing effects is more important. Of course, quantitative data may also be obtained by microdensitometer measurements on photographs, but this procedure is less accurate than direct measurement.

In order to study the angular variation of intensity at very small angles, it is essential that the incident radiation is highly parallel and the angular resolution of the detector is high. This necessitates the use of narrow slits or small pinholes so that intensities are low. Consequently, it is desirable to have an X-ray source in which the focal spot is small and of high intensity. Consequently, microfocus X-ray tubes and rotating anode generators are frequently employed. For photometric studies, data accumulation is slow so that high stability of intensity is essential. Thus, voltage and current regulation of the X-ray tube and cooling water temperature and flow regulation are recommended. It is also desirable to control the ambient temperature of the apparatus to reduce intensity variations resulting from thermal expansion of components of the collimating system.

For convenience in accumulating data over a long time period, the use of an automatic angular advance and data accumulation system is desirable.

A good general discussion of instrumentation for low angle scattering studies has been recently published.[65] Photographic cameras for orientation studies are better if they are of the pinhole rather than the slit type since pinhole optics produces less distortion of the orientation distribution. Simple cameras of this type for photographic studies have been developed by Statton[66] and Kiessig.[67] Since to achieve good collimation, fairly long path lengths (~ 40 cm) are employed, provision is made for evacuation of the cameras to minimise air scattering. With these cameras, spacings of up to 200–300 Å can be resolved. Cameras with finer pinholes to resolve longer spacings have been described[68] but these often require exposure times of over 100 h so that their use is limited.

With slit collimation, higher resolution may be obtained with much less loss of intensity. Such cameras are, for example, of the Rigaku–Denki[69] and Kratky[70] type, the latter employing an ingenious U-bar collimating system which permits resolution up to several thousand angströms. While slit cameras may not readily be used to characterise degrees of partial orientation, a compromise is possible with cameras of the Rigaku–Denki type in that both vertical and horizontal slits are provided for collimation. Since the intensity variation in the azimuthal direction is usually much more gradual than that in the θ direction, it is possible to work with slit openings in the vertical direction which are considerably greater than those in the horizontal (θ) direction so that the intensity loss is usually less than that with pinhole systems.

High collimation cameras of the Franks,[71] Elliott,[72] and Bonse–Hart[73] types make use of total reflection of the X-rays at a glancing angle for collimation. These provide higher resolution with less intensity loss than the Kratky system for studies of spacings above about 1000 Å.

For measurement of the absolute intensity of scattering it is necessary to compare the intensity of the scattered ray with that of the incident ray. Since these are usually orders of magnitude different in intensity, it is desirable to attenuate the incident beam in some way so as to make their intensities more comparable. This may be done either by using an absorbing filter (such as nickel foil for Cu-Kα radiation) or by using a rotating sector. For the former method, it is essential to use crystal monochromatised radiation because of the large dependence of absorbance of X-rays upon their wavelength. Alternatively, one may calibrate measurement against a standard sample such as a gold sol or a reference polyethylene sample of the sort supplied by Kratky's laboratory.

(C) *Slit corrections*
With slit optics of finite width, X-rays scatter over a range of scattering angle. Thus, the measured angular variation of intensity will be distorted and a correction is essential. Such correction factors have been obtained for the case of unoriented samples using (1) infinitely long and narrow slits,[74] (2) narrow slits of arbitrary height,[75,76] and (3) slits of finite breadth.[77] The application of these corrections has been greatly facilitated by computer techniques.

(D) *Scattering invariants*
Porod[78] has defined a scattering invariant as:

$$Q_s = \int_0^\infty s^2 I(s)\,\mathrm{d}s \qquad (103)$$

for pinhole collimation or:

$$Q_s = 2\int_0^\infty s I(s)\,\mathrm{d}s \qquad (104)$$

where $s = (2/\lambda) \sin(\theta/2)$†
for slit collimation. This invariant is a measure of the total scattering power of the system and is independent of its detailed structure. It is related to the mean squared fluctuation in electron density by the equation:[79]

$$\frac{\overline{(\rho - \bar\rho)^2}}{\bar\rho} = \frac{4\pi}{vD} Q_s \tag{105}$$

where ρ is the electron density at a particular position with the sample and $\bar\rho$ is its average value, $v = \lambda^2 i_e$ where λ is the wavelength expressed in angströms and i_e is the Thompson scattering constant of a free electron ($i_e = 7 \cdot 9 \times 10^{-26}$) and D is the thickness of the sample expressed in units of the number of electrons, in a volume of 1 cm² area and length equal to the sample thickness.

The quantity $\overline{(\rho - \bar\rho)^2}$ is the equivalent of $\langle \eta^2 \rangle$ introduced in Section 3.3.2. For a two-phase system of phases of volume fractions w_1 and w_2 and electron densities ρ_1 and ρ_2, it is given by:

$$\overline{(\rho - \bar\rho)^2} = (\rho_1 - \rho_2)^2 w_1 w_2 \tag{106}$$

This quantity is useful, for example, in establishing the relationship between the X-ray scattering power of a polycrystalline sample and the volume fraction of crystallites and densities of the crystalline and amorphous phases.

(E) *Dilute systems*
The X-ray scattering from dilute systems of particles follows the discussion in Section 3.3.2. For spherical particles or rods, eqns. (40) and (45) of that section apply. A general form has been obtained by expanding the $(\sin hr/hr)$ term of eqn. (66) of Section 3.3.2 in a power series to give:[80]

$$\mathscr{R} = 4\pi K_3 V \langle \eta^2 \rangle_{av} \int_{r=0}^{\infty} \gamma(r) \left(1 - \frac{1}{3!} h^2 r^2 + \frac{1}{5!} h^4 r^4 + \dots \right) r^2 \, dr \tag{107}$$

$$= \mathscr{R}(0) \left(1 - \frac{1}{3} R_g^2 h^2 + \dots \right) \tag{107}$$

† This definition, chosen for consistence with the literature differs from the definition in Section 3.3.2.

where $\mathscr{R}(0)$ is the Rayleigh ratio† at $0°$ and $\overline{R_g}^2$ is the mean squared electronic radius of gyration defined by:

$$\overline{R_g}^2 = \frac{\sum_k f_k r_k^2}{\sum_k f_k} \tag{108}$$

where f_k is the scattering factor for the kth electron and r_k is its distance from the electronic centre of gravity of the scatterer.

Thus, $\overline{R_g}^2$ can be determined from a plot of $\mathscr{R}(\theta)$ against h^2. A good approximation is:

$$\mathscr{R}(\theta) = \mathscr{R}(0)\exp(-\tfrac{1}{3}h^2 R_g^2) \tag{109}$$

which is a form of Gunier's law. The radius of gyration is really all that the scattering experiment characterises and its relationship to the dimensions of an object requires an assumption as to its shape. For example, for a sphere of radius R_s, $\overline{R_g^2} = \tfrac{3}{5}\overline{R_s^2}$, while for an ellipsoid of semi-axes a, b, and c, $R_g^2 = \tfrac{1}{5}(a^2 + b^2 + c^2)$. The effect of orientation of particles is identical with that discussed in Section 3.3.2.

(F) *Effect of interparticle interference*
From an extension of the interference equation (55) of Section 3.3.2, Zernike and Prins[81,82] have obtained an équation relating the radial distribution function of sphere centres to the scattering:

$$\mathscr{R}(\theta) = \mathscr{R}_0(\theta)\left\{1 + \frac{4\pi}{v_1}\int_0^\infty [P(r)-1]\frac{\sin hr}{hr}r^2\,dr\right\} \tag{110}$$

where $\mathscr{R}(\theta)$ is the Rayleigh ratio in the absence of interference, v_1 is the average volume available per particle and $P(r)$ is the radial distribution function of spheres separated by distance r. By making use of theories of hard sphere liquids, this equation predicts, as shown in Fig. 34, that with increasing volume fraction of spheres in a system, the intensity maximum at $\theta = 0°$ characteristic of scattering from isolated spheres decreases and a maximum builds up at some angle characteristic of the average sphere diameter. In fact, as the concentration becomes greater and the spheres

---

† While it is not customary to express low angle scattering intensities in terms of Rayleigh ratios, we have done so here for consistency with the treatment of light scattering.

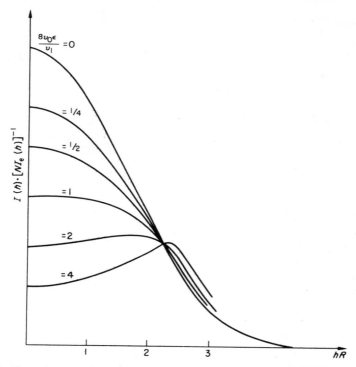

Fig. 34. The variation of relative scattered intensity with reduced scattering angle
hR *(where* h$-4\pi/\lambda$*) sin ($\theta/2$) for a non-interacting hard sphere fluid for various*
concentrations of spheres. The spheres have radius R *and volume* $v_0 = (4/3)\pi R^3$,
v_1 *is the average volume available per particle of the system and* ε *is a constant of order*
unity. (From Gunier, A. and Fournet, G. (1955). Small Angle Scattering of X-Rays,
Wiley, New York, p. 50.)

begin to pack in a regular fashion, the scattering equations pass
continuously into diffraction equations characteristic of a lattice of
spheres.

If a dense polymeric solid were perfectly homogeneous, there would be
no low angle scattering. The observation of diffuse scattering indicates the
presence of diffuse heterogeneities. The work of Hermans *et al.*[84] on the
effect of swelling has demonstrated that microvoids present in many
polymeric systems are a principal contributor to diffuse scattering. The
large electron density difference between a void and a polymer results in

even a small void being a more important contributor to scattering than that resulting from the smaller electron density difference between crystalline and amorphous regions of the polymer. These void regions occur between fibrils or lamellae in polymeric structures and depend very much upon the conditions of polymer preparation.[85]

The diffuse scattering from such polymers may be treated by the correlation function approach discussed in Section 3.3.2. The correlation distance may be related to the size of the void and the value of $\langle \eta^2 \rangle_{av}$ to the volume fraction of the void and electron density difference between the void and the polymer. If the voids can be filled with some liquid which matches the electron density of the polymer, the residual scattering may then be interpreted in terms of the electron density differences between crystalline and amorphous regions. If care is taken to eliminate the contribution of voids, then $\langle \eta^2 \rangle_{av}$ may be related to the volume fraction of crystallinity through eqn. (106). Such measurements[85] give results comparing favourably with degrees of crystallinity determined by other methods.

The diffuse scattering from oriented polymers exhibits dependence upon the equatorial scattering angle that principally reflects the orientation of the microvoids.[85] Formally, these may be treated in terms of direction dependent correlation functions as discussed in Section 3.3.2.

(G) *Discrete scattering*
As seen in Fig. 34, in systems containing appreciable concentration of structural units, scattering maxima result from the periodicity of electron density. Crystalline lamellae in polymers have thicknesses of the order of a few hundred angströms which are separated by less dense amorphous regions. This periodicity leads to scattering maxima characteristic of the inter-lamellar distance. Single crystals of polymers may be grown from dilute solution and are in the form of thin platelets.[87] These can be filtered to yield mattes where the crystal platelets lie in the plane of the matte. Low angle X-ray patterns obtained when the X-ray beam passes parallel to the plane of the matte show discrete spacings along the meridian characteristic of the intercrystalline distance.[86] If the spacing were completely regular, this distance could be obtained by the Bragg relationship. Actually, there is a distribution of crystal thicknesses, intercrystal spacings and angles between crystal planes which leads to a broadening of these diffraction maxima, both in the θ and the azimuthal direction and a decrease in intensity or even a disappearance of higher

orders. These effects have been analysed in terms of the statistics of paracrystals by Hosemann[88] and Vonk.[89]

Oriented bulk polymers show diffraction maxima characteristic of chain folded lamellae somewhat similar to those obtained with single crystal mattes.[86] The lamellae are usually oriented with their planes perpendicular to the stretching direction. These lamellae in bulk crystallised polymers differ from the solution grown single crystals in that they are separated by layers of amorphous material which may consist of tie chains as well as chain ends and folds.[87,90,91] With heat treatment there is an increase in the repeat distance which is accompanied by an increase in crystal thickness (as judged from the width of wide angle diffraction peaks). There is also usually an increase in the intensity of the scattering maxima characteristic of an increase in the density difference between the crystals and the intervening amorphous spaces. This has been interpreted as being a consequence of an increase in the regularity of chain folding.[92,93]

For unoriented samples of crystalline polymers, the meridional spacings are replaced by rings characteristic of a random distribution of inter-lamellar spacings. With stretching these rings narrow into arcs characteristic of the orientation of the lamellae. In principle, a lamellar orientation function can be obtained from the azimuthal variation of intensity in a similar manner to that described in Section 3.2 for the crystal orientation functions as determined by wide angle scattering. A comparison of the variation in wide angle and low angle scattering changes accompanying the mechanical deformation of the sample and subsequent thermal treatment serves to indicate whether the lamellae and their constituent crystals change their orientation together as an integral unit, or whether the chains tilt with respect to the lamellae plane.[94,95] Under certain conditions of orientation where the lamellae tilt at a positive and negative value of some angle with respect to the stretching direction, a so-called four-point diagram is obtained. As with wide angle scattering, it is possible to construct pole figures of orientation from the variation of the low angle scattered intensity with the Bragg and azimuthal angles.[96] Such pole figures describe the angular distribution of orientation of the lamellae in oriented samples.

3.3.4 Light scattering from solid polymers
(A) *Introduction*
The light scattering technique has been principally used in polymer science for the study of the molecular weight and size of molecules in

dilute solutions. The application of the method to solids was delayed because of the complex task of accounting for the more complex problem of interparticle and intermolecular interference in condensed systems. This extension was helped experimentally by the availability of continuous lasers which made the photographic light scattering technique practicable.

(B) *Experimental techniques*
(1) *Photographic.* While the photographic light scattering method for studying films was developed using mercury light sources,[97] modern equipment utilises laser sources. A typical set-up is shown in Fig. 35.[98] Such an apparatus permits one to obtain a scattering pattern from most systems in a fraction of a second. It is possible to see directly the patterns on a ground glass screen. Typical patterns with parallel (V_v) and perpendicular (H_v) polarisation for low density polyethylene films are shown in Fig. 36. These patterns were recorded on polaroid film using a 2 mW He–Ne laser with a 1/50 s exposure time. Weaker scattering patterns could be recorded using a more intense laser, but care must be taken in such cases to avoid sample changes produced by heating. Cells may be provided for controlling sample temperature, for immersing in liquids and for stretching.

The possibility of obtaining photographic patterns with short exposure time permits the following of rapid changes with time. This approach has been used, for example, for following the crystallisation kinetics of isotactic polystyrene.[99] By utilisation of a high speed motion picture camera, it has been possible to follow the rapid extension of samples at rates of several hundred frames per second.[100]

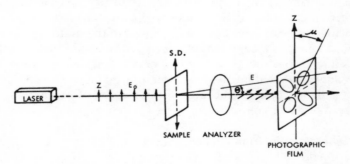

Fig. 35. A diagram of a typical low angle light scattering apparatus. (From Samuels, R. (1968). Hercules Chemist, **56**, 19.)

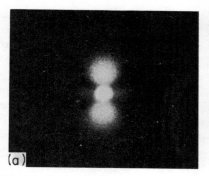 

Fig. 36. Typical V_v *(a) and* H_v *(b) light scattering patterns. (From Stein, R. S. (1969). In* Rheology *(Ed. F. Eirich), Academic Press, New York, Vol. V, Chapter 6.)*

In obtaining photographic patterns, it is desirable to eliminate surface scattering either by working with films having a very smooth surface, or by immersing the sample between glass plates using a fluid of matching refractive index which does not swell the polymer. It is important that the thickness of the film is such that excessive secondary scattering does not occur which will obscure the true scattering pattern. A good rule-of-thumb is that the sample should transmit at least 75% of the incident light.

For studies of biaxially oriented samples, the so-called XYZ technique is available where samples are studied for an incident ray propagating parallel to one of the three principal axes.[101] For this purpose, it is necessary to use thin sections of the samples cut perpendicular to these axes.

(2) *Photometric.* In photometric studies, one wishes to directly measure the intensity of scattering as a function of angle using a detector, usually a photomultiplier. The type of apparatus used for conventional solution scattering is usually not suitable because its angular resolution is not sufficient and it is not possible to make measurements at angles sufficiently close to the incident beam.

Most equipment described in the literature so far has used mercury light sources with high resolution optics.[102,103] The new generation of equipment will undoubtedly use laser sources.[104] It is advantageous to use blue lasers (argon-ion or cadmium) because of the greater scattering power at short wavelengths and because of the greater sensitivity of photomultipliers to blue light. It is desirable that both the direction of polarisation of the incident radiation and that of the analysing polariser in the scattered beam (ψ_A and ψ_P of Fig. 26 in Section 3.3.2) be adjustable. The incident beam should be parallel to within $0.1°$ and the photometer should have corresponding resolution. Because of this high resolution, the amount of room light admitted to the photometer is small so that measurements can usually be made without a light-tight enclosure. It is advantageous to chop the incident beam and use a phase-sensitive detector, or for low intensities, to use photon counting. Because of the great variation of scattering power with angle, it is essential that there is provision for insertion of neutral filters for suitable attenuation of the intensity.

The sample is usually mounted between glass cover slips using an immersion fluid and is set in a goniometer which permits tilting of the sample to the incident beam through the angle ϕ and rotating it about the sample normal (for oriented samples) through the angle Ω (Fig. 22 of Section 3.3.2).

For absolute intensity measurements, apparatus calibration is necessary. In most equipment, it is difficult to calibrate against a pure liquid because of the low scattering power and low sensitivity of solid state apparatus associated with their high resolution. Consequently, calibration is often referred to a conventional apparatus using some more highly scattering medium such as a 'Ludox' solution.

The scattering from films must be corrected for reflection, refraction and secondary scattering.[105] When making measurements on oriented samples, the scattering is complicated by the effect of birefringence.[106,107] The error is minimised by making measurements under conditions where Ω is equal or complementary to ψ_A and ψ_P.

(C) Theoretical interpretation

As discussed in Section 3.3.2, scattering may be interpreted in terms of particulate or statistical theories. The particulate theories have been most useful in interpreting photographic patterns. Such theories have been used to account for scattering from spherulitic or rodlike crystalline aggregates.

(D) Spherulite scattering

Spherulitic structure is common to many crystalline polymers. These spherulites are optically anisotropic aggregates of crystals arising from growth from a heterogeneous nucleus at their centre. In most cases, they grow until they fill the volume of the polymer and meet at planar surfaces so as to form polygonal structures (Fig. 37). The sizes of these spherulites

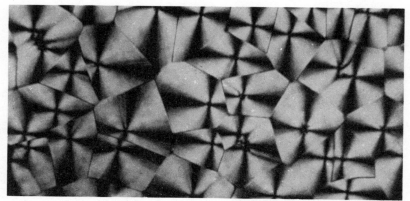

Fig. 37. A photomicrograph of polyethylene spherulites. (From Stein, R. S. (1973). In Structure and Properties of Polymer Films (Ed. R. W. Lenz and R. S. Stein), Plenum Press, New York, p. 10.)

vary from less than a micrometre to hundreds of micrometres depending upon crystallisation conditions. Their optic axis is usually along or perpendicular to their radius and the highest refractive index direction may be radial or tangential.

The scattering from such structures may be described surprisingly well by the idealised model of isolated anisotropic sphere having radial and tangential polarisabilities α_r and α_t and radius R which is imbedded in an isotropic uniform matrix of polarisability α_s. This leads to the equations:

$$I_{H_v} = K_3 V^2 \{(3/U^3)$$
$$\times (\alpha_t - \alpha_r) \cos^2(\theta/2) \sin \Omega \cos \Omega (4 \sin U - U \cos U - 3\mathrm{Si}\, U)\}^2 \qquad (111)$$
$$I_{V_v} = K_3 V^2 \{(3/U^3)(\alpha_t - \alpha_s)(2 \sin U - U \cos U - \mathrm{Si}\, U)$$
$$+ (\alpha_r - \alpha_s)(\mathrm{Si}\, U - \sin U)$$
$$- (\alpha_t - \alpha_r) \cos^2(\theta/2) \cos^2 \Omega (4 \sin U - U \cos U - 3\mathrm{Si}\, U)\}^2 \qquad (112)$$

where $V = (4/3)\pi R^3$ is the volume of the spherulite of radius R. Ω is the azimuthal scattering angle and $\mathrm{Si}\, U = \displaystyle\int_0^U (\sin x/x)\,dx$.

Plots of these equations are given in Figs. 38 and 39. The H_v pattern has a characteristic four-leaf clover appearance while the V_v pattern has

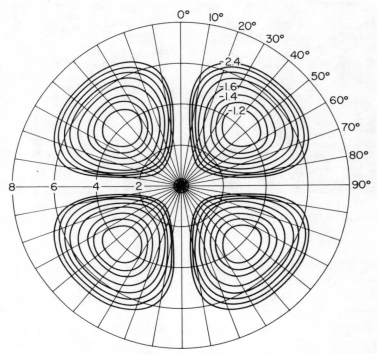

Fig. 38. A theoretical plot of an H_v *scattering pattern. (From Stein, R. S., Erhardt, P., van Aartsen, J. J. Clough, S. and Rhodes, M. (1966). J. Polymer Sci., P & C,* **13,** *1.*

two-fold symmetry. The maximum in the H_v pattern occurring at $\Omega = 45°$ is found when:

$$U_{max} = 4\cdot1 = 4\pi(R/\lambda)\sin(\theta_{max}/2) \tag{113}$$

Thus, the value of θ_{max} at which the H_v intensity is a maximum serves as a measure of R and may be used to follow the spherulite growth.[99] An automatic device has been described for scanning the H_v pattern for rapidly following the growth kinetcs.[108]

An analysis of the effect of the size distribution of spherulites demonstrates that the size obtained from the position of the maximum is highly weighted in favour of the large sizes.[109] An effect of the distribution of sizes is to 'wash out' the maxima corresponding to higher order interference.

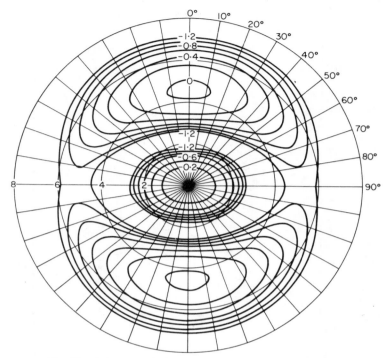

Fig. 39. A theoretical plot of a V_v *light scattering pattern.*

The appearance of the V_v pattern depends upon the polarisability of the surroundings, α_s, and has been shown by the theoretical and experimental studies of Samuels.[110] For high concentrations of spherulites, the surroundings are really other spherulites and an expression has been obtained for α_s in terms of the polarisabilities of the spherulites, that of the amorphous interspherulitic material and of the volume fraction of spherulites.[111] This approach accounts for the experimental observation that the V_v intensity passes through a maximum as the degree of crystallinity increases.

A comparison of the variation of H_v intensity with U predicted by eqn. (111) with that which is experimentally measured is shown in Fig. 40. It is seen that the experimental scattering is considerably greater than the theoretical at both large and small values of U. A part of this difference results from the spherulites not being perfect spheres, they are

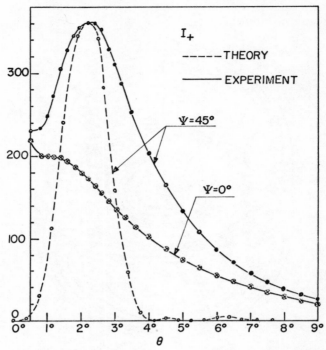

Fig. 40. A comparison of the theoretically calculated and experimentally measured variation of light scattered intensity with scattering angle θ for H_v scattering at 45° to the polariser and analyser direction. (From Hashimoto, T., Prud'homme, R. E. and Stein, R. S. (1973). J. Polymer Sci., A-2, 11, 693, 709.)

truncated by impingement with other spherulites.[112] The truncation has recently been described statistically[113] and has been shown to lead to excess scattering. The dependence of scattering upon azimuthal angle becomes less so that the intensity does not go to zero at $\mu = 0°$ and $90°$ as predicted by the theory of single perfect spherulites. The photographic scattering patterns assume the 'tennis racket' appearance described by Motegi *et al.*[114] Theoretically predicted patterns are shown in Fig. 41 for perfect and truncated spherulites.

Another cause for 'tennis racket' type patterns is the incompleteness of development of spherulites. Spherulites nucleate from bundle-like crystals which evolve into sheaves and eventually into complete spheres. Such evolution has been idealised by a model of sectors of spherulites[114,115] leading to patterns of the observed type. The evolution of patterns occurring during the early stages of growth of polyethylene terephthalate spherulites has been described by Misra and Stein[116,117] where they change progressively from rodlike through 'tennis-racket' towards spherulitic type.

A critical comparison of experimental patterns from polyethylene spherulites with those predicted for perfect truncated spherulites leads to

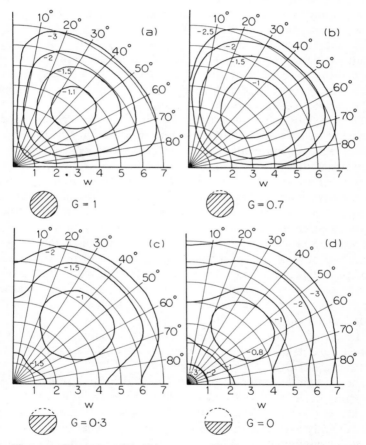

*Fig. 41. Theoretically predicted H_v light scattering patterns for perfect and truncated spherulites. The parameter G characterises the degree of truncation. (From Stein, R. S. and Picot, C. (1970). J. Polymer Sci., A-2, **8**, 3127.)*

the conclusion that the truncation can only account for a small part of the deviation from ideality. The more important cause is that of internal disorder of spherulites. A perturbation theory of deviations in which deviations from the average density within the spherulite was described in terms of a density correlation function was first offered by Stidham et al.[118] This treatment was further generalised to the case of disorder in orientation by Chu and Stein[119] and disorder in anisotropy by Hashimoto and Stein.[120] Because of mathematical limitations, it was only possible to consider either disorder in the radial or angular direction separately with this model. A more general model[121] involving an extension of a lattice model of disorder[122] has been recently developed. In these models, the optic axis direction is not fixed at a definite angle to the spherulite radial direction, but departs from it in a manner that is correlated with the departure of its neighbours. The excess scattering at large and small angles depends upon the magnitude and correlation distance of these deviations.

For arrays of spherulites, interspherulitic interference must be considered using an extension of the theory discussed in Section 3.3.2 (eqn. (50)).[123] Pairs of spherulites lead to patterns which exhibit beating of the electromagnetic waves from the individual spherulites against each other.[124] Larger spherulitic arrays give more complex interference patterns which agree with experimental patterns obtained from clusters of separated isotactic polystyrene or starch spherulites. The procedure has been extended[125] to computer generated random arrays of up to 200 spherulites for which the scattering pattern is like that of a single spherulite but modulated with a fine structure dependent upon the interspherulitic interference. The position of the H_v scattering maximum is not affected by this interference. It is postulated that this same interference effect leads to the speckling which is found in photographic scattering patterns from actual spherulitic samples.

These studies show that the theory for spherulitic scattering with its embellishments accounts quite well for most of the features of observed scattering patterns for spherulitic samples.

The change in scattering patterns upon deforming polyethylene was observed by Stein and Rhodes.[97] The analysis of these patterns was carried out for two-dimensional spherulites by Clough et al.[126] and subsequently extended to three dimensions.[127] The theory involves an affine deformation of the spherical spherulite to an ellipsoid of revolution, accompanied by the rotation of the optic axes which are assumed to be initially oriented at an angle β_0 to the radius, and at a twist angle ω

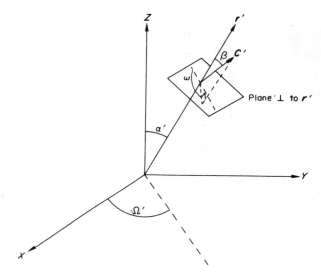

*Fig. 42. The angles β and ω defining the orientation of the optic axis (C') with respect to the spherulite radius. (From van Aartsen, J. J. and Stein, R. S. (1971). J. Polymer Sci., A-2, **9**, 295.)*

(Fig. 42) which is random. With deformation, β decreases by an amount determined by a parameter K, and ω departs from randomness by an amount described by a second parameter, η. Both of these angles change in such a way as to turn the optic axis toward the stretching direction with elongation. Some typical theoretical scattering patterns for various values of these parameters are shown in Fig. 43. These can be fitted to the observed patterns.

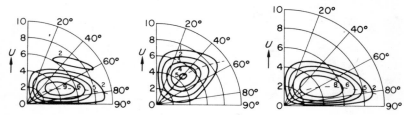

*Fig. 43. A quadrant of the theoretically predicted H_v scattering pattern for a 50% elongated sample for: (a) $K = 0, \eta = 0$; (b) $k = 0.4, \eta = 0$; (c) $K = 0, \eta = 0.50$. (From van Aartsen, J. J. and Stein, R. S. (1971). J. Polymer Sci., A-2, **9**, 295.)*

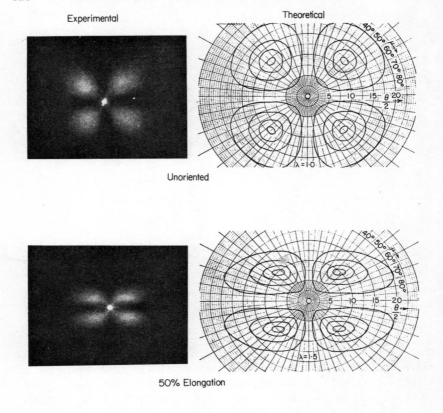

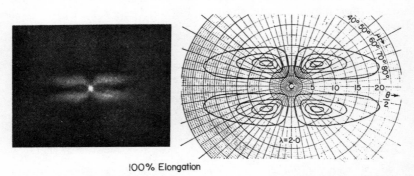

Fig. 44. The calculated H_v scattering patterns from deformed spherulites obtained using Samuels' theory. (From Samuels, R. (1968). Hercules Chemist, **56**, 19.)

Samuels has proposed a semi-empirical theory to account for the change in scattering patterns with deformation.[98] He has substituted the definition of U given by eqn. (44) of Section 3.3.2 into the Stein–Rhodes eqns. (111) and (112). While this procedure is not rigorous, it leads to predicted patterns quite similar to that for the rigorous theory and which agree quite well with experiment. This is illustrated in Fig. 44 where the experimental H_v scattering patterns for deformed polypropylene samples may be compared with calculated patterns.

Such theories may be used to determine the degree of spherulite deformation from experimental patterns, and have been used for interpreting rates of deformation from light scattering motion pictures.[100] They have been used more recently for explaining the results of dynamic light scattering measurements where samples are subjected to an oscillatory strain.[128]

The theories of spherulite deformation have been based upon the model of a perfect spherulite. The change in patterns accompanying deformation are undoubtedly affected by truncation and by internal disorder. The effect of these has been investigated.[129]

(E) Rod scattering

The theory of the scattering from rods has been generalised to include the case of anisotropic rods by Rhodes and Stein[130] who give, for example, for the H_v scattering intensity from a thin rod of length L lying in a plane perpendicular to the incident beam and tilted at an angle α with respect to the vertical direction (Fig. 45):

$$I_{H_v}(\alpha) = K_8 L\{[\sin(kaL/2)/(kaL/2)]\sin\alpha'\cos\alpha'\}^2 \qquad (114)$$

where $\alpha' = \alpha + \omega$ and ω is the angle between the rod axis and the optic axis. The anisotropy of the rod is δ, $k = 2\pi/\lambda$ (where λ is the wavelength in the medium), and

$$a = -\sin(\alpha + \Omega)\sin\theta \qquad (115)$$

For an assembly of such rods scattering independently, the total scattering is then:

$$I_{H_v} = \int N(\alpha) I_{H_v}(\alpha)\, d\alpha \qquad (116)$$

where $N(\alpha)\,d\alpha$ is the number of rods oriented in the angular interval

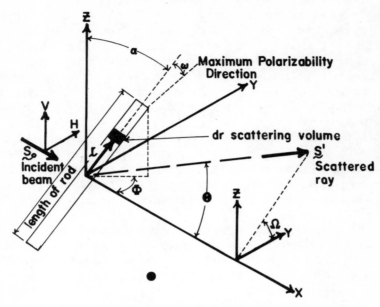

Fig. 45. The co-ordinate system for rod scattering. (From Rhodes, M. B. and Stein, R. S. (1969). J. Polymer Sci., A-2, 7, 1539.)

between α and $\alpha + d\alpha$. This distribution function depends upon the elongation of the sample. For an assumed distribution function:

$$N(\alpha) = N_0(\varepsilon^{-2}\cos^2\alpha + \varepsilon^2\sin^2\alpha)^{-1/2} \tag{117}$$

where ε is a parameter characterising the degree of orientation, scattering patterns were calculated. Some typical ones are shown in Fig. 46.

The calculations were generalised to three dimensions by van Aartsen[131] and by Hayashi and Kawai,[132] to give results which are qualitatively similar to the results in two dimensions.

Rod scattering patterns differ from spherulitic patterns in that the scattering is a maximum at $\theta = 0$ and decreases with increasing scattering angle. Hence, the size of the rod cannot be simply obtained from the position of a scattering maximum as in the case of spherulitic scattering, but must be determined from a quantitative measure of the rate of fall-off of intensity with θ.

Recent discussion and application of rod scattering theory to the scattering from hydroxypropyl cellulose has been given by Samuels[133] and from collagen by Chang and Chien[134] and Wilkes.[135]

These theories for rod scattering are for isolated rods in a uniform isotropic medium. The previously discussed treatment for interparticle

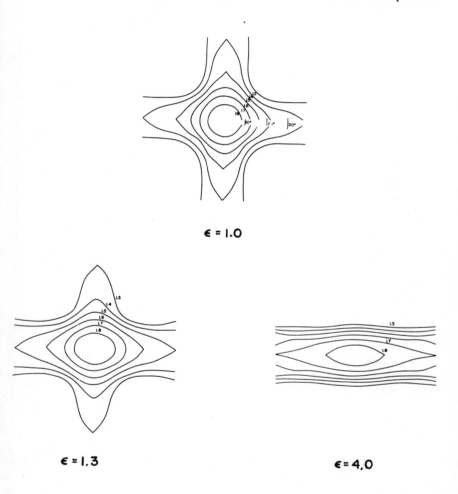

Fig. 46. Calculated H_v rod scattering patterns for anisotropic rods for which $\omega = 45°$ at elongations characterized by $\varepsilon = 1·0$, $1·3$ and $4·0$. (From Rhodes, M. B. and Stein, R. S. (1969). J. Polymer Sci., A-2, 7, 1539.)

interference may be extended to interference among rods.[136] In this case, a distribution function must be specified which describes the relative orientation as well as the relative location of the rods. Such a function may be formulated in terms of statistical parameters of the sort described in Section 3.3.2. It is found that unlike the case of spherulitic scattering, the scattering pattern is very dependent upon interparticle interference. This distribution function is dependent upon sample deformation. Thus the change in scattering patterns from rodlike systems depends not only upon the orientation of the individual rods but also upon the changing arrangement of the rods with respect to each other.

(F) *Applications of statistical theories of scattering from anisotropic systems*
As pointed out in Section 3.3.2, anisotropic scattering units which are not organised into aggregates having a definite shape scatter in a manner that is best treated by statistical theories. Random orientation correlation leads to scattering with V_v and H_v polarisation described by eqns. (75) and (76) of Section 3.3.2. Such scattering is independent of azimuthal angle, μ, and leads to patterns which are circularly symmetrical about the incident beam.

In this case, the correlation functions may be obtained by Fourier inversions of the intensity data, or correlation distances may be calculated from suitable plots of functions of intensity against angle if the correlation functions are gaussian or exponential as shown in eqns. (68) or (70) of Section 3.3.2.

In many cases, however, azimuthally dependent patterns are obtained as, for example in annealed polytetrafluoroethylene film. Such scattering may be treated in terms of the previously described non-random orientation correlation theories. An alternate approximate approach is that proposed by Keijzers et al.[103] in which a scattering pattern intermediate between that of an azimuthally independent random orientation correlation pattern and a highly azimuthally dependent spherulite pattern is represented by a summation of patterns of the two types.

$$I(\theta, \mu) = \phi_s I_s(\theta, \mu) + \phi_R I_R(\theta) \tag{118}$$

where ϕ_s and ϕ_R are the fractional contributions of the spherulitic and random contributions and $I_s(\theta, \mu)$ and $I_R(\theta)$ are the scattering functions for perfect spherulites and of random structures as given by eqns. (111) and (112) and by eqns. (75) and (76) of Section 3.3.2. In such cases, the parameters which must be specified are ϕ_s, ϕ_R, R, α_t, α_r, α_s, $\langle \eta^2 \rangle$, δ^2, $\gamma(r)$ and f(r). An application of this approach to the study of quenched poly-propylene film has been described by Keijzers et al.[103] A limitation of this approach is that the random contribution to the scattering is usually not a separate phase scattering independently from the spherulitic part, but rather it is usually part of the spherulitic structure. Consequently, cross-correlation between the spherulitic and random scattered amplitudes is undoubtedly important, so that the addition of intensities is an approximation.

The extension of the random orientation correlation theory to the description of the scattering from oriented polyethylene films was described by Stein and Hotta.[56] This application is not strictly correct since it was made prior to the development of non-random orientation correlation theory and it is now realised that random theory is not adequate for the description of such films. In fact, it is now felt that random correlations in oriented systems are quite unlikely and that the degree of non-randomness increases with orientation. While the non-random correlation theory has been generalised to permit the description of oriented systems,[113] it is still rather complex so that it has not been actually applied. In principle eqn. (118) could be used to describe oriented systems, by using the van Aartsen–Stein[127] expression for I_s and the Stein–Hotta[56] expression for I_R in the oriented state. It is likely that ϕ_s and ϕ_R may also vary with orientation. So far, this has not been done.

3.4 SONIC TECHNIQUES

3.4.1 Introduction

One method used for studying the orientation of polymers which does not make use of electromagnetic radiation is that of sonic velocity or pulse propagation. This technique is particularly suited to the characterisation of orientation in fibres (or samples having rodlike geometry) and, although the technique suffers somewhat from not having a sound theoretical basis, it is of particular usefulness in instances where an 'orientation index' or parameter is desirable for relative comparisons.

3.4.2 Propagation of a sonic pulse in a solid

Figure 47(a) shows a schematic diagram where the polymer chains are oriented at some average angle θ to the stretch axis while in Figs. 47(b) and (c), θ has values of 0° and 90°, respectively. Simple physical con-

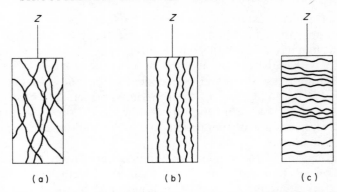

Fig. 47. Hypothetical examples of chain orientation in an amorphous polymer (a) partial parallel orientation along Z, (b) perfect parallel orientation, (c) perfect perpendicular orientation.

siderations suggest that sonic velocity (in the Z direction) is orientation dependent since in Fig. 47(b) the mechanism of propagation occurs along the polymer chain (intramolecular bond stretching), while in Fig. 47(c) the mechanism must depend on the propagation across the chains, i.e. via van der Waals bonds (intermolecular bond stretching) or other chain–chain interactions such as hydrogen bonding. Based on these different mechanisms, theoretical calculations utilising the force constants of bond stretching have predicted that the rate of propagation will be much greater when directed along the chain, i.e. along chemical bonds, than when directed perpendicularly to the chain.[137,138] Utilising the above, Moseley[139] proposed a semi-empirical treatment to relate the sonic velocity to orientation. He considered two different cases: (a) a series addition of force constants and (b) a parallel addition of force constants. The former results in eqn. (119)

$$\frac{1}{C^2} = \frac{1 - \langle \cos^2 \theta \rangle}{C_\perp^2} + \frac{\langle \cos^2 \theta \rangle}{C_\parallel^2} = \frac{\langle \sin^2 \theta \rangle}{C_\perp^2} + \frac{\langle \cos^2 \theta \rangle}{C_\parallel^2} \qquad (119)$$

where C, $C_\perp$, and $C_\parallel$ represent the sonic velocity for the sample having an average orientation angle θ, 90° (perpendicular orientation), and 0° (parallel orientation), respectively. The parallel addition case, (b), will not be considered further for it was demonstrated by Moseley that it leads to invalid results when correlated with experimental observations. We note in eqn. (119) that it was assumed that only the second moment ($\langle\cos^2\theta\rangle$) of the orientation distribution is influential in determining the velocity. At this point we will assume this to be true, but we will return to discuss this later on in the text.

To utilise eqn. (119) requires knowledge of the values for $C_\perp$ and $C_\parallel$ which may be obtained from theoretical calculations or experimentally.[140] This approach however, is usually not taken since the theoretical values are not easily calculated (and may not be highly accurate) and furthermore, limited experimental data exists. Rather, further simplification of eqn. (119) can be made as follows. If we consider the case of a randomly oriented (isotropic) sample, i.e. $\langle\cos^2\rangle = 1/3$, then eqn. (119) becomes

$$\frac{1}{C_u^2} = \frac{2}{3C_\perp^2} + \frac{1}{3C_\parallel^2} \tag{120}$$

where C_u is the sonic velocity for an unoriented sample which can be obtained experimentally. Rearranging eqn. (120) gives

$$C_\perp^2 = 2C_u^2 C_\parallel / 3C_\parallel^2 - C_u^2 \tag{121}$$

Since typical values of $C_\parallel$ and C_u are of the order of 14 km/s and 1·5 km/s respectively, we can assume $3C_\parallel^2 \gg C_u^2$ and thus

$$C_\perp^2 = \frac{2C_u^2}{3} \tag{122}$$

Substituting this for $C_\perp^2$ in eqn. (119) gives

$$\frac{1}{C^2} = \frac{3(1-\langle\cos^2\theta\rangle)}{2C_u^2} + \frac{\langle\cos^2\theta\rangle}{C_\parallel^2} \tag{123}$$

If we assume that $C_\parallel^2$ is considerably larger than C_u^2 so that the last term of eqn. (123) will be small (except at very high orientation where θ is small), then by rearrangement we obtain

$$\langle\cos^2\theta\rangle = 1 - \frac{2}{3}\frac{C_u^2}{C^2} \tag{124}$$

where the measured quantities C_u and C can be measured thus allowing determination of the orientation. Knowing $\langle\cos^2\theta\rangle$ one can then calculate the Herman's orientation function (see Section 1.5.2). The reader wishing to note the degree of validity of the assumptions and the details of the above derivation should consult Ref. 139.

The sonic velocity is related to the sonic modulus, E, $i.e.$ Young's modulus measured at high frequency, via the expression

$$E = \rho C^2 \tag{125}$$

where ρ is the density of the material. Hence eqn. (124) and those previous may be rewritten in terms of moduli rather than velocity.

Before the applicability of eqn. (124) is considered in more detail, it must be pointed out that eqn. (124) has also been arrived at by Ward[141] via a different approach. His model assumes that the solid can be considered as an aggregate of structural units whose elastic constants are identical to those of the highly oriented material. For the case of continuity of stress, Ward has shown that

$$\frac{1}{E} = \frac{\langle \sin^4 \theta \rangle}{E_\perp} + \frac{\langle \cos^4 \theta \rangle}{E_\parallel} + \langle \sin^2 \theta \cos^2 \theta \rangle \left(\frac{1}{G} - \frac{2\gamma}{E_\parallel} \right) \tag{126}$$

where G refers to the torsional modulus of the perfectly oriented material and γ is its Poisson's ratio. We note that E is not only a function of the second moments of the orientation distribution, but is also dependent on the fourth moments as well. Since γ is of the order of 0·5, the term of $\gamma/E_\parallel$ will be small as will $1/E_\parallel$ relative to $1/G$ or $1/E_\perp$. In this case, eqn. (126) reduces to

$$\frac{1}{E} = \frac{\langle \sin^4 \theta \rangle}{E_\perp} + \frac{\langle \sin^2 \theta \cos^2 \theta \rangle}{G} \tag{127}$$

If we make the further assumption that $E_\perp \approx G$, which is not too unreasonable for moderate orientation, then eqn. (127) becomes

$$\frac{1}{E} = [\langle \sin^4 \theta \rangle + \langle \cos^2 \theta \sin^2 \theta \rangle] \frac{1}{E_\perp} \tag{128a}$$

$$= \langle \sin^4 \theta + \cos^2 \theta \sin^2 \theta \rangle \frac{1}{E_\perp} \tag{128b}$$

$$= \langle \sin^2 \theta \rangle \frac{1}{E_\perp} \tag{128c}$$

$$= \frac{1 - \langle \cos^2 \theta \rangle}{E_\perp} \tag{128d}$$

$$= \frac{1 - \langle \cos^2 \theta \rangle}{C_\perp^2} \tag{128e}$$

which is identical to eqn. (119) if one again assumes the second term of eqn. (119) is negligible, i.e. ($C_\parallel \gg C_\perp$).

We therefore find that Moseley's semi-empirical equation can be obtained theoretically but only under a number of restricting conditions. Even so, the method has proven useful and correlates quite well with a number of the others discussed in this chapter.

The sonic technique suffers from one of the same difficulties as the birefringence method in that only an *average* orientation of the total system is obtained. Thus, if one has a multicomponent system (*e.g.* semi-crystalline, block copolymer, etc.), one cannot separate component orientation by this technique alone. It is noteworthy that the orientation measured by the sonic method can be correlated directly with birefringence data on the same material.[142] This ease of correlation has been suggested to lead to values of the intrinsic birefringence by extrapolation of the sonic data to perfect orientation.[142,143]

Samuels[143] has extended the treatment of Moseley to consider a two phase system—specifically that of a semi-crystalline polymer. His treatment results in the following equation

$$\frac{3}{2}(\Delta E)^{-1} = \frac{(1 - \beta)f_1}{E_1} + \frac{\beta f_2}{E_2} \tag{129}$$

where β is the volume fraction of component 2 while f_1 and f_2 are the respective Herman's orientation functions for components 1 and 2. E_1 and E_2 are the modulus values of the perfectly oriented components and

$$(\Delta E^{-1}) = \left(\frac{1}{E_u} - \frac{1}{E} \right) \tag{130}$$

where E_u and E are the respective moduli for an unoriented sample and the sample being measured. Clearly this equation can be rewritten in terms of sonic velocities by the use of eqn. (120). It is pointed out that in the derivation of eqn. (129) a number of assumptions and restrictions are involved. The first is that densities and compressibilities are assumed to be additive. This assumption has been shown to be reasonable in many instances.[144,145] It is also assumed that the Poisson's ratio is 0·33 for *each component*. This suggests that the measurements should be made near or below the glass transition temperature rather than in the rubbery region.

The usefulness of eqn. (129) becomes apparent when at least one of the *f* values can be determined by another method, *e.g.* X-ray diffraction for the crystalline orientation. Wide applicability of eqn. (129) is limited, however, since it requires *knowledge* of both E_1 and E_2 which is generally not available.

3.4.3 Measurements of sonic velocity

One of the advantages of the sonic method is its ease of measurement. Figure 48 shows a schematic of a typical experimental set-up where one simply has a piezoelectric crystal which rests on the sample. This crystal, when pulsed, transmits a sonic wave through the sample and this propagating wave is detected by a second piezoelectric crystal. Knowing the propagation distance and incremental time between pulse and signal detection allows the calculation of the velocity. Sample geometry is also important in that the length-width ratio must be of the order of ten or more, *i.e.* the sample must be rodlike in order for eqn. (120) to apply. The importance of the geometry has been discussed by Ballou and Silverman.[146]

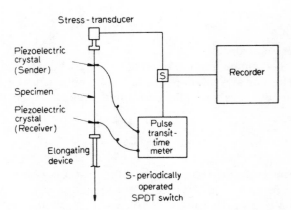

Fig. 48. Schematic of sonic apparatus for measuring orientation (after Moseley[139]).

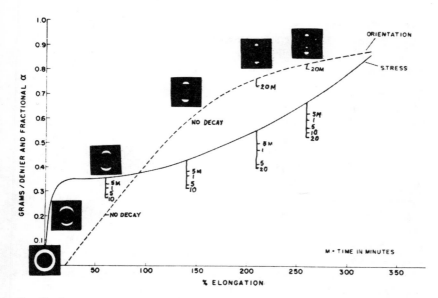

Fig. 49. Stress–strain and sonic orientation–strain curves for undrawn Nylon 66. Wide angle patterns included were taken at the corresponding elongations (from Charch and Moseley[147]).

Because the pulse frequency is very high ($\sim$ 10 kHz), one can utilise the method as an 'on-line' system to monitor orientation induced during fibre processing operations such as fibre drawing, spinning, etc.[139,147] Generally the technique is applied in conjunction with other methods. For example, in Fig. 49 data of Charch and Moseley[147] are given on a drawn Nylon 66 fibre. Note that X-ray diffraction as well as sonic velocity data have been combined with the stress strain behaviour of the filaments.

In conclusion the usefulness of the sonic technique has found merit when applied to uniaxially deformed materials that can be fabricated to give rodlike geometry. It seems that its application to biaxially deformed material, however, hardly has desirability, particularly if one is attempting to determine planar orientation.

3.5 GENERAL DISCUSSION—INTER-RELATION AND LIMITATIONS OF METHODS

3.5.1 General comments

Most polymer systems are sufficiently complex such that a single method cannot completely characterise the system, and the co-ordinated application of a number of the methods is of greatest value. We should like to discuss a few such applications.

3.5.2 Orientation of polyethylene

The orientation of the crystals of polyethylene is most completely described in terms of the pole figure as discussed in Section 3.2. Such pole figures have been published for polyethylene for example by Lindenmeyer and Lustig,[148] Desper and Stein[149] and Desper[150] as previously discussed. Orientation functions have also been calculated for these samples by Desper *et al.* and represented as points on the triangular orientation diagram. It should be emphasised that the orientation function is a second moment average over the orientation distribution and does not uniquely characterise it. It is possible to have different pole figures for samples having the same orientation functions.

It is desirable to account for the orientation change of samples in terms of detailed mechanisms. For example, Wilchinsky first attempted to account for the change in crystal orientation upon stretching polyethylene in terms of a model involving the affine deformation of spherulites.[151] Refinements of this model were proposed by Stein and coworkers[152] as well as Kawai and coworkers[153] in which various mechanisms of re-orientation of crystals were postulated which take place to different degrees in different parts of the spherulite as is experimentally indicated. The ability to predict the proper second moment orientation function is usually not a sufficiently rigorous test of a deformation model to assure uniqueness. In addition, it would be better to also attempt prediction of the fourth order orientation functions (see, for example, Ref. 155), or even better, the complete angular distribution of X-ray intensities or their pole figures.

A spherulitic sample possesses rather complex morphology for the complete analysis of the orientation mechanisms in that lamellae are distributed over a range of angles with respect to the stretching direction. Also a singularity exists at the centre of the spherulite where the deformation mode is quite complex. Consequently, measurements made on samples having 'single texture' are preferred[155] in which by processes of rolling, stretching and annealing, lamellae are arranged at a uniform angle throughout the sample. In this way, measurements may be made for which the direction of strain is at some fixed angle with respect to the lamellae direction. The results of such studies may then be used for the interpretation of changes in the more complex spherulite morphology.

As has been pointed out, a number of methods have been used for separating the orientation of polyethylene into crystalline and amorphous components. While the results of comparisons of such methods are in relatively good agreement for the specification of crystalline orientation, critical comparison has not been made for amorphous orientation. In cases where the infra-red results have been compared with measurements made by a combination of X-ray diffraction and birefringence, a different degree of amorphous orientation has been obtained by the infra-red method. This is now believed to be related to the observation that differing degrees of amorphous orientation are obtained using differing infra-red bands,[156] and this is believed due to the conformational sensitivity of infra-red absorption[157-159] as previously discussed. This conformational dependence has been verified in recent studies by Read and Hughes[160] on the dichroism of the 720 cm^{-1} band of amorphous cross-linked polyethylene. The band is shown to originate from a sequence of *trans*-conformations[157] which should theoretically[158,159] exhibit greater orientation than configurational sequences involving *gauche* bonds. Thus, the amorphous orientation as measured from the dichroism of this band represents that of a selected portion of the amorphous phase whereas the birefringence is more a function of the entire amorphous phase.

There has been some comparison of the amorphous orientation of linear polyethylene as measured by the X-ray–birefringence combination with that obtained from a Congo Red dyed sample with good agreement.[161]

There have been extensive measurements of the light scattering from oriented polyethylene as previously discussed.[162] The change in the photographic low-angle scattering pattern upon stretching can be qualitatively accounted for quite well on the basis of the assumption of affine deformation of spherulites. A quantitative comparison of the scattered intensity of the H_v pattern of unstretched samples with the theory gives reasonably good agreement provided that proper account is taken of the effects of truncation, internal disorder and interspherulitic interference.[163] In principle, if one can fit the X-ray data using a model for spherulite deformation, the same type of theory could be used to predict the local contribution of crystals to the spherulite birefringence at any position, and hence the change in scattering with deformation should be predictable. Two complications are (1) the effects of truncation

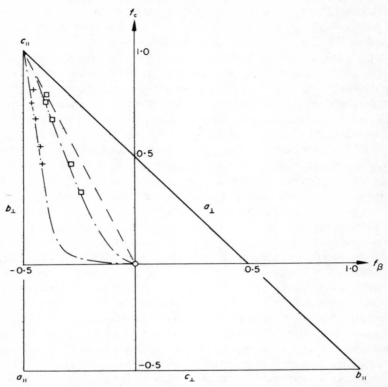

Fig. 50. Orientation function triangle diagram for the b and c-axes for isotactic polypropylene: (□) hot drawn samples (110°C); (+) melt spun fibres; (---) cold drawn polypropylene (data of Hoshino et al.[165]); (—·—) extrapolation (from Samuels[164]).

and disorder which changes appreciably upon deformation, and (2) the local amorphous orientation contribution to the spherulite birefringence is appreciable and changes with deformation.

3.5.3 Orientation of polypropylene
Samuels has extensively studied the orientation of polypropylene by a number of techniques.[164]

Typical orientation diagram plots obtained by X-ray diffraction for hot drawn film melt spun fibres, and cold drawn films of isotactic polypropylene are shown in Fig. 50. It is noted that the crystal orientation change is dependent upon the temperature at which the sample is drawn. With the cold drawn sample, there is no preferential orientation of the a or b-axis as the c-axis aligns parallel to the stretching direction, while for hot drawing, there is a tendency for the b-axis to align more perpendicularly to the stretching direction than the a-axis.

The orientation of the crystals was also followed by infra-red dichroism using the 1220 cm^{-1} crystal band. The absolute orientation could not be independently established from these measurements because of the uncertainty of the angle that the transition moment makes with respect to the crystal axes. This angle may, however, be determined by calibrating the infra-red dichroism measurements against X-ray diffraction measurements of crystal orientation. These measurements by Samuels[164] established that the transition moment for this band makes an angle of 72° with respect to the crystal c-axis.

The use of birefringence for the establishment of the relative amount of crystalline and amorphous orientation was limited by the lack of knowledge of the values of the intrinsic birefringence of the crystalline and amorphous phases. It is not possible to obtain isomorphous low molecular weight large single crystals as with polyethlene. Samuels[164] resorted to the use of the combination of sonic modulus and X-ray diffraction for calibrating his birefringence measurements and made use of eqn. (119) to obtain the amorphous orientation function using the values of the intrinsic moduli given by Samuels of $E^o_{t,cr} \cong 3 \cdot 0 \times 10^{10}$ dyn/cm^2 and $E^o_{t,am} = 1 \cdot 06 \times 10^{10}$ dyn/cm^2 and the degree of crystallinity ϕ_{cr} from density, and the crystalline orientation function by X-ray diffraction.

This procedure leads to values of $\Delta^o_{cr} = 33 \cdot 1 \times 10^{-3}$ and $\Delta^o_{am} = 46 \cdot 8 \times 10^{-3}$. It is noted that the intrinsic birefringence of the amorphous phase is greater than that of the crystalline. The difference is probably due to the anisotropic internal field of the crystal and is smaller than the corresponding difference believed to exist with polyethylene.[161] This value of Δ^o_{cr} is in reasonable agreement with the values of 26×10^{-3} obtained from studies by Wilchinsky,[167] and of 30×10^{-3} by Padden and Keith.[168] It is intermediate between the extreme values of 67×10^{-3} and 15×10^{-3} which are theoretically calculated by Keedy et al.[169] using the Denbigh[170] and the Bunn–Daubeny values of bond polarisability, respectively. This is reasonable since the internal field within a polypropylene crystal might be expected to be intermediate between that in a normal paraffin crystal (Bunn–Daubeny) and a normal paraffin vapour (Denbigh).

Once the intrinsic birefringence values are established the combination of birefringence and X-ray diffraction may then be used to resolve the

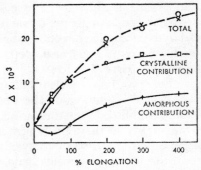

Fig. 51. Crystalline and amorphous contributions to the birefringence of deformed isotactic polypropylene (from Samuels[164]).

birefringence of polypropylene into its crystalline and amorphous components as is shown by Samuel's result in Fig. 51. It is of interest to note the negative amorphous birefringence at low elongations. Similar results were found by Hoshino et al.[165] for low density polyethylene. This may indicate that in the initial unoriented structure, there is a tendency for the

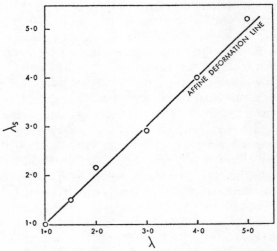

Fig. 52. Relationship between the extension ratio of the spherulite (λ_s) of isotactic polypropylene as measured by light scattering and the sample extension ratio (λ) (from Samuels[164]).

local orientation of the amorphous chains to be perpendicular to that of the crystals, so that as the crystalline c-axis orients parallel to the stretching direction, the amorphous chains orient perpendicularly.

Similar measurements have been made by Samuels[172] on as-spun and drawn polypropylene fibres and they are correlated with crystalline orientation functions, light and low angle X-ray scattering studies.

Samuels has also observed[164,172] spherulite deformation for these polypropylene films and fibres using the low-angle light scattering method and he has demonstrated, as shown in Fig. 52 that the spherulite deforms affinely so that the extension ratio of the spherulite is equal to that of the sample.

Similar analyses have also been reported by Samuels[173] for hydroxypropylcellulose. This differs from polypropylene in that the morphology is rodlike rather than spherulitic and presents a good example of the application of the theory of scattering from rodlike structures.

REFERENCES

1. E. Leitz Inc, General Catalogue, Rockleigh, N.J.
2. Tanaka, H., Masuko, T. and Okajima, S. (1972). *J. Appl. Polym. Sci.*, **16**, 441.
3. Stein, R. S. and Clough, S. (1963). ONR Tech. Rept. No. 67, Project NR 356–378, Contract No. 3357(01), Univ. Mass.
4. Erhardt, P. and Stein, R. S. (1967). *Appl. Polym. Symp. 5* (Ed. R. D. Andrews, Jr. and F. R. Eirich), Interscience, New York, Vol. 1, p. 113.
5. Wilkes, G. L. and Stein, R. S. (1969). *J. Polymer Sci. A-2*, **7**, 1525.
6. Wahlstrom, E. (1951). *Optical Crystallography*, Wiley, New York.
7. Edsell, J. T., Rich, A. and Goldstein, M. (1952). *Rev. Sci. Instrum.*, **23**, 695.
8. Hartshorne, N. H. and Stuart, A. (1970). *Crystals and the Polarizing Microscope*, Amer. Elsevier Co., New York.
9. Gurnee, E. F. (1954). *J. Appl. Phys.*, **25**, 1232.
10. Stein, R. S. (1957). *J. Polymer Sci.*, **24**, 383.
11. Bunn, C. W. and Daubeny, R. (1954). *Trans. Faraday Soc.*, **50**, 1173.
12. Wiener, O. (1912). *Abhandlgn. d. Sachs. Ges. D. Wiss., Math-Phys. Kl.*, 32.
13. Folkes, M. J. and Keller, A. (1971). *Polymer*, **12**, 222.
14. Ambronn, H. and Frey, A. (1926). *Das Polarisationmi–Kroskop*, Akademische Verlagsgesellschaft M.B.H., Leipzig.
15. Bettelheim, F. A. and Stein, R. S. (1958). *J. Polymer Sci.*, **27**, 567.
16. Kuhn, W. and Grün, F. (1942). *Kolloid-Z.*, **101**, 248.
17. Shindo, Y. and Stein, R. S. (1969). *J. Polymer Sci. A-2*, **7**, 2115.
18. Flory, P. J. and Abe, Y. (1969). *Macromolecules*, **2**, 335.
19. Nagai, K. (1971). *Prog. Polym. Sci., Japan*, 1 (Ed. M. Imoto and S. Onogi), Kodansha, Tokyo.
20. Treloar, L. R. G. (1947). *Trans. Faraday Soc.*, **43**, 277; (1954). **50**, 881; (1958). *The Physics of Rubber Elasticity*, Oxford, 2nd edn, Chap. 10.
21. Saunders, D. W. (1957). *Trans. Faraday Soc.*, **53**, 860.
22. Gent, A. N. (1961). *Macromol.*, **2**, 262.

23. Fukuda, M., Wilkes, G. L. and Stein, R. S. (1971). *J. Polymer Sci., A-2*, **9**, 1417.
24. Ishikawa, I. and Nagai, K. (1970). *Polymer J., Japan*, **1**, 116.
25. Ishikawa, I. and Nagai, K. (1969). *J. Polym. Sci., A-2*, **7**, 1123.
26. Poddubnyi, I. Y., Erenburg, G. and Yeremina, M. A. (1968). *Vysokomol. Soedin.*, **AID**, 1381.
27. Phillipoff, W. (1970). Private Communication.
28. Hoshino, S., Powers, J., LeGrand, D. G., Kawai, H. and Stein, R. S. (1962). *J. Polymer Sci.*, **58**, 185.
29. Kotani, T. and Sternstein, S. S. (1971). *Symp. on Highly Crosslinked Polymer Networks*, Plenum Press, New York.
30. *Read, B. E.* (1963). *Techniques of Polymer Science*, S. C. I. Monograph No. 11, Soc. Chem. Industry, London.
31. Stein, R. S., Onogi, S. and Keedy, D. A. (1962). *J. Polymer Sci.*, **57**, 801.
32. Alexander, L. E. (1969). *X-ray Diffraction Methods in Polymer Science*, Wiley, New York.
33. Bunn, C. W. (1961). *Chemical Crystallography*, 2nd Edn. Clarendon Press, Oxford.
34. Klug, H. P. and Alexander, L. E. (1954). *X-ray Diffraction Procedures*, Wiley, New York.
35. Bunn, C. W. (1939). *Trans. Faraday Soc.*, **35**, 482.
36. Bunn, C. W. and Garner, E. V., (1947). *Proc. Roy. Soc. (London)*, **A189**, 39.
37. Hay, I. L. and Keller, A. (1966). *J. Mater. Sci.*, **1**, 41.
38. Stein, R. S. and Norris, F. H. (1956). *J. Polymer Sci.*, **21**, 381.
39. Decker, B. F., Asp, E. T. and Harker, D. (1948). *J. Appl. Phys.*, **19**, 388.
40. Roe, R. J. and Krigbaum, W. R. (1964). *J. Chem. Phys.*, **40**, 2608.
41. Jetter, L. K., McHargue, C. J. and Williams, R. O. (1958). *J. Appl. Phys.*, **27**, 368.
42. Krigbaum, W. R., Adachi, T. and Dawkins, J. V. (1968). *J. Chem. Phys.*, **49**, 1532.
43. Desper, C. R. and Stein, R. S. (1966). *J. Appl. Phys.*, **37**, 3990.
44. Wilchinsky, Z. W. (1963). *J. Appl. Polym. Sci.*, **7**, 923.
45. Kawaguchi, T., Ito, T., Kawai, H., Keedy, D. and Stein, R. S. (1967). *ONR Tech. Rept. No. 97*, Project NR056-378, Contract No. 3357(01), Univ. Mass.
46. Oda, T. and Stein, R. S. (1972). *J. Polymer Sci. A-2*, **10**, 685.
47. Born, M. (1959). *Principles of Optics*, Pergamon Press, New York.
48. van der Hulst, H. (1957). *The Scattering of Light by Small Particles*, Wiley, New York.
49. Kerker, M. (1969). *The Scattering of Light and Other Electromagnetic Radiation*, Academic Press, New York.
50. Roess, L. S. and Shull, C. G. (1959). *J. Appl. Phys.*, **30**, 1479.
51. Gunier, A. and Fournet, G. (1955). *Small Angle Scattering of X-Rays*, Wiley, New York, p. 20.
52. Stein, R. S. and Rhodes, M. B. (1969). *J. Polymer Sci., A-2*, **7**, 1539.
53. Stein, R. S. and Wilson, P. R. (1962). *J. Appl. Phys.*, **33**, 1914.
54. Goldstein, M. and Michalik, E. R. (1955). *J. Appl. Phys.*, **26**, 1450.
55. Debye, P. and Bueche, A. M. (1949). *J. Appl. Phys.*, **20**, 518.
56. Stein, R. S. and Hotta, T. (1964). *J. Appl. Phys.*, **35**, 2237.
57. Stein, R. S., Keane, J. J., Norris, F. H., Bettelheim, F. A. and Wilson, P. R. (1959). *Ann. N.Y. Acad. Sci.*, **83**, Art. I, 37.
58. Norris, F. H. and Stein, R. S. (1958). *J. Polymer Sci.*, **27**, 87.
59. Stein, R. S., Erhardt, P., Clough, S. and Adams, G. (1966). *J. Appl. Phys.*, **37**, 3980.
60. van Aartsen, J. J. (1971). In *Polymer Networks, Structural and Mechanical Properties*. (Ed. A. J. Chompff and S. Newman), Plenum Press, New York, p. 307.
61. Gouda, J. H. and Prins, W. (1970). *J. Polymer Sci., A-2*, **8**, 2029.
62. Hashimoto, T. and Stein, R. S. (1970). *J. Polymer Sci., A-2*, **8**, 1503.
63. Scherrer, P. (1918). *Gottinger Nachrichten*, **2**, 98.

64. Buerger, M. J. (1960). *Crystal Structure Analysis,* Wiley, New York.
65. Alexander, L. E. (1969). *X-Ray Diffraction Methods in Polymer Science,* Wiley-Interscience, New York, p. 104.
66. W. H. Warhus Co., 406 Rowland Park Blvd, Barrcroft, Wilmington, Delaware 19803.
67. Kiessig, H. (1942). *Kolloid-Z.,* **98,** 213; (1957). *Ibid.,* **152,** 62; (Camera available from Richard Seifert & Co., Hamberg 13, Germany).
68. Boldvan, O. E. A. and Bear, R. S. (1949). *J. Appl. Phys.,* **20,** 983.
69. Rigaku-Denki Co. Ltd., 9-8, 2-Chome, Sotokanda, Chiyoda-ku, Tokyo.
70. Kratky, O. (1954). *Z. Electrochem.,* **58,** 49; (1955). *Kolloid-Z.,* **144,** 110; (1958). *Z. Electrochem.,* **62,** 66; (1963). *Progress in Biophysics,* Vol. 13, Pergamon Press, New York, pp. 139–141.
71. Franks, A. (1955). *Proc. Phys. Soc.,* London, **B68,** 1054; (1958). *British J. Appl. Phys.,* **9,** 349.
72. Elliott, A. (1965). *J. Sci. Instrum.,* **42,** 312; (1968). *Hilger J.,* **11,** 38.
73. Bonse, U. and Hart, M. (1966). *Z. Phys.,* **189,** 151; Koffman, D. M. (1968). *Adv. in X-Ray Analysis,* **11,** 332.
74. Gunier, A. and Fournet, G. (1955). *Small-Angle Scattering of X-Rays,* Wiley, New York, pp. 116–120.
75. Kratky, O., Porod, G. and Kahovec, L. (1951). *Z. Electrochem.,* **55,** 53.
76. Schmidt, P. W. and Hight, R. Jr. (1960). *Acta Cryst.,* **13,** 480.
77. Kratky, O., Porod, G. and Skala, Z. (1960). *Acta Phys. Austraiaca,* **13,** 76.
78. Porod, G. (1951). *Kolloid-Z,* **124,** 83.
79. Luzzati, V. (1960). *Acta Cryst.,* **13,** 939; (1963). In *X-Ray Optics and X-Ray Microanalysis.* (Ed. H. H. Pattee, V. E. Coslett and A. Engstrom), Academic Press, New York, pp. 133–156.
80. Gunier, A. (1937). *Compt. Rendues,* **204,** 1115; (1939). *Ann. Phys., Paris,* **12,** 161; (1943). *J. Chim. Physique,* **40,** 113.
81. Zernike, F. and Prins, J. A. (1927). *Z. Phys.,* **41,** 184.
82. Gunier, A. and Fournet, G. (1955). *Small-Angle Scattering of X-Rays,* Wiley, New York, pp. 42–46.
83. Born, M. and Green, H. S. (1946). *Proc. Roy. Soc., London,* **A188,** 10; (1947). **A190,** 455; Green, H. S. (1947). **A189,** 103.
84. Hermans, P. H., Heikens, D. and Weidinger, A. (1959). *J. Polymer Sci.,* **35,** 145.
85. Kratky, O. and Schwartzkopf-Shier, K. (1963). *Monatsh. Chem.,* **94,** 774; Brumberger, H., Kratky, O. and Mittelbach, P. (1964). *Monatsh. Chem.,* **95,** 1599.
86. Statton, W. O. (1962). *J. Polymer Sci.,* **58,** 205; (1964). In *Newer Methods of Polymer Characterization.* (Ed. B. Ke), Interscience, New York, Chap. 6.
87. Geil, P. H. (1963). *Polymer Single Crystals,* Wiley, New York.
88. Hosemann, R. and Bagchi, S. N. (1962). *Direct Analysis of Diffraction by Matter,* North Holland, Amsterdam.
89. Vonk, C. G. and Kortleve, G. (1967). *Kolloid-Z.,* **220,** 19; (1968). Kortleve, G. and Vonk, C. G. *ibid.,* **225,** 129.
90. Bonart, R. and Hosemann, R. (1962). *Kolloid-Z.,* **186,** 16.
91. Kavesh, S. and Schultz, J. M. (1971). *J. Polymer Sci.,* *A-2* **9,** 85.
92. Dismore, P. F. and Statton, W. O. (1964). *J. Polymer Sci., B,* **2,** 1113.
93. Schultz, J. M., Robinson, W. H. and Pound, G. M. (1967). *J. Polymer Sci., A-2* **5,** 511.
94. Hay, I. L. and Keller, A. (1965). *Kolloid-Z.,* **204,** 43.
95. Heffelfinger, C. J. and Lippert, E. L. Jr. (1971). *J. Appl. Polym. Sci.,* **15,** 2699.
96. Point, J. J., Homes, G. A., Gezovich, D. and Keller, A. (1969). *J. Mater. Sci.,* **4,** 908; Gezovich, D. and Geil, P. H. (1971). *J. Mater. Sci.,* **6,** 509.
97. Stein, R. S. and Rhodes, M. B. (1960). *J. Appl. Phys.,* **31,** 1873.
98. Samuels, R. (1966). *J. Polymer Sci., C.* **13,** 37.
99. Picot, C., Weill, G. and Benoit, H. (1968). *J. Polymer Sci., C,* **16,** 3973.

100. Erhardt, P. F. and Stein, R. S. (1967). *Appl. Polymer Symp., High Speed Testing*, Vol. VI, *The Rheology of Solids*, **5**, 113.
101. Wilkes, G. L. (1971). *J. Polymer Sci., A-2*, **9**, 1531.
102. Stein, R. S., Norris, F. H. and Plaza, A. (1957). *J. Polymer Sci.*, **24**, 455.
103. Keijzers, A. E. M., van Aartsen, J. J. and Prins, W. (1968). *J. Am. Chem. Soc.*, **90**, 3167.
104. Chu, W., private communication.
105. Stein, R. S. and Keene, J. J. (1955). *J. Polymer Sci.*, **17**, 21.
106. Stein, R. S. and Stidham, S. N. (1966). *J. Polymer Sci., A-2*, **4**, 89.
107. Stein, R. S., Erhardt, P. F. and Chu, W. (1969). *J. Polymer Sci., A-2*, **7**, 271. Stein, R. S. and Chu, W. (1970). *J. Polymer Sci., A-2*, **8**, 489.
108. Van Antwerpen, F. (1971). Thesis, *Kinetics of Crystallization Phenomena of Spherulites in Poly(ethylene Terephthalate)*, Delft Technical University, Delft, Netherlands.
109. Stein, R. S., Stidham, S. N. and Wilson, P. R. (1961). ONR Technical Report No. 30, Project: 356–378, Contract No. 3357(01), University of Massachusetts, Amherst, Massachusetts.
110. Samuels, R. (1971). *J. Polymer Sci., A-2*, **9**, 2165.
111. Yoon, D. and Stein, R. S. (1974). *J. Polymer Sci., A-2*, **12**, 735.
112. Picot, C. and Stein, R. S. (1970). *J. Polymer Sci., A-2*, **8**, 2127.
113. Prud'homme, R. and Stein, R. S. (1973). *J. Polymer Sci., A-2*, **11**, 1683.
114. Motegi, M., Oda, T., Moritani, M. and Kawai, H. (1970). *Polymer J.*, **1**, 209.
115. Stein, R. S., Picot, C., Motegi, M. and Kawai, H. (1970). *J. Polymer Sci., A-2*, **8**, 2115.
116. Misra, A. and Stein, R. S. (1972) *J. Polymer Sci., B*, **10**, 473.
117. Stein, R. S. and Misra, A. (1973). *J. Polymer Sci., A-2*, **11**, 109.
118. Stidham, S. N., Wilson, P. R. and Stein, R. S. (1963). *J. Appl. Phys.*, **34**, 46.
119. Chu, W. and Stein, R. S. (1970). *J. Polymer Sci., A-2*, **7**, 1137.
120. Hashimoto, T. and Stein, R. S. (1971). *J. Polymer Sci., A-2*, **9**, 1747.
121. Yoon, D. Y. and Stein, R. S. (1974). *J. Polymer Sci., A-2*, **12**, 763.
122. Stidham, S. N. and Stein, R. S. (1964). *J. Appl. Phys.*, **35**, 42.
123. Picot, C. and Stein, R. S. (1970). *J. Polymer Sci., A-2*, **8**, 1955.
124. Picot, C., Stein, R. S., Marchessault, R. H., Borch, J. and Sarko, A. (1971). *Macromolecules*, **4**, 467.
125. Prud'homme, R. E. and Stein, R. S. (1973). *J. Polymer Sci., A-2*, **11**, 1353.
126. Clough, S., van Aartsen, J. J. and Stein, R. S. (1965). *J. Appl. Phys.*, **36**, 3072.
127. van Aartsen, J. J. and Stein, R. S. (1971). *J. Polymer Sci., A-2*, **9**, 295.
128. Hashimoto, T., Ph.D. Thesis, University of Massachusetts, Amherst, Massachusetts; Hashimoto, T., Prud'homme, R. E. and Stein, R. S. (1973). *J. Polymer Sci., A-2*, **11**, 693, 709.
129. Yoon, D. Y., Prud'homme, R. E. and Stein, R. S., in preparation.
130. Rhodes, M. B. and Stein, R. S. (1969). *J. Polymer Sci., A-2*, **7**, 1539.
131. van Aartsen, J. J. (1960). *European Polymer Journal*, **6**, 1095.
132. Hayashi, N. and Kawai, H. (1972). *Polymer J., Japan*, **3**, 180.
133. Samuels, R. J. (1969). *J. Polymer Sci., A-2*, **7**, 1197.
134. Chang, E. P. and Chien, J. S. C. W. (1972). *Macromolecules*, **5**, 610.
135. Wilkes, G. L. (1972). *Molecular Crystals and Liquid Crystals*, **18**, 165.
136. Stein, R. S. and Prud'homme, R. E. *J. Polymer Sci., A-2*, in press.
137. Meyer, K. H. and Lotmar, W. (1936). *Helv. Chem. Acta.*, **19**, 68.
138. Lyons, W. J. (1958). *J. Appl. Phys.*, **29**, 1429.
139. Moseley, W. W. Jr. (1960). *J. Appl. Polym. Sci.*, **3**, 266.
140. Holliday, L. and White, J. W. (1971). *Pure Appl. Chem.*, **26**, 545.
141. Ward, I. M. (1964). *Tex. Res. J.*, **34**, 806.
142. Morgan, H. M. (1962). *Tex. Res. J.*, **32**, 866.

143. Samuels, R. J. (1965). *J. Polymer Sci. A-2*, **3**, 1741.
144. Urick, R. J. (1947). *J. Appl. Phys.*, **18**, 983.
145. Waterman, H. A. (1963). *Kolloid-Z.*, **192**, 9.
146. Ballou, J. W. and Silverman, S. (1944). *Tex. Res. J.*, **14**, 282.
147. Charch, W. H. and Moseley, W. W. Jr. (1959). *Tex. Res. J.*, **29**, 525.
148. Lindenmeyer, P. H. and Lustig, S. (1965). *J. Appl. Polym. Sci.*, **9**, 227.
149. Desper, C. R. and Stein, R. S. (1966). *J. Appl. Phys.*, **37**, 3990.
150. Desper, C. R. (1969). *J. Appl. Polym. Sci.*, **13**, 169.
151. Wilchinsky, Z. (1964). *Polymer*, **5**, 271.
152. Sasaguri, K., Hoshino, S. and Stein, R. S. (1964) *J. Appl. Phys.*, **35**, 47; Sasaguri, K., Yamada, R. and Stein, R. S. (1964). *J. Appl. Phys.*, **35**, 3188.
153. Fujino, K., Kawai, H., Oda, T. and Maeda, H. (1965). *Proc. Fourth Int. Congress on Rheology*. (Ed. E. H. Lee and A. L. Copley), Interscience, New York, Pt. III, p. 501; Oda, T., Normura, S. and Kawai, H. (1965). *J. Polymer Sci., A*, **3**, 1993; Oda, T., Sakaguchi, N. and Kawai, H. (1966). *J. Polymer Sci., C*, No. 15, 223; Nomura, S., Anasuma, A., Suehiro, S. and Kawai, H. (1971). *J. Polymer Sci., A-2*, **9**, 1991.
154. Hay, I. L. and Keller, A. (1965). *Kolloid-Z.*, **204**, 43; (1966). *J. Mater. Sci.*, **1**, 41.
155. Ward, I. M. (1962). *Proc. Phys. Soc.*, **80**, 1176; (1967). *Brit. J. Appl. Phys.*, **18**, 1165.
156. Read, B. E. (1968). *Macromolecules*, **1**, 116.
157. Snyder, R. G., (1967). *J. Chem. Phys.*, **47**, 1316.
158. Nagai, K. (1971). In *Progress in Polymer Science, Japan* (Ed. M. Inoto and S. Onogi), Kodensha, Tokyo.
159. Flory, P. J. and Abe, Y. (1969). *Macromolecules*, **2**, 335.
160. Read, B. E. and Hughes, D. A. (1972). *Polymer*, **13**, 495.
161. Stein, R. S. (1966). *J. Polymer Sci., C*, **15**, 185.
162. Norris, F. H. and Stein, R. S. (1956). *J. Polymer Sci.*, **20**, 209; Kratky, O. (1954). *Z. Electrochem.*, **58**, 49; (1955). *Kolloid-Z.*, **144**, 110; (1958). *Z. Electrochem.*, **62**, 66; (1963). In *Progress in Biophysics*, Vol. 13, Pergamon Press, New York, pp. 139–141.
163. Prud'homme, R., Natarajan, R. and Stein, R. S., unpublished work.
164. Samuels, R. J. (1965). *J. Polymer Sci., A*, **3**, 1741.
165. Hoshino, S., Powers, J., LeGrand, D. G., Kawai, H. and Stein, R. S. (1962). *J. Polymer Sci.*, **58**, 185.
166. Stein, R. S. (1969). *J. Polymer Sci., A-2*, **7**, 1021.
167. Wilchinsky, Z. W. (1960). *J. Appl. Phys.*, **31**, 1969.
168. Padden, F. J. and Keith, H. D. (1959). *J. Appl. Phys.*, **30**, 1479.
169. Keedy, D. A., Powers, J. and Stein, R. S. (1960). *J. Appl. Phys.*, **31**, 1911.
170. Denbigh, K. G. (1940). *Trans. Faraday Soc.*, **36**, 936.
171. Bunn, C. W. and Daubeny, R. (1954). *Trans. Faraday Soc.*, **50**, 1173.
172. Samuels, R. J. (1968). *J. Polymer Sci., A-2*, **6**, 2021.
173. Samuels, R. J. (1969). *J. Polymer Sci., A-2*, **7**, 1197.

CHAPTER 4

ULTRA-VIOLET, VISIBLE AND INFRA-RED DICHROISM

B. E. READ

4.1 ABSORBANCE AND DICHROIC RATIO

When a beam of polarised monochromatic radiation is transmitted through a polymer sample it may, depending on its wavelength λ, be partially absorbed by specific molecular groups within the material. Absorption processes generally involve some periodic relative displacement of atomic nuclei or electrons which yield a dipole moment change, or so-called transition moment M, within the molecule.[1,2] A transition of the molecule may then occur between two states of energy difference ΔE when the frequency of the radiation, $v = c/\lambda$, is equal to $\Delta E/h$. Here c and h are the velocity of the radiation and Planck's constant respectively. Since a given molecule may exhibit various absorption mechanisms, each associated with a different ΔE, radiation may be absorbed at several different wavelengths or wavelength regions. An experimental plot of transmitted intensity against wavelength of the radiation will then exhibit minima at wavelengths of maximum absorption. As is well-known, such a plot is termed the absorption spectrum of the material. Figure 1 shows schematically an absorption spectrum and illustrates the regions of wavelength corresponding to the ultra-violet (UV; $\lambda \approx 0.1$– 0.4 μm), the visible ($\lambda \approx 0.4$–0.8 μm) and the infra-red (IR: $\lambda \approx 1$– 50 μm). A given absorption peak may be characterised not only by its wavelength or frequency, which yields the energy of the molecular transition, but also by its intensity or peak absorbance A. The absorbance is proportional to the square of the transition moment magnitude and may be defined according to the Beer–Lambert law,

$$A = \log_{10}(I_0/I) \tag{1}$$

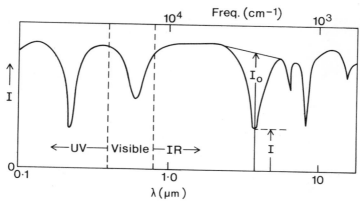

Fig. 1. Schematic absorption spectrum.

where I is the transmitted intensity at the peak and I_0 may be taken as the incident intensity. In order to correct for errors arising from partial scattering of the radiation and for the overlap of neighbouring absorption peaks, a baseline may be constructed, as indicated in Fig. 1, and I_0 determined as the transmitted intensity at the baseline.

The observed absorbance A at wavelength λ is equal to the sum of absorbance contributions from those structural units within the polymer which contain groups participating in the absorption. Furthermore, the contribution from each group depends on the angle which its transition moment vector M makes with the electric vector E of the polarised radiation. If ω is the angle between M and E then the group absorbance a_E, measured in the direction of E, is given by,[3]

$$a_E = K(M.E)^2 = K(ME)^2 \cos^2 \omega \qquad (2)$$

where K is a proportionality constant and M and E are the magnitudes of M and E respectively. If the structural units and their associated transition moments are non-randomly oriented within the polymer, it follows that the observed net absorbance will vary with the direction of E. This effect is generally referred to as dichroism, and the ratio of sample absorbances measured with the electric vector successively polarised in two mutually perpendicular directions is termed a dichroic ratio.

Let us consider (Fig. 2) three orthogonal reference axes X_1, X_2 and X_3 within an oriented polymer and regard X_3 as the draw direction for uniaxially oriented specimens and the axes X_3 and X_2 as the two stretch-

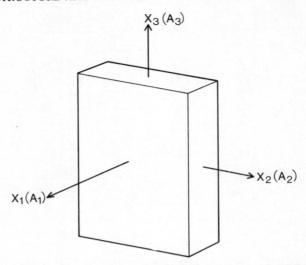

Fig. 2. Sample reference axes and principal absorbance components.

ing directions in the case of samples which are biaxially deformed. We may then define absorbances A_1, A_2 and A_3 measured with the electric vector polarised along X_1, X_2 and X_3 respectively. We note that the symmetry of the orientation distribution of structural units normally allows us to regard X_1, X_2 and X_3 as principal absorbance axes so that the absorbance in any direction of E is determined from a knowledge of A_1, A_2 and A_3. In the more general case of biaxial orientation, for which A_1 differs from A_2, we may define three dichroic ratios (see Zbinden[4]),

$$D_{32} = \frac{A_3}{A_2} \quad D_{31} = \frac{A_3}{A_1} \quad \text{and} \quad D_{21} = \frac{A_2}{A_1} \tag{3}$$

only two of which are independent since $D_{31} = D_{32} D_{21}$. If we take X_1 as the direction normal to the sample plane it will be observed that A_1 refers to the absorbance determined with the electric vector in the sample thickness direction. For thin specimens the direct determination of A_1 is thus difficult, if not impossible. Although A_1 can be evaluated from measurements on samples tilted with respect to the direction of the incident beam,[4] relatively few accurate results have yet been obtained by this method. We will thus be mainly concerned here with uniaxially oriented polymers for which $A_1 = A_2$ and for which a single dichroic ratio D_{32} suffices for the characterisation of orientation by the dichroic method.

The significance, and importance, of the dichroic ratio is that it may be quantitatively related to certain orientation functions characteristic of the average orientation of structural units within the polymer. It must be emphasised, however, that the use of such relationships requires a knowledge of the orientation of the transition moments relative to defined axes, such as the local polymer chain axis or the crystallographic axes, within the structural units. Prior to discussing these relationships in more detail (Section 4.3), we will thus consider the origins of dichroic effects in different wavelength regions, giving particular regard to the structure of absorbing groups and their transition moment directions.

4.2 MOLECULAR ORIGIN OF DICHROIC EFFECTS IN DIFFERENT SPECTRAL REGIONS

4.2.1 Ultra-violet and visible wavelengths

In the UV and visible regions, absorption processes involve electronic displacements within the molecule[2] and usually require unsaturated groups, which may either form part of the polymer molecule or may be added in the form of dye molecules. The absorption characteristics of such groups are generally specified in terms of their absorbance components along certain defined axes rather than the transition moment directions (which are mainly used to characterise infra-red absorption processes). In the case of polyvinyl chloride (PVC), absorbing sequences of conjugated double bonds, or polyene segments, can be 'built into' the polymer chain by a chemical dehydrochlorination reaction.[5] Depending on the length of the polyene segment, it will absorb either at UV or visible wavelengths, the absorption arising from the transition of π electrons between molecular orbitals of different energy. For the all-*trans* isomer, illustrated in Fig. 3, the transition moment to the lowest electronic state is almost parallel to the segment axis.[2] Consequently the absorption intensity (represented by $a_{\parallel}^{o}$) is large when the electric vector is parallel to this direction, and the small perpendicular component of absorption ($a_{\perp}^{o}$) is generally negligible. Hence, dichroism measurements at wavelengths associated with the polyene absorptions can be confidently employed to study the average orientation of polyene segment axes within the oriented polymer.

Added dye molecules, which absorb at visible wavelengths, may attach

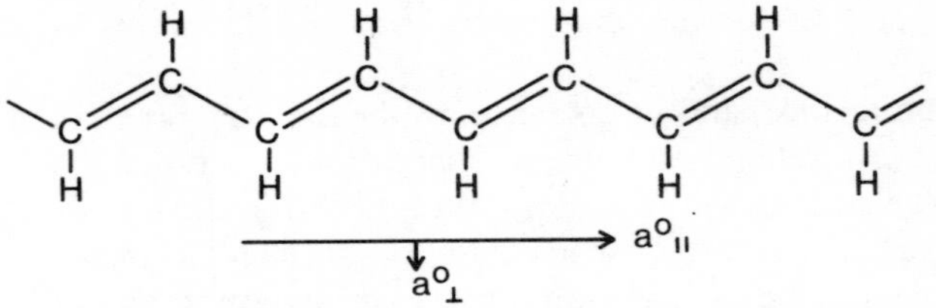

Fig. 3. A trans *polyene segment and associated absorbances.*

themselves either chemically or physically to the polymer chains. Dye molecules generally have rather complex conjugated ring structures,[6] as exemplified by the Congo-red molecule (Fig. 4) which has been used in orientation studies on several polymers. Such a molecule would be expected to exhibit maximum absorption when the electric vector lies approximately along its length in the plane of the benzene rings. Further, when the polymer is deformed, it should orient by mechanical forces with its long dimension inclined toward the polymer chain axis. Since, for partially crystalline polymers, the dye molecules are exclusively located in the amorphous phase, dichroism measurements are capable of characterising the average chain orientation within these regions. However, owing to the structural complexities of dye molecules, the transition moment direction or principal absorption intensities are seldom known with accuracy (although dichroism studies on dye crystals could help to overcome this problem), nor is the inclination of the dye molecule known with respect to the polymer chain. These factors tend to limit the usefulness of dye dichroism investigations in the precise determination of amorphous chain orientation.

4.2.2 Infra-red region
Most dichroism studies of oriented polymers have so far been made in the infra-red region, where the frequencies of absorption peaks measured

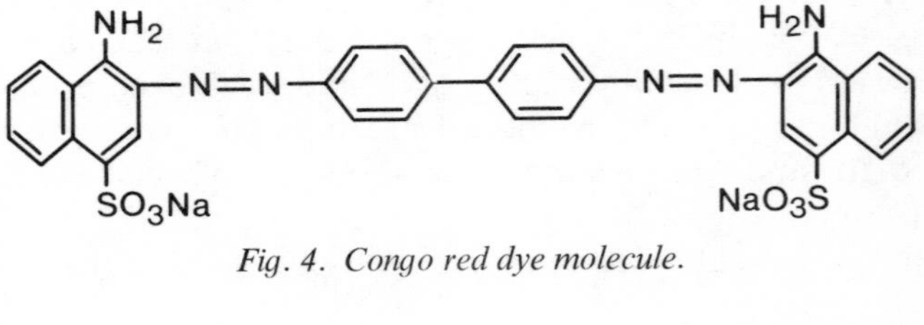

Fig. 4. *Congo red dye molecule.*

in wavenumbers, $\bar{v} = \lambda^{-1}$, typically range from about 6000 cm^{-1} to about 200 cm^{-1}. The mechanism of infra-red absorption by polymers can generally be described in terms of the normal resonance vibrations of groups of atomic nuclei which result in an increase in their vibrational energy.[4,7] Although the co-operative nature of these oscillations may generally be complex, the vibrational energy is frequently localised within certain chemical bonds or groups, and in such cases transition moment directions may be defined with little difficulty. For example, owing to the relatively small mass of the hydrogen atom, C—H stretching vibrations are frequently highly localised and the transition moment is expected to be along the chemical bond. In the case of CH$_2$ groups, which also occur widely in polymers, the two C—H bond vibrations are coupled, but the transition moment directions for various types of vibration involving the group as a whole are fairly well defined,[4] as illustrated in Fig. 5. Unsaturated chemical bonds also tend to yield localised stretching modes owing to their relatively high force con-

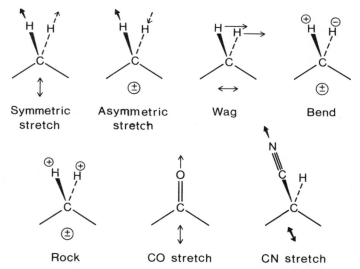

Fig. 5. Illustration of some vibrational modes of the CH$_2$ group and the stretching modes of the carbonyl and CN bonds. The plane of the CH$_2$ group is perpendicular to the plane of the page. $\oplus$ and $\ominus$ indicate movements of H atoms perpendicular to the plane of the paper in positive and negative directions respectively. Transition moment directions are indicated beneath each diagram.

stants, and again the transition moment direction should coincide closely with the bond direction. Examples within this category include the C=O bond stretching vibration in oxidised polyethylene and ethylene–carbon monoxide copolymers, and the C≡N stretching mode in poly-acrylonitrile (see Fig. 5). However, some caution must usually be exercised when considering transition moment directions. In the case of polyamides, for example, vibrations of bonds within the CONH group are strongly coupled and the transition moment for the C=O stretching mode forms an angle of about 10–20° with the bond direction.[4] Similarly, the C—Cl stretching mode in polyvinyl chloride involves significant co-operation with skeletal vibrations of the polymer chain backbone,[7] so that the precise transition moment direction is in some doubt.

For partially crystalline polymers, infra-red absorption peaks can arise from vibrations either within the crystalline phase or the amorphous phase or both. Dichroic measurements on crystalline absorption peaks can be employed to study the orientation of certain crystallographic axes, providing that the direction of the transition moment with respect to the axes is known. Within the crystal the chains are stabilised in a uniform conformation, usually planar zig-zag (all *trans*) or helical, and transition moment directions may either be rigidly fixed or may vary periodically with respect to certain crystal axes. If the crystal unit cell contains more than one polymer chain, the crystalline field may serve to weakly couple the vibrations in neighbouring chains, an effect which results in a splitting of the crystalline absorption into two peaks. This effect is particularly important from the dichroic point of view, since the two components may be polarised along different crystal axes. A well-known example concerns the CH_2 rocking vibrations in the poly-ethylene crystal (Fig. 6) which yield absorptions at about 720 cm^{-1} and 730 cm^{-1} respectively. The two components correspond to in-phase and out-of-phase rocking modes respectively, the 720 cm^{-1} component having a transition moment along the b crystal axis and the 730 cm^{-1} component being polarised along the a-axis.[7] Dichroic studies on the two absorption peaks have been much used to study the orientation of the two crystal axes in polyethylene, although the presence of an amorphous absorption contribution at 720 cm^{-1} complicates the analysis of this peak and requires additional information on the orientation of the amorphous phase.

In the case of amorphous polymers, and the amorphous regions of partially crystalline polymers, the polymer chain conformation is no longer uniform or repetitive along its length. Details of the local chain

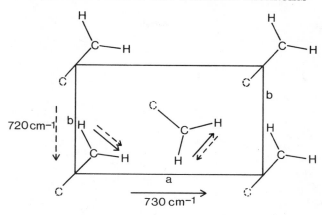

Fig. 6. *Cross-section of the unit cell of the polyethylene crystal, indicating the in-phase and out-of-phase rocking motions of the two CH₂ groups, and the net transition moments along the respective* a- *and* b-*axes (after Krimm[3]).*

conformation can, however, be specified in terms of the rotational isomeric states of consecutive skeletal bonds. For many polymers, including polyethylene, the rotational state of a given chain bond can be designated by *trans* (t), *gauche* (g) or *gauche'* (g') depending on whether its associated angle of internal rotation is 0°, +120° or −120° respectively. Local conformations can then be represented by sequences such as t, g, gg, gtg, gtg', tttt etc., and recent work of Snyder[8] has shown that many amorphous infra-red absorptions arise from vibrational modes localised within specific sequences of this type (see Fig. 7 and Table 1). Observed dichroic ratios will thus depend on the relative orientation of the absorbing conformations and their transition moment directions, and it is not surprising that appreciable differences in dichroism are observed for different amorphous peaks.[9] If transition moment directions are known, then, with the aid of appropriate statistical theories (Section 4.3), stress-dichroism measurements can be used to study the nature of the orienting segment, and hence aid in the assignment of infra-red bands of amorphous origin. Dichroism measurements on amorphous bands of partially crystalline polymers might also be used to estimate the stress on the amorphous phase if the stress-dichroism coefficients are known for the polymer in the wholly amorphous state (*e.g.* as a cross-linked rubber above the melting point). It should be added, however, that dichroic effects observed for amorphous infra-red peaks are generally of small

TABLE 1

THEORETICAL G_2^* VALUES FOR VARIOUS CONFORMATIONS IN AMORPHOUS POLYETHYLENE AND PROPOSED IR ASSIGNMENTS

Conformation	Transition moment direction	G_2^*	Proposed IR frequency (cm^{-1})	Assignment	Ref.
Independent	b″	−1·67	2850	Symmetric CH_2 stretch	8, 19
	c	−1·45	2924	Asymmetric CH_2 stretch	
g	c′	−2·10	1078	Skeletal C − C stretch (mainly g) some CH_2 wag	8, 18
tgt	c′	−2·58			
gg	a′	1·12	1352	CH_2 wag[a]	8, 18
gtg	a	−1·43	1368	CH_2 wag[a] 1368 cm^{-1}—sym. w.r.t. centre of trans bond, mainly gtg	8, 18
	b	3·69			
	a′	1·87	1303		8, 19
	a	−2·39		1303 cm^{-1}—assym. w.r.t. centre of trans bond, gtg and gtg′	
gtg′	b	2·73			8, 18
	a′	0·89			8, 19
	c	−3·26	720	CH_2 rock g(t)$_m$ g*[b], $m \geqq 4$	
			1463	CH_2 bend g(t)$_m$ g*[b], m large	8, 18
tttt	a′	6·71	2016[c]	Combination involving 720 cm^{-1} fundamental.	

[a] See Fig. 7.
[b] g* denotes either g or g′.
[c] Experimental studies indicate that this band involves *trans* sequences containing about 7 chain bonds[38] (see text).

magnitude, and sensitive experimental methods are required for precise measurements (Section 4.4). In addition, the amorphous peaks may be broadened by the overlap of neighbouring peaks from different conformational isomers. Despite these problems, and also the uncertainties in transition moment directions, the infra-red dichroism method potentially provides one of the few ways for directly studying amorphous chain orientation in detail.

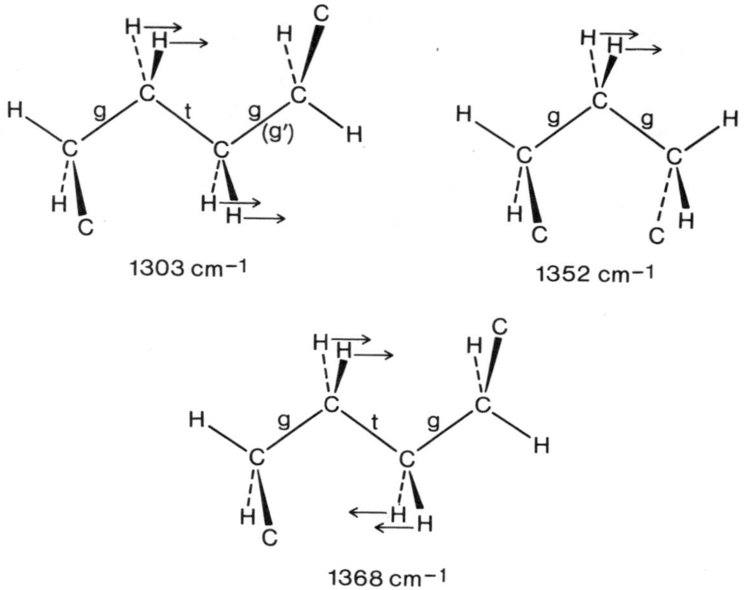

1303 cm^{-1}

1352 cm^{-1}

1368 cm^{-1}

Fig. 7. Examples of some conformationally dependent CH_2 wagging modes in polyethylene. Frequencies of IR absorptions are given under each diagram (after Snyder[8]).

4.3 RELATIONSHIPS BETWEEN DICHROIC RATIOS AND STRUCTURAL ORIENTATION FUNCTIONS

General quantitative relations between observed dichroic ratios and structural orientation functions have been considered by several authors. These include Fraser,[10] Beer,[11] Patterson and Ward,[12] Stein,[13,14] Chappel,[15] Kawai and Stein,[16] and Nomura et al.[17] More recently, Flory and Abe,[18] and also Nagai,[19] have carried out statistical calculations relating specifically to the stress-dichroism coefficients of amorphous polymer networks. Some results of these calculations will be summarised below. First, however, it is informative to derive some general relationships between the dichroic ratio and appropriate orientation functions.

4.3.1 General relationships
We first consider within the polymer some structural unit characterised

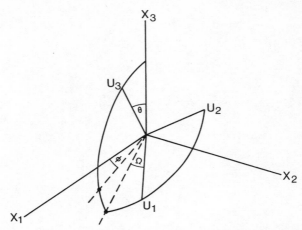

Fig. 8. Illustration of angles used to specify the orientation of a structural unit $(U_1 U_2 U_3)$ *within the sample reference axes* $(X_1 X_2 X_3)$.

by the orthogonal axes U_1, U_2, and U_3 as shown in Fig. 8. These axes may be taken to coincide, for example, with the crystal unit cell axes for a crystalline polymer or, for an amorphous polymer, we may take U_3 as the local direction of the polymer chain axis. In order to specify the orientation of the structural unit within the sample frame $X_1X_2X_3$, three independent (Eulerian) angles θ, ϕ and Ω may be defined (Fig. 8). Here θ and ϕ are the polar and azimuthal angles, respectively, of U_3 within the $X_1X_2X_3$ frame and Ω specifies the rotation of the unit around its own U_3 axis. The direction cosines t_{ik} between the X_i and U_k axes $(i, k = 1, 2, 3)$ can be written in matrix form (Read and Dean[20]),

$$[t_{ik}] = \begin{bmatrix} \cos\theta\cos\phi\cos\Omega & -\cos\theta\cos\phi\sin\Omega & \sin\theta\cos\phi \\ -\sin\phi\sin\Omega & -\sin\phi\cos\Omega & \\ \cos\theta\sin\phi\cos\Omega & -\cos\theta\sin\phi\sin\Omega & \sin\theta\sin\phi \\ +\cos\phi\sin\Omega & +\cos\phi\cos\Omega & \\ -\sin\theta\cos\Omega & \sin\theta\sin\Omega & \cos\theta \end{bmatrix} \quad (4)$$

The orientation distribution of all structural units within the sample may now be given in terms of a normalised probability function $p(\theta,\phi,\Omega)$ where $p(\theta,\phi,\Omega) \sin\theta\,d\theta\,d\phi\,d\Omega$ gives the fraction of units having axes lying in the angular range $\theta \rightarrow \theta+d\theta$, $\phi \rightarrow \phi+d\phi$, $\Omega \rightarrow \Omega+d\Omega$. Any function $f(\theta,\phi,\Omega)$ averaged over the orientations of all units is then,

$$\overline{f(\theta,\phi,\Omega)} = \int_{\Omega=0}^{2\pi} \int_{\phi=0}^{2\pi} \int_{\theta=0}^{\pi} f(\theta,\phi,\Omega)\, p(\theta,\phi,\Omega)\, \sin\theta\, d\theta\, d\phi\, d\Omega \tag{5}$$

We next consider (Fig. 9) an absorbing group within the structural unit and assume that its absorption anisotropy can be represented by a rotational ellipsoid characterised by absorbances $a_{\parallel}^{o}$ and $a_{\perp}^{o}$ in directions parallel and perpendicular, respectively, to the rotational axis. The two angles α and β (Fig. 9) are then adequate for specifying the orientation of the absorbing group within the structural unit, and in the special case where the rotational axis of the absorption ellipsoid corresponds to the transition moment direction, we note that $a_{\perp}^{o} = O$ and $a_{\parallel}^{o} = K(ME)^2$. In order to account for possible variations in the angles α and β, averages of any function $f(\alpha,\beta)$ may be defined in terms of a normalised distribution $p(\alpha,\beta)$ which is assumed to be independent of $p(\theta,\phi,\Omega)$,

$$\overline{f(\alpha,\beta)} = \int_{\beta=0}^{2\pi} \int_{\alpha=0}^{\pi} f(\alpha,\beta)\, p(\alpha,\beta)\, \sin\alpha\, d\alpha\, d\beta \tag{6}$$

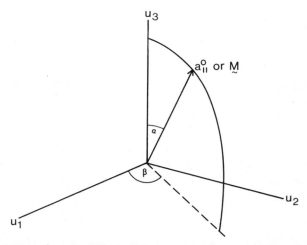

Fig. 9. Orientation of a principal absorption axis $(a_{\parallel}^{o})$ or transition moment (M) within a structural unit.

On the basis of eqn. (2), absorbance components A_{ij} for the bulk polymer can now be related to the average absorbances $\overline{a_{kl}}$ of the structural unit, where subscripts i, j refer to X_i, X_j and k, l denote U_k, U_l. The relationship represents a 2nd order tensor transformation

$$A_{ij} = N \sum_k \sum_l \overline{a_{kl}} \, \overline{t_{ik} t_{jl}} \tag{7}$$

where N is the total number of contributing units and the averages over $t_{ik} t_{jl}$ correspond to eqn. (5). The $\overline{a_{kl}}$ are given by,

$$
\begin{aligned}
\overline{a_{33}} &= (a_\parallel^o - a_\perp^o) \overline{\cos^2 \alpha} + a_\perp^o \\
\overline{a_{22}} &= (a_\parallel^o - a_\perp^o) \overline{\sin^2 \alpha \sin^2 \beta} + a_\perp^o \\
\overline{a_{11}} &= (a_\parallel^o - a_\perp^o) \overline{\sin^2 \alpha \cos^2 \beta} + a_\perp^o \\
\overline{a_{13}} &= \overline{a_{31}} = \tfrac{1}{2}(a_\parallel^o - a_\perp^o) \overline{\sin 2\alpha \cos \beta} \\
\overline{a_{23}} &= \overline{a_{32}} = \tfrac{1}{2}(a_\parallel^o - a_\perp^o) \overline{\sin 2\alpha \sin \beta} \\
\overline{a_{12}} &= \overline{a_{21}} = \tfrac{1}{2}(a_\parallel^o - a_\perp^o) \overline{\sin^2 \alpha \sin 2\beta}
\end{aligned}
\tag{8}
$$

in which the average signs correspond to eqn. (6).

For specimens uniaxially drawn in direction X_3, the orientation distribution of structural units is random with respect to ϕ. Hence $p(\theta, \phi, \Omega) = p(\theta, \Omega)/2\pi = p(\pi - \theta, \Omega)/2\pi$ and it follows from eqns. (4)–(8) that $A_{11} = A_{22}$ and all A_{ij} for $i \neq j$ vanish. We also obtain, putting $A_{ii} = A_i$ and $D = D_{32}$,

$$\frac{D-1}{D+2} = \frac{A_3 - A_2}{A_3 + 2A_2} = \frac{a_\parallel^o - a_\perp^o}{a_\parallel^o + 2a_\perp^o} \times$$

$$\left[f_\alpha f_\theta + \frac{3}{4} \overline{\sin^2 \alpha \cos 2\beta} \, \overline{\sin^2 \theta \cos 2\Omega} - \frac{3}{4} \overline{\sin^2 \alpha \sin 2\beta} \, \overline{\sin^2 \theta \sin 2\Omega} \right] \tag{9}$$

where f_θ and f_α are second moment orientation functions given by,

$$f_\theta = \frac{\overline{3 \cos^2 \theta} - 1}{2}, f_\alpha = \frac{\overline{3 \cos^2 \alpha} - 1}{2} \tag{10}$$

If the principal absorption axis in Fig. 9 is randomly oriented with respect to β, as is often assumed in the case of added dye molecules, eqn. (9) becomes,

$$\frac{D-1}{D+2} = \frac{a_\parallel^o - a_\perp^o}{a_\parallel^o + 2a_\perp^o} f_\alpha f_\theta = \frac{D_0 - 1}{D_0 + 2} f_\theta \tag{11}$$

in which D_0 is the value of D corresponding to perfect orientation of structural units ($f_\theta = 1$). When the principal absorption axis corresponds to the transition moment direction (*i.e.* $a_\perp^0 = 0$) it follows from (10) and (11) that $D_0 = \overline{2\cot^2\alpha}$, a result similar to that given by Fraser.[10]

For the particular cases in which the transition moment is directed along either U_1, U_2 or U_3 we may take θ_k ($k = 1$, 2 or 3) to be the angle between axis U_k and X_3 and, noting that $\theta_3 \equiv \theta$, define orientation functions,

$$f_{\theta_k} = \frac{\overline{3\cos^2\theta_k - 1}}{2} \tag{12}$$

Together with the orthogonality relation,

$$\sum_k f_{\theta_k} = 1 \tag{13}$$

it then follows from (4) and (9) that,

$$\frac{D-1}{D+2} = f_{\theta_k} \tag{14}$$

The above general analysis is readily extended to the case of biaxially oriented samples, as shown by Nomura *et al.*[17] If the orientation distribution is symmetrical with respect to the $X_1 X_2$, $X_1 X_3$ and $X_2 X_3$ planes, then biaxial orientation functions of the form $\overline{\sin^2\theta\cos2\phi}$ may be obtained from measurements of $(A_2 - A_1)/(A_3 + A_2 + A_1)$. Assuming that the distributions over θ and ϕ are independent, then functions such as $\overline{\cos2\phi} = \overline{2\cos^2\phi} - 1$ can subsequently be obtained, as shown by Stein.[13,14] Owing to experimental difficulties, however, no applications of these analyses seem to have yet been made. For both uniaxial and biaxial orientation,

$$A_3 + A_2 + A_1 = 3A_0 = N(\bar{a}_{33} + \bar{a}_{22} + \bar{a}_{11}) \tag{15}$$

The sum of absorbance components is thus independent of orientation and proportional to the concentration of absorbing groups.

Equations (9)–(14) illustrate that the determination of structural orientation functions by the dichroism method requires a knowledge of the average orientation of the absorbing axis within the structural unit. The analysis also shows that the orientation functions, like those derived from birefringence data, are second moments (mean square functions) of the orientation distribution.[21]

4.3.2 Amorphous polymer networks

As mentioned above (Section 4.2), infra-red absorption bands of amorphous polymers frequently arise from vibrational modes localised within specific conformational sequences of the polymer chain. In such cases, dichroic studies on extended polymers at different frequences can be used to investigate the relative orientation of different chain bonds or bond sequences with respect to the chain end-to-end vector. The theoretical aspects of this problem have been considered by Flory and Abe,[18] and independently by Nagai,[19] using methods based on an extension of their respective treatments of strain birefringence. The methods employed in the two studies are similar, and the results obtained are essentially in agreement. Each method considers a polymer chain whose unperturbed average conformation is determined by the rotational isomeric states of its n skeletal bonds. The absorbance tensor for a given bond i, or for an associated group of bonds, is represented by components $(m_k m_l)_i$. Here m_k and m_l are the transition moment components along local k and l axes of the ith group for absorption at a specified frequency. For the polymer molecule, the absorbance tensor is then formulated by summing the constituent group tensors, each weighted by the probability of finding the group in the conformational state necessary for absorption. The anisotropy of molecular absorbance, averaged over all chain conformations consistent with a given end-to-end distance, is evaluated as a function of the chain extension, and subsequently of the average absorbance components derived for the polymer chain network. For long chains, the results of each theoretical study may be expressed in terms of the amorphous transition moment orientation function referred to the sample reference axes (compare eqn. 14),

$$f_{am} = \frac{D-1}{D+2} = \frac{G_2^*}{2n} \left(\frac{V}{V_0}\right)^{2/3} (\lambda_e^2 - \lambda_e^{-1}) \qquad (16)$$

$$= \frac{G_2^* M_0 t}{2\rho RT} \qquad (17)$$

Here λ_e and t are the extension ratio and tensile stress respectively, V is the volume of the strained specimen and V_0 a reference volume which may usually be taken as the unstrained volume. $\rho, R, T,$ and M_0 are, respectively, density, gas constant, absolute temperature and repeat unit molecular weight. The parameter G_2^*, defined by Flory and Abe, provides a measure of the average correlation of local transition moment directions with respect to the chain end-to-end vector. Positive values of G_2^* denote a tendency for the transition moments to orient in line with the end-to-end vector and negative G_2^* values are characteristic of

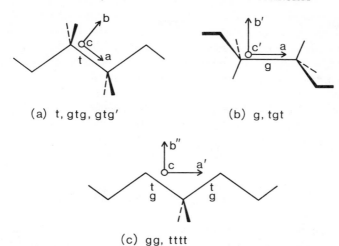

(a) t, gtg, gtg'   (b) g, tgt

(c) gg, tttt

Fig. 10. Local orthogonal axes defined by Flory and Abe[18] for the conformations indicated.

perpendicular orientation. In Nagai's calculations an equivalent orientation parameter, $f_k^* = G_2^*/2$, is defined.

Values of G_2^* (or f_k^*) were evaluated for various conformational sequences and transition moment directions in polyethylene. Table 1 lists some selected values which are considered, on the basis of Snyder's assignments,[8] to be appropriate to the major amorphous infra-red absorptions. Values taken from Flory and Abe's work correspond to transition moment directions along local axes chosen to conform with symmetry features for the various conformations, as illustrated in Fig. 10 (a), (b) and (c). The values obtained from Nagai each refer to the axes specified in Fig. 10(c). All values correspond to a temperature of 140°C and to conformational energies $E_g = 500$ cal/mol and $E_{gg'} = 2500$ cal/mol, E_g and $E_{gg'}$ being the energies of g and gg' states, respectively, relative to reference *trans* states of zero energy.

From Table 1 we observe that the symmetric and asymmetric CH_2 stretching modes are regarded, to a good approximation, as being independent of local conformation and are polarised along axes which orient on average perpendicular (negative G_2^*) to the chain. The negative G_2^* values for the g and tgt conformations with transition moment along c', reflect the strong tendency for the *gauche* bond (a-axis) to orient parallel to the chain vector. On the other hand a *trans* bond situated between two *gauche* bonds exhibits a transverse orientation, as evidenced by the negative G_2^* values for the *a*-axis in both gtg and gtg′ conformations. Long sequences of *trans* bonds show a very strong parallel orientation, as shown by the high positive G_2^* value for the *a*′-axis in the tttt conformation. The latter result is consistent with an analysis by Shindo and Stein[22] of model chains comprising freely jointed segments of varying length. This analysis yielded the result that the segment orientation function is proportional to the square of the segment length. Further discussion of Table 1 is reserved for Section 4.5 below.

4.4 EXPERIMENTAL METHODS

Only a brief account of experimental techniques will be given here. Details of individual methods and sources of error are obtainable from original references, and for a recent review the reader is referred to Wilkes.[23] In the UV and visible regions, single beam optical methods have been developed. The sample is placed in an optical beam from an appropriate source (*e.g.* mercury–xenon), the beam having been rendered parallel with a condensing lens, monochromatised with filters or monochromator, and polarised with a suitable polariser (polaroid or prisms). At UV wavelengths, quartz optical components are required for adequate transmission, and special polarisers are available. The intensity of radiation transmitted through the specimen is conveniently detected with a photomultiplier. For uniaxially oriented polymers, the dichroic ratio is obtained from the measured intensities with the polarisation direction successively parallel and perpendicular to a sample reference axis (*e.g.* the draw direction). As is evident from eqns. (1) and (3), intensities I_0 are also required for the two polarisation directions. These may be measured with the sample absent or, in order to correct for possible scattering, from the baseline method illustrated in Fig. 1. Details of equipment of the above type are given by Shindo *et al.*[5] For monofilaments, methods based on a polarising microscope have been described by Patterson and Ward[12] and by Chappel.[15]

In the infra-red region, dichroic measurements are conveniently made on double beam instruments, the sample and polariser each being placed in the usual sample beam, and the spectra recorded with the polariser orientation successively parallel and perpendicular to the sample reference axis. If wire grid polarisers are employed. errors due to polariser inefficiency are usually negligible, and errors due to polarisation effects of the monochromator are eliminated by orienting the polariser and sample at $\pm 45°$ to the monochromator entrance slit. The effects of scattering and overlap of neighbouring peaks may be corrected for by suitable choice of baseline (Section 4.1). Other sources of error arise from insufficient spectrometer resolution, beam convergence and spectral dilution. A detailed discussion of the various errors and correction procedures is given by Zbinden.[4] For biaxially oriented specimens, absorbance measurements on tilted films are required in order to determine the component A_1. The principles of this method and the problems and errors encountered are discussed by Schmidt,[24] Zbinden[4] and Koenig and coworkers.[25–27]

For the measurement of small dichroic effects, as are encountered with the absorption peaks of amorphous origin, a differential technique may be used. In this method a polariser is placed in each of the two beams after splitting, the two polarisers being crossed and, to eliminate machine polarisation, each oriented at $+45°$ or $-45°$ to the monochromator slit. The specimen is similarly inclined, and is placed in the common beam either before splitting or after the two beams have been recombined. From the spectrum subsequently recorded, the dichroic difference $A_3 - A_2$ is directly obtained, and the dichroic ratio determined from the additional measurement of the average absorbance using unpolarised radiation. This method is considerably more sensitive than the conventional method and also enables changes in dichroism to be measured during the continuous elongation and subsequent relaxation processes. One disadvantage, however, concerns the decrease in available energy, a problem which may require relatively wide slit widths and large corrections for resolution. Gotoh and coworkers[28–30] have employed this technique to study infra-red dichroism changes during the stress relaxation of several polymers, and Wilkes et al.[31] have described a method of this kind capable of investigating the rapid changes of UV, visible or infra-red dichroism during high speed stretching. Read and coworkers[32] have also developed a differential infra-red technique, and have quantitatively discussed the sensitivity of the method, as well as correction procedures for resolution and for sample emission in high .temperature measurements.

Measurements of dichroic ratios at fundamental absorption frequencies usually require very thin specimens or films (typical thickness range 2 μm to 0·1 mm) in order to avoid complete absorption of the radiation. Such specimens require careful preparation using either melt pressing or solvent casting techniques.[7,9] Thicker samples may be used at infra-red wavelengths if measurements are restricted to overtone or combination bands, or if attenuated total reflectance methods[33] are developed. When using the differential method, an increase in sensitivity is obtained by increasing the sample thickness (and hence the average absorbance), providing that sufficient transmitted energy remains for accurate operation.[32]

4.5 RESULTS FOR SELECTED POLYMERS

A comprehensive discussion of the many dichroic studies so far reported for individual polymers is beyond the scope of this chapter, the main aim being to illustrate the nature of the orientational and structural effects which are capable of investigation by the dichroic method. The most extensive investigations have so far been made on polyethylene and its copolymers, and it is instructive to consider the results of these studies in some detail. Regarding other polymer systems, a brief summary of results obtained for dehydrochlorinated PVC should serve to indicate the value of combining dichroic data at UV, visible and infra-red wavelengths with complementary birefringence results. References to studies made on a range of other polymers will finally be collected in Table 2.

4.5.1 Polyethylene and copolymers
The dichroic studies so far reported for polyethylene are conveniently grouped into two categories. In the first category we include those investigations concerned with the structural rearrangements during the initial stages of deformation and up to about 100% elongation. The majority of such studies have been made on low density (branched) polyethylene, and of particular concern are the deformation processes within, or destruction of, the spherulite structure. In the second category we include those studies made on predrawn samples at high draw ratios, where the spherulitic morphology no longer exists. Most studies of this type have been concerned with the linear, high density, polymer.

Regarding the dichroic behaviour at low extensions, we first consider data relevant to the crystalline phase.[9] Figure 11 illustrates the dependence on extension ratio of the second moment orientation functions f_c, f_a and f_b for low density polyethylene. These functions were determined from wide-angle X-ray measurements and characterise the

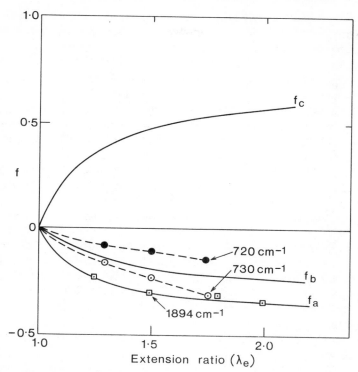

Fig. 11. Crystalline orientation functions f_c, f_a, and f_b (determined from X-ray diffraction) plotted against extension ratio for low density polyethylene. Also shown are the orientation functions $(D-1)/(D+2)$ determined from the infra-red bands at the frequencies indicated (Read and Stein[9]).

orientation of crystal c, a and b axes, respectively, relative to the stretching direction. Also shown are values of $(D-1)/(D+2)$ from the infrared peaks at 720, 730 and 1894 cm^{-1} respectively. We recall from Section 4.2 that the 730 cm^{-1} crystalline band and the crystalline component of the 720 cm^{-1} band have transition moments along the crystal a- and b-axes respectively. If we ignore the amorphous contribution to the 720 cm^{-1} peak, and regard the a and b axes as coinciding with the U_1 and U_2 axes, respectively, in Fig. 8, then it follows from eqn. (14) that the $(D-1)/(D+2)$ values for the 730 and 720 cm^{-1} bands should be equal to f_a and f_b respectively. The discrepancies observed in Fig. 11 between values of f_a and f_b estimated in this way and the corresponding

X-ray values may partly be due to the fact that the samples employed in the X-ray and infra-red studies were not identical.[9] Other possible reasons for the discrepancies include the close overlap of the two peaks, the presence of a broad amorphous component centred at 720 cm^{-1} which may contribute to both peaks, and the necessity for using thin films which may cause the orientation distribution to deviate somewhat from uniaxial symmetry. Some doubt must therefore exist about the quantitative accuracy of crystal orientation functions evaluated from the 720 and 730 cm^{-1} peaks. Further X-ray and infra-red studies on identical samples would help to resolve the apparent discrepancies.

Figure 11 illustrates a good correspondence between the X-ray values of f_a and the $(D-1)/(D+2)$ values from the 1894 cm^{-1} band. The 1894 cm^{-1} band originates entirely from the crystalline phase and involves a combination between the Raman active fundamental at 1168 cm^{-1} and the 720–730 cm^{-1} bands.[7] The transition moment is perpendicular to the c-axis but there is some doubt as to its precise direction relative to the a- and b-axes. The correspondence in Fig. 11 suggests that the polarisation direction may be close to the a-axis, a conclusion consistent with the B_{1u} symmetry mode listed by Krimm.[7] However, the dichroic studies of Desper[34] on extruded polyethylene suggest that the transition moment makes an angle of about 39° with the a-axis, and further studies of this problem should be worthwhile. The relatively low intensity of the 1894 cm^{-1} band, which eliminates the necessity for employing thin films, and the absence of any amorphous component, make this band particularly convenient for investigating the crystalline orientation.

The main conclusions resulting from Fig. 11 are that the crystal c-axes orient parallel and the a- and b-axes orient perpendicular to the stretching direction. Since f_a has larger negative values that f_b, it follows from eqns (4), (12) and (14) that $\sin^2\theta\cos^2\Omega$ is less than $\sin^2\theta\sin^2\Omega$ and hence that the crystal orientation distribution is non-random with respect to Ω. According to Onogi and coworkers,[35–37] this result shows that the initial stages of deformation mainly involve rotations around the b crystal axes, corresponding to a twisting of crystal lamellae about the spherulite radii. For a more detailed discussion of dichroic data in terms of lamellae or spherulitic deformation processes the reader is referred to Onogi and Asada [37] and Gotoh and coworkers.[29]

With regard to the amorphous orientation during the initial stages of deformation, Fig. 12 shows some $(D-1)/(D+2) = f_{am}$ values for the same low density material as that used to obtain the data in Fig. 11.

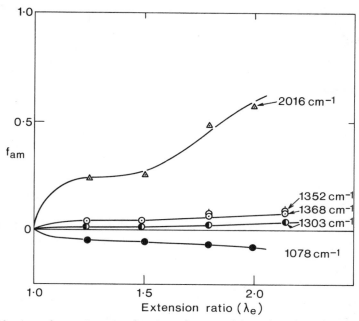

Fig. 12. *Amorphous orientation functions, determined from the infra-red peaks at the indicated frequencies, plotted against extension ratio for low density polyethylene (Read and Stein[9]).*

The peaks at 1078, 1303, 1352 and 1368 cm^{-1} originate entirely from the amorphous phase and the respective $(D-1)/(D+2)$ values were obtained directly. For the 2016 cm^{-1} peak, which contains both an amorphous and a crystalline component, the $(D-1)/(D+2)$ values are appropriate to the *amorphous* component only. They were obtained from the measured net D values for the 2016 cm^{-1} peak, in combination with X-ray data and dichroic data on the 1894 cm^{-1} crystalline band. Details of the method used for resolving the amorphous orientation function are given by Read and Stein.[9]

A comparison of Figs. 12 and 11 shows that the amorphous orientation functions are generally lower than the crystal orientation functions at low extensions. Figure 12 also reveals a qualitative correspondence between the experimental f_{am} values and the theoretical $G_{\frac{*}{2}}$ values in Table 1, as might be expected if eqn. (16) is approximately valid for the amorphous regions of the polymer. The negative f_{am} values for the 1078

cm^{-1} band are consistent with the negative G_2^* values predicted for skeletal C—C stretching vibrations for *gauche* bonds having a transition moment along c' (Fig. 10). The contribution to this band from *trans* bond stretching is likely to be small owing to the centrosymmetry of the bond and correspondingly small dipole moment change. A possible contribution from CH_2 wagging, however, has been emphasised,[8] although it is difficult to comprehend how this mode could give rise to the observed dichroism. The very small positive values of f_{am} for the 1303 cm^{-1} band are also consistent with the CH_2 wagging assignment for gtg and gtg' conformations. These modes should have transition moments directed between the a- and b-axes (Fig. 10) and the appropriate G_2^* values are small and positive, particularly for the gtg' conformation. At first sight the relatively large f_{am} values for the 1368 cm^{-1} peak, compared with the values for the 1303 cm^{-1} peak, seem surprising in view of the similar assignments proposed for the two absorptions. However, as Snyder has suggested,[8] the 1368 cm^{-1} band probably has its origins in gtg rather than gtg' conformations since, for the latter, symmetric CH_2 wagging (Fig. 7) would tend to be infra-red inactive owing to the local inversion centre on the *trans* bond. The relatively high G_2^* values for the gtg conformation (Table 1) might then explain the higher f_{am} values for the 1368 cm^{-1} band. For the 1352 cm^{-1} band, which is assigned to CH_2 wagging within gg sequences (Fig. 7), small positive f_{am} values are also to be expected, and values close to those observed for the 1368 cm^{-1} band seem reasonable (Table 1). It is not clear, however, to what extent the close overlap of the 1352 and 1368 cm^{-1} bands might have affected this result.

The most striking feature of Fig. 12 is the high f_{am} values obtained from the 2016 cm^{-1} peak. The 2016 cm^{-1} absorption is a combination band which partly involves the 720 cm^{-1} fundamental and which has a transition moment parallel to the chain axis. Since the amorphous component of the 720 cm^{-1} band is assigned to extended *trans* sequences comprising four or more *trans* bonds, a similar conformational sequence might be responsible for the 2016 cm^{-1} band. The high experimental values of f_{am} lend support to this view since, as illustrated in Table 1 and discussed in Section 4.3, extended *trans* sequences should be highly oriented and have correspondingly high G_2^* values. Results of recent stress-dichroism studies on cross-linked amorphous polyethylene at 170°C (Read and Hughes[38]) add further support to the above views. When analysed by means of eqn. (17) these results yielded a G_2^* value between 10·1 and 12·6, depending on the choice of baseline. A tentative extrapolation of Flory and Abe's theoretical G_{2}^{*} values suggested that *trans* sequences of about 7 chain bonds are involved. It is not clear to what extent this result may be influenced by possible correlations between the conformations adopted by neighbouring chain segments, or by the postulated 'local order' in amorphous polymers. Such factors, which are not accounted for in the theory, are apparent from the effects of swelling agents on stress-birefringence data,[19,39] and further dichroic studies on swollen amorphous samples (and at different wavelengths) could be illuminating.

From Fig. 12 it is apparent that, apart from the results on the 2016 cm^{-1} band, the amorphous orientation functions are much smaller than the crystal orientation functions at each extension. Low estimates of amorphous orientation have also been obtained from visible dichroism data on polyethylene containing the Congo red dye molecule[40] (Fig. 4) and from the dichroic ratio of the IR $C \equiv N$ stretching band in ethylene–acrylonitrile copolymers. In the latter study (Read and Stein[9]) the copolymer contained about 2 wt. % acrylonitrile. This was employed as a tracer for the study of amorphous orientation on the grounds that the $C \equiv N$ group is too large to be accommodated within the polyethylene crystal lattice and that the transition moment is perpendicular to the local amorphous chain axis. Similar dichroic studies of the $C = O$ stretching band in oxidised polyethylene[9] and in ethylene–carbon monoxide copolymers[41] yielded relatively high orientation functions. These results were undoubtedly influenced by the partial incorporation of $C = O$ groups within the crystals.

Time dependent dichroic measurements during the stress relaxation of low density polyethylene have been reported by Gotoh et al.[29] and more recently by Fukui et al.[35] and by Uemura and Stein.[42] Fukui and coworkers determined both crystalline and amorphous orientation functions at a constant strain between 2·5 and 5%. At room temperature, values of $-f_a$ and $-f_b$, estimated from data on the 730 and 720 cm^{-1} bands respectively, increased with time towards steady values which were attained after about 10^3 s. These time dependencies were considered to largely determine the observed time dependence of the strain-optical coefficient. Values of f_{am} from the 1352 cm^{-1} band showed little time dependence, but the amorphous orientation was estimated to contribute the larger amount to the magnitude of the strain-optical coefficient.

Dichroic investigations of highly drawn samples of linear polyethylene have been reported by Koenig et al.[25] and more recently by Glenz and Peterlin.[43–45] In the former study samples were drawn by about 800%

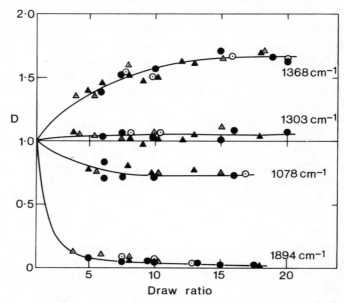

Fig. 13. Dichroic ratio v. draw ratio for various infra-red bands of linear polyethylene drawn at 60°C. Frequencies in wavenumbers are indicated for each curve. (⊙), slowly cooled Marlex 6050; (●), quenched Marlex 6050; (△), slowly cooled Marlex 6002; (▲), quenched Marlex 6002 (from Glenz and Peterlin[43]).

at temperatures ranging from 25°C to 85°C. The crystal orientation, as evidenced by the dichroic ratios of the bands at 1894 and 2016 cm^{-1}, was found to be very high, and independent of the initial draw temperature. The much lower amorphous orientation, detected from the bands at 1303 and 1352 cm^{-1}, decreased slightly with increasing draw temperature, a result attributed to the higher mobility of amorphous chain segments at the higher temperatures. From observations of the average absorbance $A_0 = (A_3 + 2A_2)/3$ on the respective crystalline and amorphous peaks, no changes in crystallinity were detected but an apparent small increase in *gauche* content was reported after drawing. The latter results appear to be at some variance with the data of Glenz and Peterlin.[43]

Glenz and Peterlin[43–45] have made an extensive study of the dichroic properties of linear polyethylene drawn at 60°C to draw ratios between 5 and 25. Figure 13 shows the dichroic ratios determined from the bands at 1894, 1368, 1303 and 1078 cm^{-1} for Marlex 6050 ($M_w \approx 80\,000$, $\overline{M_n} \approx 9000$) and Marlex 6002 ($\overline{M_w} \approx 150\,000$, $\overline{M_n} \approx 13\,000$). Samples of each polymer were studied after slow cooling and also after quenching from the melt at 160°C. The observed dichroic ratios are independent of molecular weight, within the range studied, and of thermal treatment, and are quantitatively consistent with the results outlined above for low density polyethylene at lower strains (Figs. 11 and 12). Comparing the data on the crystalline 1894 cm^{-1} band with the results on the three amorphous bands, it is seen that the crystals are more highly oriented, and approach their maximum orientation at lower draw ratios, than the amorphous chain segments. This result is further demonstrated in Fig. 14, where the relative orientation function,

$$f_{rel} = \frac{D-1}{D+2} \frac{D_{max}+2}{D_{max}-1} \tag{18}$$

is plotted against draw ratio for the bands at 1894 and 1368 cm^{-1} for a linear sample (LU 0·5) prepared with a Phillips catalyst and containing less than 0·5 methyl groups per 1000 CH$_2$ groups. The quantity D_{max} in eqn. (18) is the dichroic ratio observed at the highest draw ratio or at

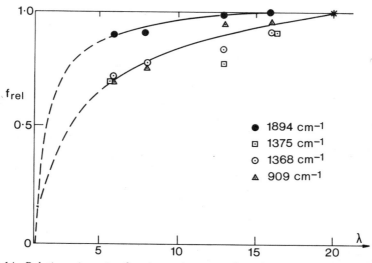

Fig. 14. *Relative orientation function* v. *draw ratio for infra-red bands at 1894* (●), *1368* (⊙), *1375* (▯) *and 909 cm^{-1}* (△) *for LU 0·5 linear polyethylene (from Glenz and Peterlin[44]).*

saturation (compare eqn. (11)). Also included in Fig. 14 are the relative orientation functions for sample LU 0·5 derived from the bands at 1375 and 909 cm^{-1}, which arise from the methyl and vinyl end groups, respectively. The f_{rel} values for the 1368, 1375 and 909 cm^{-1} bands show a similar dependence on draw ratio. The above results thus provide direct evidence for a two-phase structure of the drawn polymer in which the crystalline and amorphous regions are segregated and in which the chain ends are exclusively located in the amorphous phase (Fig. 15).

Glenz and Peterlin[44] have also observed the changes in dichroic ratio for both crystalline and amorphous bands during the annealing of oriented ethylene–butene–copolymer samples (LU 7, containing 7 ethyl side groups per 1000 CH$_2$ groups). The samples were initially drawn to a draw ratio of 10 and subsequently unclamped and allowed to shrink whilst being annealed for 5 h at different temperatures. Figure 16 shows the macroscopic shrinkage $S = (l_d - l)/(l_d - l_0)$, where l_d, l_0 and l are lengths of the drawn sample, the undrawn sample and the annealed and shrunk sample, respectively, as a function of annealing temperature. Also shown are the values of f_{rel} for the bands at 1894, 1375, 1368 and 909 cm^{-1}. These were calculated from eqn. (18) using, instead of D_{max}, the value of D before annealing. It is seen that the amorphous chain

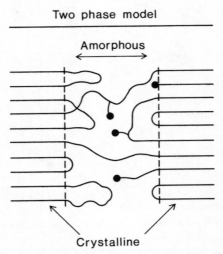

Fig. 15. Illustration of two-phase model proposed by Glenz and Peterlin[44] for drawn polyethylene.

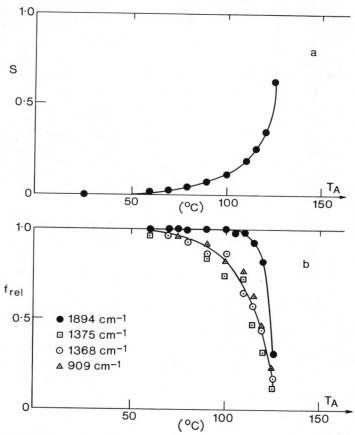

Fig. 16. (a) Macroscopic shrinkage S *and (b) relative orientation function* f$_{rel}$ *plotted against annealing temperature* T$_A$ *for oriented ethylene-butene-1 copolymers LU7. Frequencies of infra-red bands in wavenumbers are indicated in (b) (Glenz and Peterlin[44]).*

segments (1368 cm^{-1} band) and the methyl and vinyl end groups (1375 and 909 cm^{-1} bands), begin to disorient at the annealing temperature at which sample shrinkage is first observed. The disorientation of crystals (1894 cm^{-1} band) sets in at a significantly higher annealing temperature. We may note that these results provide further evidence for the two-phase structure (Fig. 15) and that the disorientation of the amorphous phase may be partly responsible for the decrease in longitudinal Young's modulus frequently observed after the annealing of highly drawn polyethylene.

Assuming that the ethyl side groups in copolymer LU 7 are exclusively located in the amorphous phase, and that the orientations of the backbone groups to which the side groups are attached are representative of the mean orientation of amorphous chain axes, Glenz and Peterlin have evaluated, from their data on the 1375 cm^{-1} bands for LU 7 and LU 0·5, an average orientation function for the amorphous phase. At high draw ratios (15–20) they obtained $f_{am} = 0·35$–$0·57$ compared with a value of about 1·0 for the crystal orientation function. Although involving rather extensive computations (owing to uncertainties in the conformations adopted by the ethyl side chains) this analysis attempted to overcome the problem arising from the conformational dependence of most amorphous bands, and provided additional evidence for the relatively low average amorphous orientation in the drawn polymer. In the writer's view, however, it must be questionable as to whether the orientation of chain segments adjacent to the side groups is representative of the average orientation of all types of amorphous chain segment. Later results have been reported by Glenz and Peterlin[45] relevant to the orientation of methyl and vinyl end groups in the linear polymer.

In addition to the orientational information gained from the analysis of observed dichroic ratios, Glenz and Peterlin's results have also yielded valuable information concerning changes in the degree of crystallinity and chain conformation upon drawing.[43] The latter information comes from the evaluation of the so-called 'reduced structural factor' (A') given by

$$A' = A_0/A_0 \text{(ref)} \tag{19}$$

where A_0 (cf. eqn. (15)) is the average absorbance for a particular peak and equal to $(A_3 + 2A_2)/3$ for uniaxially oriented samples. A_0 (ref) is the value of A_0 for a reference peak, usually taken as the 909 cm^{-1} vinyl band in the case of polyethylene. As discussed above, A_0 is independent of orientation and proportional to the concentration of absorbing groups, whilst division by A_0 (ref) makes A' independent of the sample thickness and density. Glenz and Peterlin have evaluated the weight fraction crystallinity α from measurements of A' for the 1894 cm^{-1} crystalline peak,

$$\alpha = \frac{A'(1894 \text{ cm}^{-1})}{A'(1894 \text{ cm}^{-1})_{\alpha=1}} \tag{20}$$

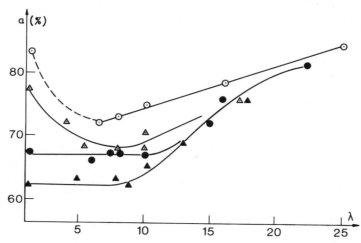

Fig. 17. Mass fraction crystallinity v. draw ratio for linear polyethylene drawn at 60°C. Point symbols as in Fig. 13 (from Glenz and Peterlin[43]).

where $A'_{\alpha=1}$ is the reduced structural factor for a completely crystalline sample and was obtained by extrapolation of A' values for unoriented samples of different crystallinity, as evaluated by wide-angle X-ray measurements. It will be observed from Fig. 17 that appreciable changes of crystallinity occur on drawing. Furthermore at draw ratios up to about 10, the crystallinity changes are dependent on the initial heat treatment, whereas at higher draw ratios the crystallinity changes in all specimens are similar. These results were ascribed to the transition from spherulitic to fibre morphology, the final fibre structure being essentially independent of the initial morphology.

From the values of A' and α, Glenz and Peterlin have also calculated values of $A_{rel} = A'/(1-\alpha)$ for a number of amorphous peaks. Such values are proportional to the concentration of absorbing groups per unit mass of amorphous component. Significant decreases were observed in the values of A_{rel} for the 1368, 1352 and 1303 cm^{-1} bands with increasing draw ratio whereas A_{rel} for the 1078 cm^{-1} band was essentially unchanged. These results are indicative of a decrease in *gauche* content and a corresponding increase in *trans* bonds upon drawing, and were attributed to an increase in the number of tie molecules linking the lamellae at the expense of chain loops within the inter-lamellar regions. This interpretation is consistent with the observed increases in longitudinal modulus and tensile strength upon drawing. The observed crystallinity and conformational changes also show that the drawing process cannot be explained solely in terms of the orientation of crystals but must also involve considerable structural changes in the intercrystalline regions.

Miller et al.[46] have recently reported infra-red dichroic data obtained for high density polyethylene crystallised under the orientation and pressure effects of a pressure capillary viscometer. Their data for a number of crystalline bands (including the 1894 cm^{-1} absorption) showed that the crystal c-axes were almost perfectly oriented ($f_c \approx 1$) in the initial extrusion direction. The amorphous orientation functions were generally lower, but corresponded to an extension ratio between 2 and 7 when compared with the above results of Read and Stein and of Glenz and Peterlin. Further evidence was also obtained for the relatively high orientation of the amorphous component of the 2016 cm^{-1} band ($f_{am} = 0.66$–0.72).

4.5.2 Dehydrochlorinated polyvinyl chloride

As discussed in Section 4.2, polyene segments (Fig. 3) can be incorporated into the polyvinyl chloride molecule by a dehydrochlorination reaction, and the orientation of these segments can be studied by means of UV or visible dichroism. In order to prevent the formation of cross-links arising from the intermolecular removal of HCl, the dehydrochlorination reaction is preferably carried out in solution rather than the solid state. Shindo et al.[5] have described a solution method in which an alcoholic KOH solution is added to a solution of PVC in tetrahydrofuran, and the reaction allowed to proceed in a controlled manner at 7°C. Portions of the mixed solution were removed after reaction times of 5, 10, 20, 40, 80 and 160 min, and were then neutralised and precipitated with dilute HCl. The dehydrochlorinated polymers were quantitatively characterised from a comparison of their UV and visible absorption spectra in solution with corresponding spectra for model polyene solutions. As the reaction time increased from 5 to 160 min it was found that the number of PVC chains containing polyene segments increased from about 2 to 7% and the average number of conjugated double bonds in the polyene segments increased from about 5 to 9.

Measurements of birefringence Δn, and the UV or visible dichroic ratio D were made on films cast from tetrahydrofuran solution as a function of extension (up to 200%) at 80°C. At this temperature the polymer behaves as a rubberlike material, effectively cross-linked by the small number of crystallites present. Both the birefringence and dichroic ratio were observed to increase with reaction time at a given extension. These results were quantitatively consistent with theoretical calculations which demonstrated that an increase in length of the polyene segment caused an increase in the anisotropy of the segment polarisability and absorbance (at the maximum wavelength) and also a marked increase in the average segment orientation.

Also consistent with theoretical predictions was the observation that the ratio $\Delta n/f_D$ was essentially constant, independent of extension and reaction time. Here $f_D = (D-1)/(D+2)$ is the polyene segment orientation function which follows from eqn. (11), noting that the absorbance is a maximum in the direction of the segment axis and that the small perpendicular component of absorbance is negligible. Assuming that the constant value $\Delta n/f_D = k$ may also be applied to the pure PVC, we may write,

$$f_{PVC} = \Delta n_{PVC}/k \qquad (21)$$

where Δn_{PVC} and f_{PVC} are the birefringence and average orientation function, respectively, for the PVC chain segments. We note that k provides an estimate of the intrinsic birefringence of fully oriented PVC and that at 80°C the derived value of 27×10^{-3} is less than the value of 85×10^{-3} estimated from stress-optical data. The experimental points shown in Fig. 18 correspond to f_{PVC} values obtained using eqn. (21). The triangular points represent other estimates of f_{PVC} obtained by extrapolating the dichroic f_D values to zero reaction time.

Also shown in Fig. 18 are amorphous orientation functions derived from infra-red dichroism data on selected absorption peaks. The curves marked IR(—C=C—C=C—) and IR(—C=C—) were obtained from bands at 1610 cm^{-1} and 1675 cm^{-1}, which were considered to arise from C=C stretching modes of conjugated and unconjugated bonds, respectively. Since the transition moment should lie approximately along the respective segment axes for these absorptions, the orientation functions were evaluated from $f = (D-1)/(D+2)$ according to eqn. (14). The curves marked IR(C—Cl Trans) and IR(C—Cl Gauche) were obtained from the respective peaks at 615 cm^{-1} and 690 cm^{-1}. The 615 cm^{-1} band is thought to be largely influenced by a component at 610 cm^{-1} arising from C—Cl stretching modes in syndiotactic segments of extended *trans* conformation.[7,47] The 690 cm^{-1} peak also originates from C—Cl stretching vibrations, but in this case the local absorbing sequence contains both *trans* and *gauche* bonds.[47] Since the transition

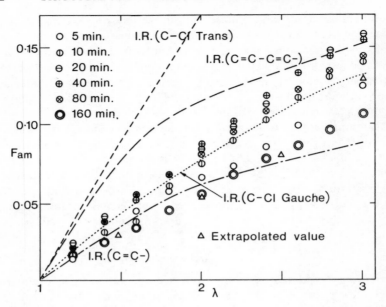

Fig. 18. Amorphous orientation functions v. extension ratio for PVC elongated at 80°C. Circular points were obtained from birefringence measurements calibrated against UV dichroism on dehydrochlorinated samples of specified reaction time (see key and text). Triangles (△) obtained by extrapolating UV orientation functions to zero reaction time. Broken lines from different IR absorption bands (Shindo et al.[5]).

moments are considered to lie approximately along the C—Cl bonds, orientation functions from the 615 cm^{-1} and 690 cm^{-1} bands were calculated using $f = -2(D-1)/(D+2)$ (see eqn. (11) with $D_0 = 2 \cot^2 \alpha = 0$).

The most significant fact to emerge from Fig. 18 is that the orientation functions estimated by means of the birefringence–UV–visible–dichroism method correlate closely with those obtained from the infra-red dichroism of bands originating from *gauche* containing sequences and unconjugated C=C bonds. The infra-red orientation functions derived for extended *trans* sequences are appreciably higher, and do not appear representative of the average orientation of amorphous PVC chain segments. We recall that similar results were obtained for low density polyethylene (Fig. 12).

4.5.3 References to other polymers

The dichroism technique has been employed in structural and orientation studies on a wide range of polymers, and a detailed discussion of the many results would require considerably more space than is appropriate for this chapter. In Table 2 we have collected together a fairly comprehensive list of references to recent quantitative dichroic studies on individual polymers, copolymers and polymer blends. No references are given to the numerous qualitative dichroic results reported, although it should be mentioned that these have provided valuable information concerning the structure of polymers and their spectral assignments.

TABLE 2

REFERENCES TO DICHROIC STUDIES ON VARIOUS POLYMERS

Polymer	Method[a]	Authors	Ref.	(year)
Polypropylene	IR (U, B)	Schmidt	48	(1963)
	IR (U)	Samuels	49	(1965)
	IR (U)	Onogi and Asada	37	(1971)
	IR (U)	Trott	33	(1970)
	V (U)	Nakayama et al.	50	(1971)
Polybutene-1	IR (U)	Onogi and Asada	37	(1971)
Polyethylene terephthalate	V (U)	Patterson and Ward	12	(1957)
	IR (B)	Schmidt	24	(1963)
	IR (B)	Koenig and Cornell	27	(1967)
	V (U)	Okajima et al.	51	(1970)
Nylon 66	V (U)	Chappel	15	(1960)
Nylon 6	V (U)	Yamada and Stein	52	(1964)
	V (U)	Yamada et al.	53	(1966)
	IR (U, B)	Sibilia	54	(1971)
Polyvinyl alcohol	V (U)	Yamada et al.	53	(1966)

TABLE 2—*continued*

Polymer	Method[a]	Authors	Ref.	(year)
Polyacrylonitrile	IR (U)	Bohn et al.	55	(1961)
	IR (U)	Zbinden	4	(1964)
	IR (U, B)	Koenig et al.	26	(1969)
			56	(1970)
Polystyrene	IR (U)	Zbinden	4	(1964)
	IR (U)	Milagin et al.	57	(1970)
Polycarbonate	IR (U)	Yannas and Lunn	58	(1971)
Natural rubber	IR (U)	Gotoh et al.	28	(1965)
Polychloroprene	IR (U)	Takenaka et al.	30	(1970)
Silicone	IR (U)	LeGrand	59	(1965)
Polyethylene + poly-propylene blends	IR (U)	Onogi and Asada	37	(1971)
Polypropylene + ethylene–propylene rubber blends	IR (U)	Onogi et al.	60	(1969)
		Onogi and Asada	37	(1971)
Ethylene–propylene block copolymers	IR (U)	Onogi and Asada	37	(1971)
Segmented polyurethanes	IR (U)	Ishihara et al.	61	(1970)
	IR (U)	Estes et al.	62	(1971)
Salts of ethylene–methacrylic acid copolymers	IR (U)	Uemura et al.	63	(1971)

[a] IR and V denote infra-red and visible respectively. (U) and (B) denote uniaxial and biaxial deformation, respectively.

REFERENCES

1. Barrow, G. M. (1962). *Introduction to Molecular Spectroscopy*, McGraw-Hill, New York.
2. Murrell, J. N. (1963). *The Theory of the Electronic Spectra of Organic Molecules*, Methuen (London) and John Wiley & Sons (New York).
3. Krimm, S. (1963). 'Infra-red spectra of solids: dichroism and polymers,' in *Infra-red Spectroscopy and Molecular Structure*, (Ed. Mansel Davies) Elsevier, Amsterdam.

4. Zbinden, R. (1964). *Infra-red Spectroscopy of High Polymers,* Academic Press, New York, London.
5. Shindo, Y., Read, B. E. and Stein, R. S. (1968). *Die Makromol. Chemie,* **118,** 272.
6. Venkataraman, K. (1952). *The Chemistry of Synthetic Dyes,* Academic Press, New York.
7. Krimm, S. (1960). *Fortschr, Hochpolym.—Forsch.,* **2,** 51.
8. Snyder, R. G. (1967). *J. Chem. Phys.,* **47,** 1316.
9. Read, B. E. and Stein, R. S. (1968). *Macromolecules,* **1,** 116.
10. Fraser, R. D. B. (1953). *J. Chem. Phys.,* **21,** 1511.
11. Beer, M. (1956). *Proc. Roy. Soc. (London),* **A236,** 136.
12. Patterson, D. and Ward, I. M. (1957). *Trans. Faraday Soc.,* **53,** 1516.
13. Stein, R. S. (1958). *J. Polymer Sci.,* **31,** 327, 335.
14. Stein, R. S. (1961). *J. Polymer Sci.,* **50,** 339.
15. Chappel, F. P. (1960). *Polymer,* **1,** 409.
16. Kawai, K. and Stein, R. S. (1964). ONR Tech. Rept. No: 68, Project NR 356–378, Contract NONR 3357–(01) Polymer Res. Inst., Univ. of Mass.
17. Nomura, S., Kawai, H., Kimura, I. and Kagiyama, M. (1967). *J. Polymer Sci., A-2,* **5,** 479.
18. Flory, P. J. and Abe, Y. (1969). *Macromolecules,* **2,** 335.
19. Nagai, K. (1971). In *Progress in Polymer Science, Japan,* Vol 1, (Ed. M. Imoto and S. Onogi), Kodansha, Tokyo.
20. Read, B. E. and Dean, G. (1970). *Polymer,* **11,** 597.
21. Stein, R. S. and Read, B. E. (1969). *Applied Polymer Symposia,* No. 8, 255.
22. Shindo, Y. and Stein, R. S. (1969). *J. Polymer Sci., A-2,* **7,** 2115.
23. Wilkes, G. L. (1971). *Fortschr. Hochpolym.—Forsch.,* **8,** 91.
24. Schmidt, P. G. (1963). *J. Polymer Sci., A,* **1,** 1271.
25. Koenig, J. L., Cornell, S. W. and Witenhafer, D. E. (1967). *J. Polymer Sci., A-2,* **5,** 301.
26. (a) Koenig, J. L., Wolfram, L. and Grasselli, J. (1969). *ACS Polymer Preprints,* **10,** 959.

   (b) Wolfram, L., Grasselli, J. and Koenig, J. L. (1970). *Appl. Spect.,* **24,** 263.
27. Koenig, J. L. and Cornell, S. W. (1967). *J. Macromol. Sci., Phys.,* **B1** (2), 279.
28. Gotoh, R., Takenaka, T. and Hayama, N. (1965). *Kolloid-Z.,* **205,** 18.
29. Gotoh, R., Takenaka, T., Umemura, J. and Hayashi, S. (1966). *Bull. Inst. Chem. Res. Kyoto Univ.,* **44,** 286.
30. Takenaka, T., Shimura, Y. and Gotoh, R. (1970). *Kolloid-Z.,* **237,** 193.
31. Wilkes, G. L., Uemura, Y. and Stein, R. S. (1971). *J. Polymer Sci., A-2,* **9,** 2151.
32. Read, B. E., Hughes, D. A., Barnes, D. C. and Drury, F. W. M. (1972). *Polymer,* **13,** 485.
33. (a) Trott, G. F. (1970). *J. Appl. Polymer Sci.,* **14,** 2421.
   (b) Fluornoy, P. A. and Schaffers, W. J. (1966). *Spectrochim. Acta,* **22,** 5.
34. Desper, C. R. (1969). *J. Appl. Polymer Sci.,* **13,** 169.
35. Fukui, Y., Asada, T. and Onogi, S. (1972). *Polymer Journal (Japan),* **3,** 100.
36. Onogi, S. and Asada, T. (1967). *J. Polymer Sci.,* Part C, No. 16, 1445.
37. Onogi, S. and Asada, T. (1971). In *Progress in Polymer Science, Japan,* Vol. 2, (Ed. M. Imoto and S. Onogi), Kodansha, Tokyo.
38. Read, B. E. and Hughes, D. A. (1972). *Polymer,* **13,** 495.
39. Liberman, M. H., Abe, Y. and Flory, P. J. (1972). *Macromolecules,* **5,** 550.
40. Stein, R. S. (1966). *J. Polymer Sci.,* Part C, No. 15, 185.
41. Phillips, P. J., Wilkes, G. L., Delf, B. W. and Stein, R. S. (1971). *J. Polymer Sci., A-2,* **9,** 499.
42. Uemura, Y. and Stein, R. S. (1972). *J. Polymer Sci., A-2,* **10,** 1691.
43. Glenz, W. and Peterlin, A. (1970). *J. Macromol. Sci., Phys.,* **B4** (3), 473.
44. Glenz, W. and Peterlin, A. (1971). *J. Polymer Sci., A-2,* **9,** 1191.

45. Glenz, W. and Peterlin, A. (1971). *Die Makromolek, Chem.*, **150,** 163.
46. Miller, P. J., Jackson, J. F. and Porter, R. S. (1972). *ACS Polymer Preprints,* **13,** 335.
47. Krimm, S. (1964). *J. Polymer Sci.,* **C.7,** 3.
48. Schmidt, P. G. (1963). *J. Polymer Sci., A,* **1,** 2317.
49. Samuels, R. J. (1965). *J. Polymer Sci., A,* **3,** 1741.
50. Nakayama, K., Okajima, S. and Kobayashi, Y. (1971). *J. Appl. Polymer Sci.,* **15,** 1453.
51. Okajima, S., Nakayama, K., Kayama, K. and Kato, Y. (1970). *J. Appl. Polymer Sci.,* **14,** 1069.
52. Yamada, R. and Stein, R. S. (1964). *J. Polymer Sci., B,* **2,** 1131.
53. Yamada, R., Hayashi, C. and Onogi, S. (1966). *Intern. Symp. Macromol. Chem., Tokyo,* **8,** 160.
54. Sibilia, J. P. (1971). *J. Polymer Sci., A-2,* **9,** 27.
56. Koenig, J. L., Wolfram, L. E. and Grasselli, J. G. (1970). *J. Macromol. Sci., Phys.,* **B4,** 491.
57. Milagin, M. F., Gabarayeva, A. D. and Shishkin, I. I. (1970). *Polymer Sci., U.S.S.R.,* **12,** 577.
58. Yannas, I. V. and Lunn, A. C. (1971). *J. Polymer Sci., B,* **9,** 611.
59. LeGrand, D. G. (1965). *J. Polymer Sci., A,* **3,** 301.
60. Onogi, S., Asada, T. and Tanaka, A. (1969). *J. Polymer Sci., A-2,* **7,** 171.
61. Ishihara, H., Kimura, I., Saito, K. and Ono, H. (1970). *Rep. Prog. Polymer Phys., Japan,* **13,** 409.
62. Estes, G. M., Seymour, R. W. and Cooper, S. L. (1971). *Macromolecules,* **4,** 452.
63. Uemura, Y., Stein, R. S. and MacKnight, W. J. (1971). *Macromolecules,* **4,** 490.

CHAPTER 5

POLARISED FLUORESCENCE AND RAMAN SPECTROSCOPY

D. I. BOWER

5.1 INTRODUCTION

If fluorescent molecules are dispersed in a solid polymer which has been oriented by drawing, the fluorescent light emitted when the polymer is suitably illuminated is usually found to be partially polarised, with the maximum intensity observable when the electric vector of the light transmitted by the analyser is parallel to the draw direction. This may be understood qualitatively by considering the way in which an individual fluorescent molecule absorbs and emits light.

A fluorescent molecule absorbs light of a certain wavelength, or range of wavelengths, provided that there is a component of the electric vector of the incident light parallel to a certain axis fixed in the molecule, which is called the absorption axis. The rate of absorption is proportional to $\cos^2 A$, where A is the angle between the polarisation vector of the incident light and the absorption axis. After a short time it emits light of a longer wavelength or wavelengths and this light is emitted from an electric dipole parallel to an emission axis fixed in the molecule. The intensity, I, observed is thus proportional both to the rate of absorption and to $\cos^2 B$, where B is the angle between the emission axis and the polarisation axis of the analyser. Thus

$$I \propto \cos^2 A \cos^2 B \qquad (1)$$

The emission and absorption axes usually either coincide or are separated by a small angle. If some mechanism exists which causes the emission axes of the fluorescent molecules to be aligned preferentially parallel to the polymer chains, then the observation of a maximum fluorescent intensity when the analyser is parallel to the draw direction is readily understood.

Morey[1] appears to have been the first to make use of this phenomenon to obtain information about the distribution of orientations of the structural units of the polymer. Most of his observations were made using fluorescent molecules incorporated in natural fibres such as ramie, cotton and flax, but some measurements were made on rayons. He used 'unpolarised' ultra-violet light to excite the fluorescence and expressed his results in terms of a percentage orientation defined as

$$p_M = \frac{I_{max} - I_{min}}{I_{max} + 2I_{min}} \times 100 \tag{2}$$

where I_{max} and I_{min} were the maximum and minimum values of the intensity when this was observed as a function of the angle between the polarisation vector of the analyser and the fibre axis. Morey concluded (not quite correctly) that p_M was linearly related to $\overline{\cos^2 \theta_e}$, with θ_e equal to the angle between the emission axis of a fluorescent molecule and the direction corresponding to I_{max} and the bar denoting the average over the distribution of orientations. An overall correlation was observed between the value of p_M and mechanical properties such as the tensile strength of the fibres and, in addition, good agreement was found between the percentage orientation of the fluorescent molecules and the orientation of the crystalline regions of the fibres as deduced from X-ray scattering. The latter observation suggested that the fluorescent molecules interacted with the polymer molecules in a well-defined way, and Morey suggested that although the fluorescent molecules probably did not penetrate the crystallites they might become attached to their surfaces and so give information about their orientations. In most of the more recent work the exactly contrary hypothesis has been made, that the fluorescent molecules are dispersed in the non-crystalline regions of the polymer and that their orientations are determined by those of the polymer chains in these regions.

The information obtained in Morey's work was very closely similar to that obtainable from measurements of the absorption dichroism of dye molecules dispersed in the polymer or of the infra-red dichroism of the polymer itself (see Chapter 4). In dichroism experiments the absorption of polarised light is measured as a function of the angle between the electric vector of the incident light and a chosen axis in the specimen and information is obtained about $\overline{\cos^2 \theta_a}$, where θ_a is the angle between the chosen axis and the absorption axis of a dye molecule. Nishijima et al.[2] pointed out that if polarised exciting light is used in the fluorescence experiment the anisotropy of both the absorption and emission processes can be exploited and additional information obtained. It is clear, for instance, that if the emission and absorption axes coincide and if both exciting and analysed fluorescent light are polarised parallel to the chosen axis in the polymer, then $A = B = \theta_a$ in eqn. (1) and the intensity observed will be proportional to $\overline{\cos^4 \theta_a}$. The ability of this method to give, in principle at least, information about what are often called the second and fourth moments of the distribution of orientations of polymer molecules has been the primary reason for interest in it, since information about the fourth moment is not easily obtained by other methods, particularly for the polymer molecules in the noncrystalline regions. It should, how-ever, be remembered that the information obtained relates directly to the fluorescent molecules, and information about the distribution of orienta-tions of the polymer molecules can only be deduced if the orientation of the fluorescent molecules with respect to the polymer molecules is known. This will be considered again later.

Another phenomenon which involves an optical anisotropy with respect to both incident and emitted light is the Raman effect, and it was suggested by Stein and Read[3] and by Cornell and Koenig[4] that measurements of the intensity of polarised Raman scattering might be used to obtain directly information about the distribution of molecular orientations in a solid polymer. Raman scattering is similar to fluorescence in that it involves irradiating the sample with light of one wavelength and observing radiation emitted at a longer wavelength. The process may be pictured classically[5] as the scattering of radiation incident with frequency ω_0 by a molecule vibrating with a normal-mode frequency ω_m. The polarizability of the molecule varies at the frequency ω_m and thus the electric vector of the incident light induces a dipole which oscillates at the frequency ω_0 with an amplitude which varies, at the frequency ω_m, by an amount proportional to the derivative of the polarisability with respect to the normal co-ordinate. This is equivalent to the sum of three oscillations with frequencies $\omega_0 - \omega_m$, ω_0 and $\omega_0 + \omega_m$. Quantum-mechanically, the incident photon of energy $\hbar\omega_0$ either increases the excitation of the molecule by one quantum of vibrational energy, $\hbar\omega_m$, and is scattered with energy $\hbar(\omega_0 - \omega_m)$ or it decreases the excitation by one quantum and is scattered with energy $\hbar(\omega_0 + \omega_m)$. These correspond to the Stokes and anti-Stokes lines, respectively. Since at normal temperatures the ground state is the only one with any appreciable population, only the Stokes line is observed with appreciable intensity.

It should be clear that if the molecule were polarisable only along one axis fixed within it, the information obtainable about the orientation of the polymer molecules from the Raman effect would be identical with that obtainable about the orientation of the molecules of a fluorescent additive from polarised fluorescence (assuming coincident emission and absorption axes). The polarisability is, however, not a vector but a second rank tensor (see Section 1.1.1). This means that any component, P_i, of the polarisation induced in the molecule by an applied electric field, E, is given by an expression of the form

$$P_i = \sum_j p_{ij} E_j \tag{3}$$

where p_{ij} is a component of the polarisability tensor $[p_{ij}]$. This expression applies for any orientation of the reference axes within the molecule, but there is a particular choice of axes, called the principal axes, for which $p_{ij} = 0$ unless $i = j$. The non-zero components, which may be called p_1, p_2 and p_3, are the principal components of the tensor. The tensor nature of the polarisability has the effect both of making the theory of the Raman method more complex than that of the fluorescence method and of offering potentially even more information about molecular orientation.

In the following section a brief account of the general principles of the theory of the two methods will be given. A further section will deal with apparatus and experimental details and with the results so far obtained. It is worth drawing attention here, however, to the important fact that both methods are suitable only for specimens of good optical quality, that is, specimens which do not appreciably scatter light which passes through them so that the polarisation of the light is not scrambled.

5.2 THEORY

The polymer solid is assumed to have at least orthotropic statistical symmerty and axes $O - x_1 x_2 x_3$ are chosen parallel to the two-fold rotation axes. If the solid has the simpler uniaxial symmetry, Ox_3 will be chosen parallel to the unique axis. The theory of the fluorescence method will be considered first, since it is somewhat simpler than that of the Raman method.

5.2.1 Fluorescence
In the treatments of the theory by Nishijima and his co-workers[2,6-13] and in the treatment by Seki,[14] calculations are made of the intensity of the fluorescence to be expected from various assumed distributions of

orientations of the fluorescent molecules. The intensity is a function of the orientation of the polarisation vectors of the exciting radiation and the analyser with respect to the symmetry axes of the specimen. Desper and Kimura[15] appear to have been the first authors to consider what parameters determine the intensities in the general case and what are the minimum number of intensities that must be measured in order to deduce them. Later work on the theory has usually adopted this type of approach and it will also be used here.

Let the absorption and emission axes of a particular fluorescent molecule have direction cosines f_i^a and f_i^e, respectively with respect to $O - x_1 x_2 x_3$, where $i = 1, 2, 3$, and let the direction cosines of the polarisation vectors of the incident and observed scattered light be l_i^a and l_i^e, respectively. It follows from eqn. (1) that the observed intensity from an aggregate of molecules is then given by

$$I = I_0 \sum \cos^2 A \cos^2 B$$

$$= I_0 \sum \left(\sum_i l_i^a f_i^a \right)^2 \left(\sum_i l_i^e f_i^e \right)^2 = I_0 \sum \left(\sum_{ij} l_i^a l_j^e f_i^a f_j^e \right)^2 \tag{4}$$

where I_0 is a constant depending on instrumental factors, the incident light intensity and the quantum efficiency of the fluorescent molecules. The summation outside the bracket is over all fluorescent molecules contributing to the observed intensity, and effects due to the birefringence of a macroscopic sample are neglected for the moment.

Equation (4) may be written

$$I = I_0 \sum_{ijpq} \left(\sum f_i^a f_j^a f_p^e f_q^e \right) l_i^a l_j^a l_p^e l_q^e$$

$$= \sum_{ijpq} F_{ijpq} \, l_i^a l_j^a l_p^e l_q^e \tag{5}$$

The quantities F_{ijpq} defined by eqn. (5), which contain information about the distribution of orientations, are in fact the components of a fourth rank tensor $[F_{ijpq}]$. It is clear from the definition of F_{ijpq} that interchange of i and j or p and q does not change its value, so that there are at most thirty-six independent components. It follows, however, from the assumed orthotropic symmetry of the specimen that only twelve of these are non-zero, nine of the form F_{iijj} and three of the form F_{ijij}. All the information available about the distribution of orientations from measurements of fluorescence intensities is thus contained in these twelve quantities.

Any one of the nine components of the form F_{iijj} can in principle be determined directly by setting the polarisation vector of the incident light parallel to Ox_i and the analyser polarisation vector parallel to Ox_j. The observed intensity is then F_{iijj}. The remaining three components can only be determined by setting both polarisation vectors away from the symmetry axes of the specimen. For such measurements the effects produced by the birefringence of the polymer sample must be considered, and this has been done by Seki,[14] Onogi and Nishijima,[16] Bower[17] and Nobbs et al.[18] In practice making such measurements usually introduces three further unknowns into the problem, because although it is possible to predict the form of the birefringence effect the absolute magnitude cannot easily be calculated. These measurements cannot in general, therefore, be used to give information about the components F_{ijij}. Fortunately, all the information about the distribution of orientations may be obtained in many cases from the nine components F_{iijj}, and the details of the birefringence correction will not, therefore, be discussed here.

If we consider the simplest possibility, that the emission and absorption axes coincide, then $f_i{}^a = f_i{}^e = f_i$ and $F_{ijij} = F_{iijj} = F_{jjii}$. Only six different components now determine any observed intensity. This was first pointed out by Desper and Kimura[15]. If the polar and azimuthal angles of a typical absorption axis are θ and ϕ then

$$f_1 = \sin\theta\cos\phi, f_2 = \sin\theta\sin\phi, f_3 = \cos\theta \tag{6}$$

Substituting these values into eqn. (5) leads to the following expressions for the six independent components, where N_0 is the total number of fluorescent molecules involved:

$$F_{1111} = I_0 N_0 \overline{\sin^4\theta\cos^4\phi}$$

$$F_{2222} = I_0 N_0 \overline{\sin^4\theta\sin^4\phi}$$

$$F_{3333} = I_0 N_0 \overline{\cos^4\theta}$$

$$F_{1122} = I_0 N_0 \overline{\sin^4\theta\sin^2\phi\cos^2\phi} \tag{7}$$

$$F_{2233} = I_0 N_0 \overline{\sin^2\theta\cos^2\theta\sin^2\phi}$$

$$F_{3311} = I_0 N_0 \overline{\sin^2\theta\cos^2\theta\cos^2\phi}$$

If the sample is uniaxially oriented around Ox_3 then for any value of θ, ϕ takes all possible values from 0 to 2π with equal probability and the functions of ϕ can be replaced by their average values. This leads to

$$F_{1111} = F_{2222} = 3F_{1122} = \tfrac{3}{8}I_0 N_0(1 - 2\,\overline{\cos^2 \theta} + \overline{\cos^4 \theta})$$

$$F_{3333} = I_0 N_0 \overline{\cos^4 \theta} \tag{8}$$

$$F_{2233} = F_{3311} = \tfrac{1}{2}I_0 N_0 \overline{(\cos^2 \theta - \cos^4 \theta)}$$

and $\overline{\cos^2 \theta}$ and $\overline{\cos^4 \theta}$ can be determined from three suitable intensity measurements, using the equation

$$(8/3)F_{1111} + 4F_{2233} + F_{3333} = I_0 N_0 \tag{9}$$

For a sample of the more general orthotropic symmetry the six equations (7) appear to contain six unknown averages and one unknown constant $I_0 N_0$. There are, however, only five independent averages, since the six functions involved in the averages can be expressed in terms of the five spherical harmonic functions $P_l^m(\cos \theta)\cos m\phi$ for $l = 2, 4$ and $m = 0, 2, 4$. The average values of the $P_l^m(\cos \theta)\cos m\phi$ are then the coefficients, v_{lm}, of an expansion of the distribution function in terms of spherical harmonics. Bower[17] has given a table which enables equations for the intensities to be written directly in terms of the five v_{lm}. The use of spherical harmonics to describe the distribution of orientations has, in addition to the advantage of dealing in terms of independent quantities, the advantage that if the axes of the fluorescent molecules have a known distribution of orientations with respect to the axes of the polymer chains, the relationship between the distribution functions of the polymer chains and of the axes of the fluorescent molecules can easily be written down using the addition theorem for spherical harmonics, as discussed by Roe,[19] by Nomura et al.[20] and by Bower.[17]

It should be noted that for a randomly oriented sample $\overline{\cos^2 \theta} = 1/3$ and $\overline{\cos^4 \theta} = 1/5$, so that if $I_{\|} = F_{1111} = F_{2222} = F_{3333}$ and $I_{\perp} = F_{1122} = F_{2233} = F_{3311}$, then $R = I_{\|}/I_{\perp} = 3$. Experimentally, R is usually found to be less than 3 and two types of explanation for this have been suggested. The first is that although the emission and absorption axes of the fluorescent molecules coincide, the molecule rotates between emission and absorption because of thermal motions. The modifications to the theory that this brings about have been discussed by Kimura et al.,[21] and by Badley et al.[22] They will not be considered here, since it seems unlikely that significant rotation of the rather long fluorescent molecules usually used occurs during the lifetime of excitation when they are incorporated in a solid polymer, and there is some experimental evidence to support this view.[18]

The second type of explanation for finding values of R less than 3 involves the assumption that the emission and absorption axes of the fluorescent molecule are not coincident. Kimura et al.[21] have considered a model in which the absorption and emission axes each have, independently, a cylindrically symmetric distribution of orientations around a third unique axis in the molecule, and Nobbs et al.[18] have considered a model which includes both this and the possibility that there is a fixed angle between the emission and absorption axes which are otherwise uniformly distributed around a third unique axis. In the more general model at least three parameters are required to specify the relationships between the directions of the emission and absorption axes and that of the unique axis of a fluorescent molecule and these are not generally known. For orthotropic symmetry, five v_{lm} are required to characterise the distribution of orientations of the unique axes and if the constant $N_0 I_0$ is included, there is a total of at least nine unknown quantities. No attempt has so far been made to evaluate these from intensity measurements on an orthotropic sample. For a uniaxial sample only two parameters, $\overline{\cos^2 \theta}$ and $\overline{\cos^4 \theta}$, are required for characterising the distribution of orientations and by making various approximations[18] the total number of unknown quantities can be reduced to six. Their evaluation then becomes a practical possibility.

5.2.2 Raman

The first treatment of the theory of the intensities and polarisation effects to be expected in the Raman scattering from an oriented polymer sample appears to be that given by Cornell and Koenig.[4] This treatment must be regarded at best as a very rough approximation, since the tensor nature of the effect is not taken into account properly. Snyder[23] has given a correct account of the theory for rather special distributions of orientations of the Raman scatterers but his work concentrates on the information that can be obtained about the Raman tensors if the orientation distribution is known. The only treatment that has considered how much information can in principle be obtained about the distribution of orientations and what measurements are necessary to obtain it is that of Bower[17] and Bower and Purvis.[24] In this treatment the similarities and differences between the theories of the fluorescence and Raman methods are apparent and an account of it follows.

The scattering from each unit (molecular segment or crystallite) of the polymer may be described, for a particular Raman line, by a second-rank tensor $[\alpha_{ij}]$ which corresponds classically to the derivative of the polarisability tensor of the unit.[5] The scattered intensity, I_R, is then given by

$$I_R = I_0' \sum \left(\sum_{ij} l_i^a l_j^e \alpha_{ij} \right)^2 \tag{10}$$

where I_0' is a constant depending on instrumental factors and the incident light intensity, l_i^a and l_j^e are direction cosines of the polarisation vectors of the incident and observed scattered radiations and α_{ij} is the ijth component of $[\alpha_{ij}]$ for a typical scattering unit, all expressed with respect to the axes $O - x_1 x_2 x_3$. It will be assumed, as is usually done, that the Raman tensor is symmetric, i.e. that $\alpha_{ij} = \alpha_{ji}$. The summation outside the bracket is again over all scattering units and effects due to the birefringence of a macroscopic sample are again neglected for the moment. Equation (10) is seen to be very similar to eqn. (4), with α_{ij} corresponding to $f_i^a f_j^e$. This last quantity may also be considered to be a second rank tensor, but it is not symmetric unless the absorption and emission axes coincide. If they coincide, $f_i^a = f_i^e = f_i$ for $i = 1, 2, 3$ and $f_i^a f_j^e = f_j^a f_i^e$. The theory of the fluorescence method is then simply a special case of the theory of the Raman method.[17]

The scattered intensity thus depends, for chosen values of the l_i^a and l_i^e, on quantities of the form $I_0' \sum \alpha_{ij} \alpha_{pq}$. These may be considered to be the components of a fourth-rank tensor for the sample as a whole, and it follows from the assumption that $\alpha_{ij} = \alpha_{ji}$ and the general symmetry restrictions on any fourth-rank tensor describing a material with orthotropic symmetry[25] that only the nine sums of the form $\sum \alpha_{ii} \alpha_{jj}$ or $\sum \alpha_{ij}^2$ are distinct and different from zero. For a material with uniaxial symmetry it can be shown that

$$\sum \alpha_{22}^2 = \sum \alpha_{11}^2, \qquad \sum \alpha_{31}^2 = \sum \alpha_{23}^2,$$

$$\sum \alpha_{11} \alpha_{22} = \sum \alpha_{11}^2 - 2 \sum \alpha_{12}^2, \qquad \sum \alpha_{33} \alpha_{11} = \sum \alpha_{22} \alpha_{33} \tag{11}$$

There remain for this symmetry only five independent quantities $I_0' \sum \alpha_{ij} \alpha_{pq}$ in terms of which all intensities can be expressed.

$I_0' \sum \alpha_{ij}^2$ is simply the intensity observed when the polarisation vector of the incident light is parallel to Ox_i and the analyser polarisation vector is parallel to Ox_j. The quantities $I_0' \sum \alpha_{ii} \alpha_{jj}$ for $i \neq j$ can only be determined by setting both polarisation vectors away from the symmetry axes of the specimen.

Just as the direction cosines f_i of the absorption axis of a particular fluorescent molecule can be expressed in terms of polar and azimuthal angles θ and ϕ, the tensor components α_{ij}, and hence the quantities $\alpha_{ij}\alpha_{pq}$, can be expressed in terms of the principal components $\alpha_1, \alpha_2, \alpha_3$ of $[\alpha_{ij}]$ and three Euler angles (θ, ϕ, ψ) defining the orientation of the principal axes with respect to $O - x_1 x_2 x_3$.† If the expressions for $\alpha_{ij}\alpha_{pq}$ in terms of θ, ϕ, ψ are expanded in terms of generalised spherical harmonics $Z_{lmn}(\xi)e^{im\phi}e^{in\psi}$, where $Z_{lmn}(\xi)$ is a generalisation of Legendre functions ($\xi = \cos\theta$), then it follows that

$$\sum \alpha_{ij}\alpha_{pq} = 4\pi^2 N_0' \sum_{lmn} v_{lmn} A_{lmn}^{ijpq} \tag{12}$$

where A_{lmn}^{ijpq} is a coefficient in the expansion of $\alpha_{ij}\alpha_{pq}$ and v_{lmn} is a coefficient in the expansion of the distribution function of the principal axes of the tensors in terms of the complex conjugate set of generalised spherical harmonics. In the summation, $l = 0, 2, 4$ and $|m|$, $|n|$ are even and $\leq l$. The limitation to even m follows from the assumption that the solid has at least orthotropic statistical symmetry, whereas the limitations on n and l follow partly from this and partly from the nature of the Raman effect. The A_{lmn}^{ijpq}, which are independent of the sign of m or n, have been tabulated[17] as functions of α_1, α_2 and α_3. The v_{lmn} are similarly independent of the sign of m and n. In the general case, the nine quantities $\sum \alpha_{ij}\alpha_{pq}$, and thus the observed Raman intensities, are determined, apart from the factor $I_0' N_0'$, by the three principal tensor components α_1, α_2 and α_3 and thirteen independent coefficients v_{lmn} (v_{000} is a constant, equal to $1/(4\pi^2\sqrt{2})$). It is thus not possible to solve for all the individual v_{lmn} even if the values of α_1, α_2 and α_3 (or their ratios) are known, which is usually not the case. Explicit expressions have been given[24] for the nine independent combinations of v_{lmn} and α_1, α_2, α_3 that can be obtained.

In the simplest possible case, where the sample has uniaxial statistical symmetry and where $\alpha_1 = \alpha_2 = \alpha_t$, the five independent $I_0' \sum \alpha_{ij}\alpha_{pq}$ are given by

$$I_0' \sum \alpha_{11}^2 = I_0' \sum \alpha_{22}^2 \tag{13}$$

$$= \tfrac{1}{8} N_0' I_0' [8\alpha_t^2 + 8\alpha_t \alpha_L + 3\alpha_L^2 - (8\alpha_t \alpha_L + 6\alpha_L^2)C_2 + 3\alpha_L^2 C_4]$$

$$I_0' \sum \alpha_{33}^2 = N_0' I_0' [\alpha_t^2 + 2\alpha_t \alpha_L C_2 + \alpha_L^2 C_4]$$

† The definition of the Euler angles is given in Section 1.5.1. The angles θ and ϕ are the polar and azimuthal angles of a chosen axis in the unit and ψ measures the rotation around this axis. In Ref. 17 and 19 ϕ and ψ are interchanged by comparison with the present use.

$$I'_0 \sum \alpha_{12}^2 = \tfrac{1}{8} N'_0 I'_0 \alpha_L^2 [1 - 2C_2 + C_4] \tag{13}$$

$$I'_0 \sum \alpha_{23}^2 = I'_0 \sum \alpha_{13}^2 = \tfrac{1}{2} N'_0 I'_0 \alpha_L^2 [C_2 - C_4]$$

$$I'_0 \sum \alpha_{22} \alpha_{33} = \tfrac{1}{2} N'_0 I'_0 [2\alpha_t^2 + \alpha_t \alpha_L + (\alpha_L^2 + \alpha_t \alpha_L)C_2 - \alpha_L^2 C_4]$$

where $C_2 = \overline{\cos^2 \theta}$, $C_4 = \overline{\cos^4 \theta}$ and $\alpha_L = \alpha_3 - \alpha_t$. If the ratio α_t/α_L is not known, all five equations must be used in order to solve for $\overline{\cos^2 \theta}$ and $\overline{\cos^4 \theta}$. It is thus necessary to make measurements in which the polarisation directions of both the incident and observed scattered radiations are not either parallel or perpendicular to the unique axis and to which corrections must therefore be made for the effects of the birefringence of the sample. For samples of high birefringence it will be very difficult[17] to determine the value $I'_0 \sum \alpha_{22} \alpha_{33}$, whereas Purvis and Bower[26] have described a method of determining it for samples of low birefringence (see Section 5.3.2). If the ratio α_t/α_L is determined from a randomly oriented sample, then $\overline{\cos^2 \theta}$ and $\overline{\cos^4 \theta}$ can be determined from any three of the other four independent $I'_0 \sum \alpha_{ij} \alpha_{pq}$. It should, however, be noted that measurements on a random sample can always be fitted by two possible values of α_t/α_L. If α_t is set equal to zero in eqns. (13) they take the same form as eqns. (8) for coincident emission and absorption axes in the fluorescence method, as expected.

So far the discussion has been concerned with the characterisation of the distribution of orientations of the principal axes of the tensors $[\alpha_{ij}]$, whereas the information of most interest is usually the distribution of orientations of the axes of the structural units, the molecular segments or crystallites. There are essentially two ways in which the problem of obtaining this information may be viewed. In the first,[17] the principal axes of the tensors are regarded as having a known distribution of orientations with respect to the axes of the structural units. This is the case when it is permissible to treat the vibrations of a particular type of group in the polymer chain as essentially independent of the vibrations of other groups and when the axes of the Raman tensor for a mode of vibration of these groups can be predicted from a knowledge of the geometry of the molecule. If M_{lmn} is now a coefficient similar to v_{lmn} but referring to the distribution of the axes of the structural units, eqn. (12) is replaced by

$$\sum \alpha_{ij} \alpha_{pq} = 16\pi^4 N'_0 \sum_{lmn\mu} [2/(2l+1)]^{1/2} M_{lm\mu} D_{l\mu n} A_{lmn}^{ijpq} \tag{14}$$

where the $D_{l\mu n}$ are the coefficients in an expansion of the distribution†
of the principal axes of the Raman tensor in terms of a series of
generalised spherical harmonics defined with respect to the axes of the
structural unit.

In the second approach,[24] the use of principal axes of the Raman
tensors is not retained, and the tensors are referred directly to the
symmetry axes of the structural units. If these axes are not the principal
axes of the Raman tensor then either more or fewer than three independent
tensor components may be needed, but in any case, components of the
type α_{ij}^o for $i \neq j$ will be introduced (in addition possibly to those of type
α_{ii}^o), where the superscript signifies reference to symmetry axes of the
structural unit. Equation (12) is now replaced by

$$\sum \alpha_{ij} \alpha_{pq} = 4\pi^2 N_0' \sum_{lmn} M_{lmn} B_{lmn}^{ijpq} \tag{15}$$

where B_{lmn}^{ijpq} is a coefficient similar to A_{lmn}^{ijpq} but with $\alpha_{ij}\alpha_{pq}$ now regarded as
a function of the Euler angles and up to six different possible α_{ij}^o. Bower
and Purvis[24] have discussed what information can be obtained from
Raman lines having non-zero components for particular values of i and
j and to what extent a previous knowledge of the values of the components
is needed if specific information is to be obtained about the distribution
of orientations.

5.3 EXPERIMENTAL WORK

5.3.1 Fluorescence

In its simplest form the exciting system will consist of a source, illumination
optics, filter and polariser, and the intensity-measuring system will consist
of an analyser, collection optics, filter and detector. The source of ultra-
violet light is usually a mercury or xenon arc and the filter on the source
side of the specimen selects from its output a narrow band of wavelengths near the peak of the absorption spectrum of the particular
fluorescent molecule in use. The illumination optics should transmit the
light as a fairly parallel beam to the specimen, so that it can be given a

---

† This 'distribution' will in many cases reduce to a delta function in the Euler angles,
since the tensor principal axes will usually be fixed in a specific orientation with respect
to the axes of the structural unit.

well-defined polarisation. The polariser should preferably be capable of being rotated so that either of two polarisation directions at right angles may be selected and the analyser should be capable of being set parallel or perpendicular to the direction of the polariser.

At its simplest, the collection optics will contain no lenses but will simply define an approximately parallel beam by means of apertures. If small areas of the sample are to be investigated a microscope system may be used, but in that case consideration must be given to the effect of the large collection angle on the polarisation of the detected light. The detection system will usually be a photomultiplier and amplifier whose output is observed either on an electrometer or a chart recorder, although an oscilloscope has been used for dynamic recording.[6,7]

It must be noted when considering the geometrical form of the specimen and the directions of the incident and fluorescent light beams that it is the direction of polarisation inside the specimen which is important. The use of certain geometries may necessitate corrections for refraction effects. It is also important for any type of system to correct for any sensitivity of the illuminating or detection systems to the polarisation vectors.

Figure 1 shows the apparatus used by Badley et al.[22] as an example of a complete system. Special features of other systems will be mentioned in discussing particular experiments.

The choice of fluorescent probe depends on a variety of factors. It has already been pointed out that what is determined directly is information which characterises the distribution of orientations of the fluorescent molecules. The ideal experiment would be one in which the polymer molecules themselves contained fluorescent groups. Stein[27] has considered the theory of the fluorescence method specifically for a uniaxially oriented fluorescent rubber but no experiments to study orientation have been reported for such a system. Nishijima et al.[28] have, however, made some qualitative observations on the polarisation of the fluorescent light from polyvinylchloride films which had been first stretched and then irradiated with light of wavelength 185 nm to produce fluorescent polyene segments.

It has usually been asserted that the fluorescent molecules are located only in the non-crystalline regions of crystalline polymers, although no clear evidence for this statement is given. If this is so, it seems reasonable to expect that, where no strong specific alignment forces exist between the polymer molecules and the fluorescent molecules, long thin fluorescent molecules will tend to align themselves parallel to the polymer chains. If, in addition, the absorption and emission axes of the fluorescent molecule

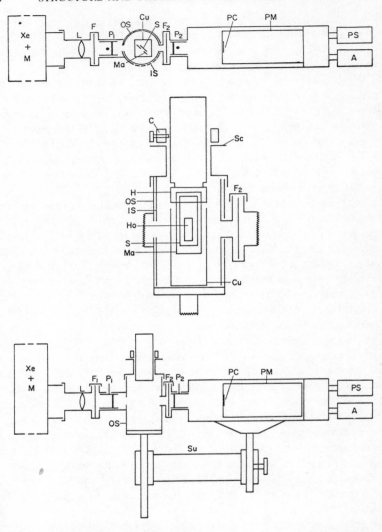

*Fig. 1. Apparatus used to make polarised fluorescence intensity measurements; from the top (a), the cell compartment (b), and from the side (c). Xenon arc (Xe), monochromator (M), lens (L), filter (F₁), polariser (P₁), outside shell (OS), inside shell (IS), cuvet (Cu), slide (S), mask (Ma), filter (F₂), polariser (P₂), photocathode (PC), photomultiplier (PM), power supply (PS), microammeter (A), locking collar (C), scale (Sc), slide holder (H), hole in mask (Ho), movable support (Su).
(Reproduced by permission from Ref. 22. Copyright American Chemical Society.)*

lie close to its geometric long axis the results will be most easily interpreted.

A molecule which appears from both theoretical and experimental evidence to satisfy these criteria and which has been used in a number of investigations is 2,2′(vinylene-di-*p*-phenylene)bisbenzoxazole) (VPBO), which is a stilbene derivative (it may be considered to be 4,4′(dibenzoxazolyl)stilbene). It has the chemical structure shown in Fig. 2(a). This compound has been incorporated into the molten polymer either after or during polymerisation. Figures 2(b) and 2(c) show the chemical structure of two of the other probes that have been used. Some probes can be diffused into a polymer from solution and this permits studies of the influence of orientation on diffusion. A large number of possible compounds might be used as probes, but it is important to note that any probe should have a high fluorescence efficiency so that it can be used in low concentration. This is important both to avoid perturbing the polymer structure and to ensure that no resonance energy transfer takes place between the fluorescent molecules, since this will affect the observed polarisation. The latter process can be detected by making experiments at different concentrations. For VPBO, for example, it is found that no such effects can be detected at concentrations of 200 ppm and this

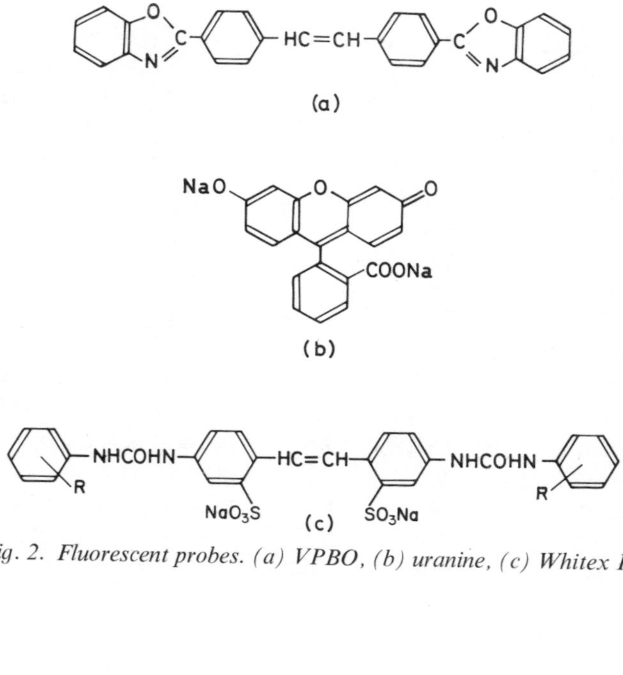

(a)

(b)

(c)

Fig. 2. Fluorescent probes. (a) VPBO, (b) uranine, (c) Whitex R.P.

concentration gives a satisfactorily strong signal from the thicknesses of specimens used.

Consideration will now be given to the experiments that have been carried out. Nishijima and his co-workers have performed a large number of experiments, most of which have, however, been of a qualitative or semi-quantitative nature. The polymers studied include polypropylene,[7,29-31] polyvinyl alcohol,[6,32-35] polyvinyl chloride[31,36] and polyethylene[31,37] and a variety of different fluorescent probes was used. In the simplest type of experiment[6,7,32,36] the specimen, which was either one-way or two-way drawn sheet, was mounted in the apparatus so that its plane contained both the exciting and fluorescent light polarisation vectors and in such a way that it could be rotated about the normal to its plane. The fluorescent light intensities $I_{\parallel}$ and $I_{\perp}$ were then recorded as a function of the angle γ between the polarisation vector of the exciting light and the draw direction of the specimen. Here $I_{\parallel}$ and $I_{\perp}$ are the intensities for parallel and perpendicular polarisation vectors of exciting and fluorescent radiations. From the observed intensities the degree of polarisation, p, was calculated, where

$$p = \frac{I_{\parallel} - I_{\perp}}{I_{\parallel} + I_{\perp}} \tag{16}$$

Polar plots of $I_{\parallel}$, $I_{\perp}$ and p were made and were compared with those predicted on the basis of simple models for possible distributions of orientations in an attempt to characterise the distribution in terms of these models. Figure 3 shows some polar plots of $I_{\parallel}$ for polyvinyl alcohol.

This procedure has several disadvantages. First, the model distributions chosen were given no theoretical justification and were in some cases extremely unlikely to correspond with the true distributions. Since, in the simplest case, the observed intensities depend only on $\cos^2 \theta$ and $\cos^4 \theta$, many different models, each having the same values of both these parameters, could have been chosen to give equally good fits. Secondly, the values of $\cos^2 \theta$ and $\cos^4 \theta$ can be determined from the intensities measured for $\gamma = 0$ and $90°$ (assuming coincident emission and absorption axes) and, in the absence of effects due to birefringence, they alone determine the complete shape of the polar plots. The effects of the birefringence may, however, be very important in modifying the shape of the polar plots. It is not permissible, for instance, to claim, without other evidence, that because $I_{\parallel}$ shows maxima for $\gamma = 0$ and $90°$ some preferred orientation takes place in the direction perpendicular to the stretching

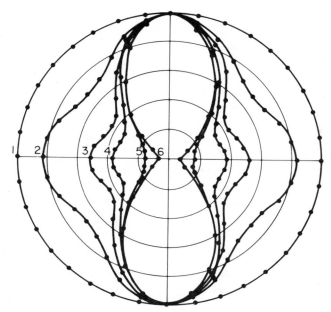

Fig. 3. Angular distributions of the polarised component of fluorescence $I_{\parallel}$, *emitted from the fluorescent compound dispersed in uniaxially stretched polyvinyl alcohol film at draw ratios of (1) 1, (2) 1·08, (3) 1·3, (4) 1·6, (5) 2·0, and (6) 5·0. (Reproduced by permission from Ref. 40.)*

axis as well as that taking place parallel to it. The maximum at $\gamma = 90°$ can occur because the birefringence effects lower the intensity observed in the region $\gamma \sim 45°$ below what it would otherwise be.[18] Finally, no attempt was made in this work to take account of the fact that for all the probes used the value of p for a random sample was <0.5, sometimes very significantly so, which means that the emission and absorption axes are not coincident.

Nishijima and his co-workers have also made measurements in which the planes of the films were tilted with respect to the polarisation directions and have compared the observed intensities with those predicted on the basis of various models in an attempt to obtain information about the full three-dimensional orientation distribution.[29,30,33,34,38] They have also described apparatus for the continuous study of molecular orientation behaviour during deformation of a specimen.[31,37,39] Similar criticisms to those above can be made about this work. More recently Onogi and Nishijima[16] have considered the effects of birefringence and have stated that corrections should be made to earlier work but none has been published and the conclusions drawn in the earlier work must therefore be treated with caution. Nishijima has written reviews[40,41] of some of this earlier work.

Kryszewski[42] has used a method very similar to that of Nishijima and his co-workers to study films of polyvinyl alcohol and polyethylene which were uniaxially stretched in the fluorescence apparatus. The PVA films were cast from aqueous solution containing uranine which acted as the fluorescent probe and the polyethylene films were cast from molten material containing anthracene as the fluorescent probe. The parameter, P, used as a measure of orientation was equal to p as defined above for $\gamma = 0$, and P/P_0 was plotted against the percentage extension, where P_0 is the value of P for the unstretched film. From the shape of the curve for PVA and the assumption that the fluorescent molecules are present only in the non-crystalline regions it is concluded that the orientation of the non-crystalline regions precedes that of the crystallites. For polyethylene the results were more complicated and dependent on whether the films were cooled quickly or slowly, since in the latter case the fluorescent probes could be incorporated in the crystallites. P/P_0 was found to increase uniformly with draw ratio in the region of reversible deformation of the polymer structure and it was concluded that elongation of the chains in the non-crystalline regions took place together with elastic deformation of spherulites. These results are again of a qualitative nature. The use of a single parameter to specify the orientation removes one of the chief advantages of the fluorescence method over the simpler measurement of absorption dichroism, particularly when, as in this case, the parameter is not simply related to those determinable by other methods.

Another study of orientation in polyvinyl alcohol by means of measurements of the degree of polarisation, p, of fluorescence has been made by Gulrajani and Padhye.[43] The specimens were films cast from aqueous solution and were drawn to draw ratios 2, 3, 4 and 5. The fluorescent probe was a reactive or a non-reactive optical brightening agent introduced by diffusion from solution. Although p was studied as a function of γ using polarised incident light, the numerical measure of orientation chosen was a single function of p determined using unpolarised incident light. The results were in qualitative agreement with those obtained from measurements of birefringence and infra-red and visible dichroism, but again full use of the information available from fluorescence was not made.

Seki[44] has studied orientation in a variety of polyamide fibres, again using a method similar to that of Nishijima's group. He has considered the effects of birefringence on the polar plots (for a film) but his conclusion[14] that the correction is only important when the birefringence is large and the non-crystalline orientation is nearly random does not agree with that of Nobbs et al.[18] The effects of spinning conditions on the molecular and crystalline orientations and on the mechanical properties are discussed, together with changes produced by drawing and heat treatments. Since his conclusions are all drawn from polar plots of $I_{\parallel}$ (γ) they must be regarded as rather tentative, particularly those concerned with a planar type of orientation perpendicular to the fibre axis.

Orientation has been studied in polyethylene terephthalate) using the fluorescence method by McGraw[45,46] and by Nobbs et al.[18] McGraw used a commercial spectrophotofluorometer (American Instrument Co. Model SPF-125) for the measurements. This instrument has two manually controlled grating monochromators which replace the filters used to select the wavelengths of the exciting and observed fluorescent light in the simplest systems. As an additional precaution a narrow-bandpass filter was placed in front of the IP21 photomultiplier which measured the intensities. The exciting and fluorescent light beams were at 90° to each other and at 45° to the surface of the rotatable samples. Glan prisms were used for polariser and analyser. No corrections were made to the data to take account of refraction at the surface of the samples nor to take account of the birefringence. Corrections for the former effect were considered unnecessary since the data were used only for comparisons among similar samples, and corrections for the latter were considered unimportant as a consequence of Seki's conclusions.[14] A correction was made for the polarisation sensitivities of the monochromators.

The measurements were made on fibres containing 200 ppm of VPBO. One series was spun in air and then drawn in steam at 140°C to draw ratios 1, 2, 3 and 4, another series was spun-drawn at spin–draw ratios from about 50 to 400 and a third series was shrunk at various temperatures. The results of the measurements were initially expressed in terms of the angular distribution of the degree of polarisation, p, but the final parameter used for comparing different specimens was the ratio of the maximum to minimum values of p, which McGraw calls the 'fluorescence orientation'. The use of this parameter is open to the following objections. First, the minimum value of p usually occurs when the polarisation vectors of both incident and observed scattered light are at approximately 45° to the draw direction, which is when the corrections for the effects of birefringence are greatest. Secondly, the use of any single parameter to describe the results removes one of the chief advantages of this method, as already mentioned.

The fibres drawn after spinning have a high birefringence, Δn, or more precisely $d\Delta n \gg \lambda$, where d is the sample thickness and λ the wavelength. Bower[17] has concluded that in this limit the observed intensities for plane-parallel specimens are independent of the value of Δn, although different from what they would be if Δn were zero. It may therefore be true that for the fibres drawn after spinning the 'fluorescence orientation' can be used to compare different specimens. McGraw's results show that this parameter first increases with draw ratio and then falls slightly above draw ratio 2 or 3 and he suggests that this may be explained as partial destruction of the crystalline regions and liberation of amorphous chains at high draw ratios. The spun-drawn fibres show values of $d\Delta n$ which are of the same order as λ and it is difficult to believe that the fluorescence orientation as defined by McGraw is then a useful parameter. The conclusion that spin-drawing can produce high chain orientation in the non-crystalline regions must thus be viewed with reservations. The bire-fringences of the fibres used in the shrinkage experiments are not given but are probably high, so that McGraw's claim that his results confirm the hypothesis that shrinkage reflects disorientation of the non-crystalline regions may be justified.

A number of criticisms has been made of the experimental work so far described and of the use of the data. The work of Nobbs et al.[18] was undertaken with a view to avoiding as many as possible of the limitations of the previous work so that the true limitations and usefulness of the fluorescence method could be established. The simplest possible type of specimen was chosen, uniaxially oriented specimens of polyethylene terephthalate in the form of tapes containing VPBO introduced during polymerisation. The tapes were produced by melt-spinning and one series was subsequently drawn at 80°C to draw ratios up to six. Another series was drawn at various temperatures to a draw ratio of 2·7. The crystallinity of the specimens was low and thus the question whether the VPBO enters predominantly the non-crystalline regions was avoided. A straight-through geometry was chosen for the system so that the incident and fluorescent light both travelled normally to the plane of the specimen, which could be rotated about this direction so that $I_{\parallel}$ and $I_{\perp}$ could be obtained as functions of γ. Monochromators were used as filters of the exciting and fluorescent light and the effects of the optical systems on the intensity of the light for different polarisation directions were corrected for.

The theory of the method was developed to include, in addition to the non-parallelism of the absorption and emission vectors and the effects of the birefringence, the effect of the finite absorption of the fluorescent molecules and its dichroism. The results were expressed in terms of $\overline{\cos^2 \theta}$ and $\overline{\cos^4 \theta}$, where θ is the polar angle of the unique axis of a typical VPBO molecule. It was also possible to calculate $\overline{\cos^2 \theta_a}$, where θ_a is the polar angle of the absorption axis of a typical VPBO molecule. The quantity $\overline{\cos^2 \theta_a}$ was also determined directly from measurements of the absorption dichroism of the VPBO and the two values were found to be in good agreement. Theory shows that, to a good approximation, the birefringence of a polymer sample should be linearly related to $\overline{\cos^2 \theta_p}$, where θ_p is the polar angle of a typical polymer chain (see Section 3.1). Figure 4 is a plot of $\overline{\cos^2 \theta}$ against birefringence for both series of tapes and shows that although the fluorescent molecules appear to have a higher

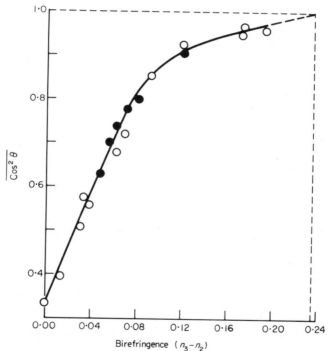

Fig. 4. $\overline{\cos^2 \theta}$ for the unique axes of fluorescent VPBO molecules dispersed in uniaxially oriented tapes of polyethylene terephthalate plotted against the birefringence $(n_3 - n_2)$, of the tapes. ○, samples drawn at 80°C to draw ratios between 1 and 6; ●, samples drawn to draw ratio 2·7 at temperatures between 65° and 90°C. (Reproduced by permission from Ref. 18. Copyright IPC Business Press Ltd.)

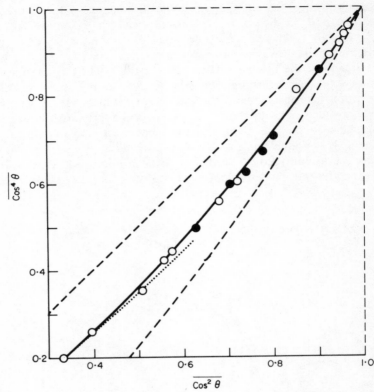

Fig. 5. $\overline{Cos^4\,\theta}$ plotted against $\overline{cos^2\,\theta}$ for the unique axes of fluorescent VPBO molecules dispersed in uniaxially oriented tapes of polyethylene terephthalate. ○, samples drawn at 80°C to draw ratios between 1 and 6; ●, samples drawn to draw ratio 2·7 at various temperatures between 65° and 90°C.
---- $\overline{cos^4\,\theta} = \overline{cos^2\,\theta}$ and $\overline{cos^4\,\theta} = (\overline{cos^2\,\theta})^2$
—— pseudo-affine aggregate deformation model
···· affine rubber elasticity model.
(Reproduced by permission from Ref. 18. Copyright IPC Business Press Ltd.)

orientation than the polymer chains the relationship between the two orientations is the same for both series.

From these results it is reasonable to conclude that the values of $\overline{cos^2\,\theta}$ and $\overline{cos^4\,\theta}$ determined for the fluorescent molecules in these experiments are probably correct, but that the fluorescent molecules do not, as often assumed, simply align themselves parallel to the polymer chains. Figure 5 shows the observed values of $\overline{\cos^4 \theta}$ plotted against the corresponding observed values of $\overline{\cos^2 \theta}$. The two dashed lines show the mathematical limits on corresponding values and the continuous and dotted curves show the predicted relationships between $\overline{\cos^4 \theta_p}$ and $\overline{\cos^2 \theta_p}$ according to the pseudo-affine aggregate deformation model and the affine rubber elasticity model, respectively (see Section 1.5.4). The closeness of the points to these lines may be to some extent fortuitous if the fluorescent molecules do not lie parallel to the polymer chains, but it does indicate that the values of $\overline{\cos^2 \theta}$ and $\overline{\cos^4 \theta}$ correspond to a distribution of orientations that is reasonable on physical grounds.

If values of $\overline{\cos^2 \theta_p}$ and $\overline{\cos^4 \theta_p}$ are required, the relationship between the two distributions of orientations must be known. This can only be established initially by comparison of the fluorescence results with those of other methods which give direct information about $\overline{\cos^2 \theta_p}$ and $\overline{\cos^4 \theta_p}$, such as infra-red and Raman spectroscopy, and studies along these lines are in progress (see Section 5.3.2). If the relationship can be satisfactorily established the fluorescence method could become a much simpler method of quantitatively characterising molecular orientation in amorphous polymers than any of the other methods that offer information about both $\overline{\cos^2 \theta_p}$ and $\overline{\cos^4 \theta_p}$.

5.3.2 Raman

The experiments by Cornell and Koenig[4] on oriented polystyrene appeared to show that orientation has a large effect on the observed intensity for given polarisations of incident and observed scattered light. Because of the approximate nature of the analysis they used, together with the fact that no attention was paid to the possibility of distortion of the intensity distribution by the effects of birefringence, no numerical data directly characterising the distribution of orientations of the polymer molecules were produced. The first such data were obtained, as far as is known, by Purvis et al.[47]

In this work uniaxially oriented specimens of polyethylene terephthalate were studied, most of which were from the first series used in the work of Nobbs et al. described in Section 5.3.1. The Raman intensity measurements were made using a Coderg PHO spectrometer and a CRL 52A argon ion laser tuned to 488 nm. The tape samples were mounted parallel to the spectrometer slit. The partially focused laser beam was incident normally on them and the scattered light was collected in directions making approximately 180° with the incident light direction. The incident and scattered light polarisation vectors could be chosen parallel or perpendicular to the length of the tapes, so that intensities proportional to $\sum \alpha_{33}^2, \sum \alpha_{11}^2, \sum \alpha_{13}^2$ and $\sum \alpha_{31}^2$ could be determined. The Raman line studied was that at 1616 cm^{-1} which is due to a benzene ring mode which, in a simple para-disubstituted benzene, would have a Raman tensor with one principal axis perpendicular to the ring plane and one of the other principal axes (chosen to correspond to α_3) parallel to the line joining the para-substituted carbon atoms.

Since only three independent intensities were measured, only two unknown quantities could be determined, and in order to reduce the number of unknowns various approximations and assumptions were made. It was first assumed that the polymer chains had no preferred orientation around their long axes. This assumption, together with the fact that the specimens had uniaxial symmetry, reduced the number of unknown orientation parameters to two, which were M_{200} and M_{400} or, equivalently, $\overline{\cos^2 \theta}$ and $\overline{\cos^4 \theta}$, where θ is the angle between a typical chain axis and the draw direction. Secondly, it was assumed that $\alpha_1 \cong \alpha_2 = \alpha_t$. The ratio $r = \alpha_t / \alpha_3$ could then be determined from measurements on a random sample. Substitution of $\overline{\cos^2 \theta} = \frac{1}{3}, \overline{\cos^4 \theta} = \frac{1}{5}$ into eqns. (13) and rearranging shows that

$$(\sum \alpha_{ij}^2)/(\sum \alpha_{ii}^2) = (1 - 2r + r^2)/(8r^2 + 4r + 3) \tag{17}$$

With these assumptions $\overline{\cos^2 \theta_R}$ and $\overline{\cos^4 \theta_R}$ could be deduced directly from eqns. (13), where θ_R refers to the orientation of the principal axis of the tensor corresponding to α_3. By assuming that this axis makes the same angle $\gamma (= 19° 12')$ with the chain axis as it would in the crystal phase it was possible to use the Legendre addition theorem to deduce M_{200} and M_{400} and hence $\overline{\cos^2 \theta}$ and $\overline{\cos^4 \theta}$. More formally, the whole calculation can be expressed in terms of the following particular example of eqn. (14):

$$\sum \alpha_{ij} \alpha_{pq} = 16\pi^4 N_0' \sum_{l=0,2,4} [2/(2l+1)]^{1/2} M_{l00} D_{l00} A_{l00}^{ijpq} \tag{18}$$

where M_{l00} and D_{l00} are simply proportional to the Legendre polynomials $P_l(\cos \theta)$ and $P_l(\cos \gamma)$, respectively, and A_{l00}^{ijpq} can be expressed in terms of r. It should be noted that if the assumption $\alpha_1 \cong \alpha_2$ had not been made then terms containing $D_{l0n} A_{l0n}^{ijpq}$ for $n = 0$ would in general be required in the above equation and the simpler method of calculation would not apply.

Figure 6 shows a plot of the values of $\overline{\cos^2 \theta}$ obtained against the birefringence. A good fit to a straight line is obtained and the indicated

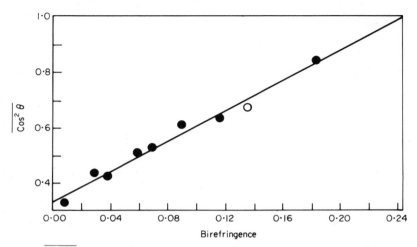

Fig. 6. $\overline{Cos^2\theta}$ plotted against sample birefringence, $(n_3 - n_2)$, for chain axes in uniaxially oriented tapes of polyethylene terephthalate. $\overline{Cos^2\theta}$ determined from Raman intensity measurements on the 1616 cm^{-1} line. (Reproduced by permission from Ref. 47. Copyright IPC Business Press Ltd.)

maximum birefringence (for $\overline{\cos^2\theta} = 1$) agrees well with that suggested by Kashiwagi et al.[48] The relationship between the values of $\overline{\cos^4\theta}$ and $\cos^2\theta$ observed experimentally was compared with those predicted by the affine rubber elasticity model and the pseudo-affine aggregate deformation model. These relationships are not very different from each other (see Fig. 5) or from that suggested by the experimental points. Measurements on three other Raman lines for the same specimens[49] gave results which were in quite good agreement with those obtained from the measurements on the 1616 cm^{-1} line. Two of these other lines correspond to other benzene ring modes and give quantitative agreement, whereas the orientation of the principal axes of the tensor for the other line was not known, so that only qualitative agreement was obtained. In these later measurements both 180° and 90° scattering geometries were used and the results were consistent with the assumption that the values of r were independent of molecular orientation.

Figure 7 shows the values of $\overline{\cos^2\theta}$ and $\overline{\cos^4\theta}$ determined from the results for the 1616 cm^{-1} line plotted against draw ratio, together with curves showing the predicted relationships according to the two deformation models already referred to. It is seen that moderately good agreement

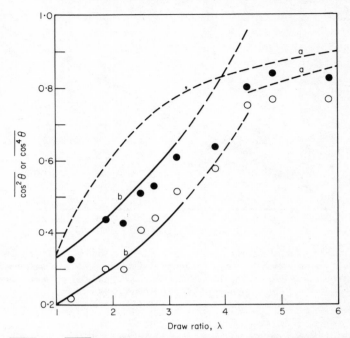

Fig. 7. $\overline{\cos^2\theta}$ and $\overline{\cos^4\theta}$ *plotted against draw ratio, λ, for chain axes in uniaxially oriented tapes of polyethylene terephthalate.* ●, $\overline{\cos^2\theta}$ *and* ○, $\overline{\cos^4\theta}$ *determined from Raman intensity measurements on the 1616 cm^{-1} line. Curves (a) pseudo-affine aggregate deformation model, curves (b) affine rubber elasticity model.*

with the rubber model is obtained for draw ratios less than about 3 and that there is some indication that for draw ratios greater than about 4·5 the pseudo-affine aggregate model may predict the values of $\overline{\cos^2\theta}$ and $\overline{\cos^4\theta}$ satisfactorily. Work is in progress on a more general model for the deformation mechanisms involved in drawing and it is hoped that by comparing this model with the results obtained by Raman, fluorescence and infra-red spectroscopy a more complete understanding of the drawing process for polyethylene terephthalate may be obtained.

In the experiments just described a rather limited number of intensities was measured and various assumptions were made to enable the orientation parameters to be calculated. A rather more detailed experimental study of the dependence of the scattered intensity on the orientation of the polarisation vectors of the incident and scattered light has been carried out by Purvis and Bower[26] on polymethylmethacrylate. In this experiment the three specimens studied were again of uniaxial symmetry and were machined from sheets of commercial PMMA drawn at various temperatures to a fixed draw ratio. Each specimen was in the form of a disc 1 cm in diameter and about 3 mm thick, with the draw direction (Ox_3) lying in the plane of the disc. It was mounted on a holder which allowed it to be rotated about the normal to the plane (Ox_1). Unfocused laser light of wavelength 488 nm passed through the edge of the disc so as to travel along a diameter and the scattered light was observed along the normal to the plane.

A set of axes $O-X_1X_2X_3$ fixed in the laboratory was defined so that the OX_1 axis coincided with Ox_1 and the OX_3 direction coincided with the direction of the incident laser beam. The polarisation vector of the polariser could be chosen parallel to OX_1 or OX_2 and that of the analyser parallel to OX_2 or OX_3. Four scattered intensities $I_{ij}(\beta)$ were defined so that $I_{ij}(\beta)$ was the intensity observed when the incident light had the polarisation vector parallel to OX_j, the analyser had the polarisation vector parallel to OX_i and β was the angle between Ox_3 and OX_3. It has already been pointed out that if either of the polarisation vectors is not parallel to a symmetry axis of the specimen then the observed intensity depends on the birefringence of the sample. The birefringence of the most oriented sample in this work was $1\cdot48 \times 10^{-3}$ (Fig. 8). It was thus possible to make the correction for birefringence small by placing a slit in front of the specimen so that only a region $0\cdot1$ mm thick, centred at a distance below the surface equal to a whole multiple of $\lambda/\Delta n$, was illuminated.

Four Raman lines were studied, at wavenumber shifts 486, 562, 604 and 1732 cm^{-1}. The first three overlapped each other and some weaker lines and were separated out by curve-fitting. Figure 8 shows the polar plots of $I_{ij}(\beta)$ for the 604 cm^{-1} line for the most oriented specimen studied. The curves are the theoretical angular distributions including the small birefringence correction. Apart from this correction, their forms can be deduced by substituting l_i^a and l_j^e in terms of β into eqn. (10). All four curves were fitted using only seven parameters, six values of $I_0' \sum \alpha_{ij} \alpha_{pq}$ and one angle to allow for experimental error in choosing the zero of β. Only five $\sum \alpha_{ij} \alpha_{pq}$ are independent for uniaxial symmetry, but this was not assumed in the fitting. The observed close equality of $I_{21}(\pi/2)$ and $I_{32}(0)$ showed immediately, however, that $\sum \alpha_{31}^2 \cong \sum \alpha_{32}^2$. The equality of $I_{21}(\beta)$ and $I_{31}(\beta + \pi/2)$ for all β confirmed the correction factors applied to the

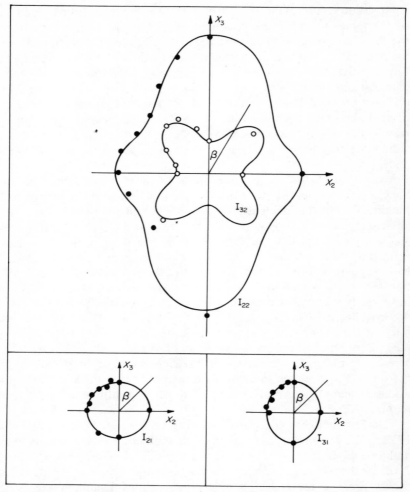

Fig. 8. Intensity of Raman scattering from a disc of uniaxially oriented polymethyl-methacrylate as a function of the orientation of the sample with respect to the polarisation vectors of the incident and scattered radiation. Results for the 604 cm^{-1} line from a sample of birefringence $(n_3 - n_2) = 1.48 \times 10^{-3}$. (Reproduced by permission from Ref. 26. Copyright IPC Business Press Ltd.)

directly measured intensities to allow for the differential polarisation sensitivity of the spectrometer.

From the values of $I_0' \sum \alpha_{ij} \alpha_{pq}$ observed for each line and for each specimen the quantity ρ_l was calculated. This quantity is the ratio of the intensity which would be found for perpendicular polarisation directions of incident and observed scattered light to that which would be found for parallel polarisation directions for a sample containing a random distribution of orientations of scatterers with the same Raman tensor. If the Raman tensors of the scatterers were independent of the distribution of their orientations this quantity would be the same for all samples. It was, however, found to increase steadily with the birefringence of the samples. This result suggested that either the nature of the structural units changed from specimen to specimen in a systematic way or that the structural units remained the same but the Raman tensors changed because of changes in the interactions between the units.

Consideration of the molecular conformations likely to be found in the polymer, together with the vibrational assignments of the Raman lines given in the literature, suggested that if the scatterers were essentially single molecules the intensities of the 604 cm^{-1} line, and possibly those of the 1732 cm^{-1} line, might be accounted for reasonably well using the assumption that one of the principal axes of the Raman tensor was parallel to the long axis of the molecule and that the principal components perpendicular to this (α_1 and α_2) were equal. The 486 and 562 cm^{-1} lines were not expected to give good fits to this axially symmetric tensor model. In practice the data for the 604, 1732 and 562 cm^{-1} lines gave similar values for $\overline{\cos^2 \theta}$ and $\overline{\cos^4 \theta}$ when fitted to eqns. (13) but the data for the 486 cm^{-1} line gave quite different values and the fits were poorer.

A more general model was also tried, in which α_1 and α_2 were not necessarily equal but α_3 remained parallel to the long axis of the molecule. This involved solving eqns. (12) by an iterative procedure to find M_{200}, M_{400} and α_1/α_3 and α_2/α_3. The results for this general tensor model also indicated that $\alpha_1 \cong \alpha_2$ for the 604 cm^{-1} line and that α_1 and α_2 were of the same sign for the 1732 and 562 cm^{-1} lines but it showed that they were of opposite sign for the 486 cm^{-1} line. A third model assumed that the molecules were helical and that each observed line was a superposition of the three lines into which the line for a group vibration is in principle split for such a molecule.

The values of $\overline{\cos^2 \theta}$ and $\overline{\cos^4 \theta}$ obtained from the general tensor and helical models were not very different. For both models the values corresponding to the different lines for a given specimen were also in good agreement and the average values for each specimen were very close to those obtained on the same or closely similar specimens of PMMA by Kashiwagi et al.[50] who used broad line nuclear magnetic resonance spectroscopy and fitted the data on the basis of a helical model for the molecule. They made measurements on a larger number of specimens within the same range of birefringence and it can be concluded that the observed relationship between $\overline{\cos^2 \theta}$ and $\overline{\cos^4 \theta}$ is in reasonable accord with that predicted on the basis of the aggregate model (see Chapter 6, Section 6.2).

A few other studies of the Raman spectra from oriented polymers have been reported. Hendra and Willis have studied the spectra from bundles of oriented fibres of polypropylene[51] and linear polyethylene.[52] They observed changes in the intensities of several lines according to whether the polarisation vector of the incident light was parallel or perpendicular to the fibre axes but they did not attempt to characterise the molecular orientation in the fibres. Carter[53] has also studied the Raman spectrum of highly oriented polyethylene fibres with a view to establishing the assignment of the B_{2g} methylene wagging mode. Derouault et al.[54] have attempted to study the variations of orientation within inhomogeneous samples of polyethylene terephthalate. They appear to be able to detect such variations but have not given any quantitative estimates of the molecular orientation.

5.4 SUMMARY

The fluorescence and Raman scattering methods of studying molecular orientation in solid polymers have been shown, both theoretically and experimentally, to be capable of providing quantitative information about the second and fourth moments of the orientation distribution functions in non-crystalline polymers, information which is difficult to obtain by other methods, except for broad line NMR. It is possible that the methods will be able to give this information directly about the non-crystalline regions of crystalline polymers in some of the cases where NMR cannot do so. For the fluorescence method it will be necessary to disperse the fluorescent molecules with certainty into the non-crystalline regions only and for the Raman method it will be necessary to identify clearly lines which are due only to molecules in these regions and to determine the orientations of the principal axes of the corresponding tensors within the molecule.

Even for single-phase materials the interpretation of the observed fluo escence intensities is not as straightforward as it promised to be in the early work of Nishijima and his co-workers and the interpretation of the Raman intensities is complicated by the possibility that it may be necessary to take account of changes in the nature of the scatterers or of the interactions between them which are dependent on the distribution of orientations. Nevertheless, the effort to deal with these problems seems justified by the promise that the methods will become useful tools for investigating the nature of the non-crystalline regions of polymers and their influence on such practically important properties as elastic modulus and shrinkage.

REFERENCES

1. Morey, D. R. (1933). *Text. Res. J.*, **3**, 325; (1934). *Text. Res. J.*, **4**, 491; (1935). *Text. Res. J.*, **5**, 483.
2. Nishijima, Y., Onogi, Y. and Asai, T. (1965). *Rep. Prog. Polym. Phys. Jap.*, **8**, 131.
3. Stein, R. S. and Read, B. E. (1969). In *International Symposium on Polymer Characterization (Appl. Polym. Symp.*, **8**) (Ed. K. A. Boni and F. A. Sliemers), Interscience, New York, p. 255.
4. Cornell, S. W. and Koenig, J. L. (1968). *J. Appl. Phys.*, **39**, 4883.
5. Herzberg, G. (1945). *Molecular Spectra and Molecular Structure*. II. *Infra-red and Raman Spectra of Polyatomic Molecules,* Van Nostrand Reinhold, New York, p. 242.
6. Nishijima, Y., Onogi, Y. and Asai, T. (1966). In *U.S.-Japan Seminar in Polymer Physics (J. Polymer Sci. C,* **15**) (Ed. R. S. Stein and S. Onogi), Interscience, New York, p. 237.
7. Nishijima, Y., Fujimoto, T., Onogi, Y. (1966). *Rep. Prog. Polym. Phys. Jap.*, **9**, 457.
8. Nishijima Y., Onogi, Y. and Asai, T. (1966). *The International Symposium on Macromolecular Chemistry,* **7**, 161.
9. Nishijima, Y., Onogi, Y. and Asai, T. (1967). *Rep. Prog. Polym. Phys. Jap.*, **10**, 461.
10. Nishijima, Y., Onogi, Y., Asai, T. and Yamazaki, R. (1967). *Rep. Prog. Polym. Phys. Jap.*, **10**, 465.
11. Nishijima, Y. and Onogi, Y. (1968). *Rep. Prog. Polym. Phys. Jap.*, **11**, 395.
12. Nishijima, Y., Onogi, Y., Asai, T. and Yamazaki, R. (1968). *Rep. Prog. Polym. Phys. Jap.*, **11**, 399, 403.
13. Nishijima, Y. and Asai, T. (1968). *Rep. Prog. Polym. Phys. Jap.*, **11**, 419, 423.
14. Seki, J. (1969). *Sen-i Gakkaishi,* **25**, 16.
15. Desper, C. R. and Kimura, I. (1967). *J. Appl. Phys.*, **38**, 4225.
16. Onogi, Y. and Nishijima, Y. (1971). *Rep. Prog. Polym. Phys. Jap.*, **14**, 533, 537, 541.
17. Bower, D. I. (1972). *J. Polymer Sci.* (Polym. Phys. Edn.), **10**, 2135.
18. Nobbs, J. H., Bower, D. I., Ward, I. M. and Patterson, D., (1974). *Polymer,* **15**, 287.
19. Roe, R-J. (1970). *J. Polymer. Sci., A-2,* **8**, 1187.
20. Nomura, S. Kawai, H., Kimura, I. and Kagiyama, M. (1970). *J. Polymer Sci., A-2,* **8**, 383.
21. Kimura, I., Kagiyama, M., Nomura, S. and Kawai, H. (1969). *J. Polymer Sci., A-2,* **7**, 709.
22. Badley, R. A., Martin, W. G. and Schneider, H. (1973). *Biochemistry,* **12**, 268.
23. Snyder, R. G. (1971). *J. Mol. Spectrosc.,* **37**, 353.
24. Bower, D. I. and Purvis, J., to be published.
25. Birss, R. R. (1966). *Symmetry and Magnetism,* North-Holland, Amsterdam, Ch. 2.

26. Purvis, J. and Bower, D. I. (1974). *Polymer*, **15**, 645.
27. Stein, R. S. (1968). *J. Polymer Sci.*, *A-2*, **6**, 1975.
28. Nishijima, Y., Yamamoto, M., Oku, S. and Umegae, M. (1966). *Rep. Prog. Polym. Phys. Jap.*, **9**, 501.
29. Nishijima, Y., Onogi, Y. and Yamazaki, R. (1968). *Rep. Prog. Polym. Phys. Jap.*, **11**, 415.
30. Yamazaki, R., Onogi, Y. and Nishijima, Y. (1969). *Rep. Prog. Polym. Phys. Jap.*, **12**, 439.
31. Asai, T., Onogi, Y. and Nishijima, Y. (1969). *Rep. Prog. Polym. Phys. Jap.*, **12**, 433.
32. Nishijima, Y., Onogi, Y. and Asai, T. (1966). *Rep. Prog. Polym. Phys. Jap.*, **9**, 461.
33. Nishijima, Y., Onogi, Y., Yamazaki, R. and Kawakami, K. (1968). *Rep. Prog. Polym. Phys. Jap.*, **11**, 407.
34. Nishijima, Y. and Onogi, Y. (1968) *Rep. Prog. Polym. Phys. Jap.*, **11**, 411.
35. Onogi, Y., Kawakami, K. and Nishijima, Y. (1970). *Kogyo Kagaku Zasshi*, **73**, 57.
36. Nishijima, Y., Onogi, Y. and Ogawa, M. (1966). *Rep. Prog. Polym. Phys. Jap.*, **9**, 465.
37. Asai, T. and Nishijima, Y. (1969). *Rep. Prog. Polym. Phys. Jap.*, **12**, 437.
38. Nishijima, Y., Onogi, Y. and Asai, T. (1968). *Rep. Prog. Polym. Phys. Jap.*, **11**, 391.
39. Nishijima, Y. and Asai, T. (1969). *Rep. Prog. Polym. Phys. Jap.*, **12**, 429.
40. Nishijima, Y. (1970). In *Polymers and Polymerization* (*J. Polymer Sci. C*, **31**) (Ed. C. G. Overberger and T. C. Fox), Interscience, New York, p. 353.
41. Nishijima, Y. (1970). *Ber. Bunsen-Gesellschaft*, **74**, 778.
42. Kryszewski, M. (1967). *Faserforsch. Textiltech.*, **18**, 189.
43. Gulrajani, M. L. and Padhye, M. R. (1971). *Ind. J. Technol.*, **9**, 211.
44. Seki, J. (1969). *Sen-i Gakkaishi*, **25**, 24.
45. McGraw, G. E. (1970). *J. Polymer Sci.*, *A-2*, **8**, 1323.
46. McGraw, G. E. (1972). In *Structure and Properties of Polymer Films* (Ed. R. W. Lenz and R. S. Stein), Plenum, New York, p. 97.
47. Purvis, J., Bower, D. I. and Ward, I. M. (1973). *Polymer*, **14**, 398.
48. Kashiwagi, K., Cunningham, A., Manuel, A. J. and Ward, I. M. (1973). *Polymer*, **14**, 111.
49. Purvis, J., Bower, D. I. and Ward, I. M., to be published.
50. Kashiwagi, M., Folkes, M. J. and Ward, I. M. (1971). *Polymer*, **12**, 697.
51. Hendra, P. J. and Willis, H. A. (1967). *Chem. Ind.*, 2146.
52. Hendra, P. J. and Willis, H. A. (1968). *Chem. Comm.*, 225.
53. Carter, V. B. (1970). *J. Mol. Spectrosc.*, **34**, 356.
54. Derouault, J. L., Hendra, P. J., Cudby, M. E. A. and Willis, H. A. (1972). *Chem. Comm.*, 1187.

CHAPTER 6

NUCLEAR MAGNETIC RESONANCE

M. J. FOLKES and I. M. WARD

6.1 THE NUCLEAR RESONANCE PHENOMENON

We will begin the discussion with a brief account of the nuclear resonance phenomenon. For a more complete treatment the reader is referred to the standard texts (for example Refs. 1, 2, 3).

Many atomic nuclei, most notably the proton, possess an intrinsic spin which can be regarded classically as a circulating electric current giving rise to a magnetic dipole moment. Analogous to a gyroscope in the earth's gravitational field, such a magnetic dipole will experience a torque when placed in a static magnetic field H_0. This gives rise to a PRECESSION of the nuclear spin about H_0 (as shown in Fig. 1) with a LARMOR frequency $\omega_0 = \gamma H_0$ where γ is the GYROMAGNETIC RATIO and is a constant for a particular nucleus.

Now consider a dilute assembly of spins so that interactions between them can be neglected. In the presence of the field H_0 the spins will all precess at the same frequency ω_0 but there will be no phase correlation between them. If a small field H_1 is applied in the xy plane (see Fig. 1), which rotates about the z-axis with angular frequency ω_0, the spins will be forced into phase with each other and with the applied field H_1, creating a macroscopic magnetic moment M, also precessing about the z-axis with the Larmor frequency. This is NUCLEAR RESONANCE and results in an absorption of power from the alternating field H_1.

Under the conditions just described, an absorption of power will occur at only *one* critical frequency ω_0 for a fixed H_0. As the spin density increases, however, dipole–dipole interactions play an increasingly important role so that the local steady field H' at any one spin will be the

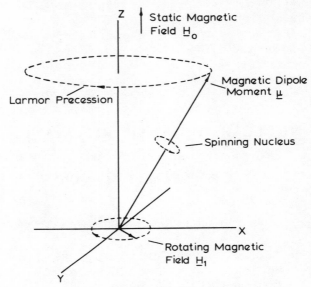

Fig. 1. The resonance phenomenon.

vector sum of the applied field H_0 and the field due to neighbouring spins. The nuclear resonance absorption line is then broadened since the precessional frequencies of the spin ensemble cover a range about the central frequency ω_0. The greater the spin interactions the broader the resonance and for solids, where the resonant nuclei are close together, the line width $\sim 5{-}10$ gauss $(5{-}10 \times 10^{-4}T)$. For this reason, NMR measurements performed on solids are conventionally referred to as BROAD-LINE NMR to distinguish them from the more widely used HIGH-RESOLUTION techniques for liquids, for which the dipole–dipole interactions average to zero.

If the field H_1 is removed very rapidly, the assembly of spins gradually lose their phase coherence. This process has a time constant T_2, the SPIN–SPIN RELAXATION TIME, which is inversely related to the spread in local field H' throughout the sample—the greater the spin interactions the smaller T_2. For solids $T_2 \sim 10^{-5}$ s while for liquids it can be of the order of seconds.

The very fact that the nuclear resonance condition can be detected by means of the power absorbed from the field H_1 implies that *continuous* absorption of power is occurring. Since the energy of the spin system cannot continue to increase indefinitely, some process must exist by which an equivalent amount of energy is being dissipated by the spins. This process is known as SPIN–LATTICE relaxation and has an associated spin–lattice relaxation time T_1. Thermal lattice vibrations produce fluctuating magnetic fields around the nuclei and these can cause a net transfer of energy from the spin system to the lattice. Since the spin–lattice relaxation process depends on the lattice vibration frequency it will in turn be very dependent on temperature. In polymeric materials, therefore, it is commonly found that T_1 v. temperature curves exhibit minima, corresponding to the onset of characteristic modes of molecular motion.

T_1 and T_2 are related to the width, shape and intensity of the NMR absorption line. These quantities can be derived from the measured absorption line or alternatively directly using pulsed NMR techniques (see e.g. Ref. 1). The latter can also be employed for measurements of $T_{1\rho}$, which is also a spin–lattice relaxation time but measured at lower frequencies ~kHz as distinct from MHz for T_1. This offers the advantage of enabling relaxations to be detected with greater resolution.

The shape of the absorption curve cannot be readily synthesised (but see Refs. 4 and 5) from a knowledge of the molecular structure except in a few instances where the nuclei form distinct groups, which are relatively isolated from each other. Consequently a measure of line shape is required which can be directly related to the structure of the material. Van Vleck[6] showed that the even moments of the line shape can in principle be calculated. For an absorption line described by $f(H)$, the nth moment is defined by the following relationship, where H^* is the applied field at the centre of resonance:

$$\langle \Delta H^n \rangle = \langle (H - H^*)^n \rangle = \frac{\int_{-\infty}^{\infty} f(H)(H - H^*)^n \, d(H - H^*)}{\int_{-\infty}^{\infty} f(H) \, d(H - H^*)}$$

Since $f(H)$ is a symmetric function, all odd numbered moments are zero. Van Vleck has also shown how the second moment $\langle \Delta H^2 \rangle$ and fourth moment $\langle \Delta H^4 \rangle$ are simply related to the structure of the material.

NMR studies of oriented polymers can be conveniently discussed under three principal headings, accepting that there must be interrelationships between these.

(1) Anisotropy of the rigid lattice response (such that the frequency of molecular motion is much less than the NMR line width in Hz, which often occurs at low temperature), usually determined from line width or second moment measurements.

(2) Anisotropy of the response where molecular motions are occurring again usually determined from line widths or second moments.

(3) Analysis of composite spectra, where the responses of the crystalline, amorphous and intermediate regions are superimposed.

In addition to these there are T_1 and more recently $T_{1\rho}$ measurements on oriented polymers. These are much less extensive and merit only a short discussion at the end of this chapter.

6.2 THE ANISOTROPY OF THE RIGID LATTICE RESPONSE AND THE MEASUREMENT OF MOLECULAR ORIENTATION

The dipole–dipole interaction for a pair of identical nuclei of spin $\frac{1}{2}$ (*i.e.* protons) gives rise to a double signal of line width splitting:

$$H = \frac{3\mu}{r^3}(\cos^2 \beta - 1)$$

where μ is the magnetic moment of the proton, r is the distance between the protons and β is the angle between the line joining the protons and H_0. In oriented polymers where the predominant dipolar interactions arise from a proton pair, *e.g.* CH_2 in polyethylene $(CH_2)_n$ and polyethylene terephthalate[7,8,9]

$$\left[-\overset{O}{\underset{\|}{C}} - \left\langle\!\!\!\bigcirc\!\!\!\right\rangle - \underset{\underset{O}{\|}}{C} \diagdown O \diagdown (CH_2)_2 \diagup O \diagdown \right]_n$$

such a simple doublet structure can sometimes be observed with some broadening due to the other internuclear dipolar interactions (Fig. 2), and this structure is very dependent on the direction of H_0. See also Ref. 4 for an earlier study of the proton pair resonance arising from the water of crystallisation in gypsum.

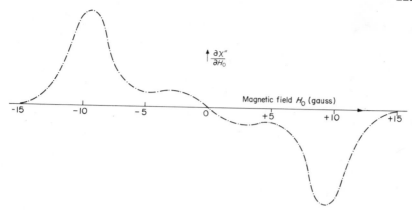

Fig. 2. Derivative wide line NMR signal $(\partial \chi'' / \partial H_0)$ for highly oriented polyethylene at $-196°C$, steady magnetic field H_0 along fibre axis (Smith, J. B. Manuel, A. J. and Ward, I. M., unpublished).

For more precise discussion we use the Van Vleck second moment equation.

$$\langle \Delta H^2 \rangle = \frac{G}{N} \sum_{j>k} r_{jk}^{-6} (3 \cos^2 \beta_{jk} - 1)^2$$

where $G = \frac{3}{4} I(I+1) g^2 \mu_n^2$ and I is the nuclear spin number $(= \frac{1}{2}$ for protons), g is the nuclear g-factor, μ_n the nuclear magneton, N the number of nuclei over which the sum is taken, r_{jk} the length of the vector joining nuclei j and k, and β_{jk} is the angle between the vector r_{jk} and direction of the external magnetic field H_0. (We have ignored any interactions with non-resonant nuclei.)

Now consider the case of a uniaxially oriented polyethylene film or fibre, with full orientation, where the chain axes are parallel to the draw direction or fibre axis and consider only the proton pair interaction arising from the CH_2 groups. The CH_2 groups are now such that the proton pairs lie in a plane perpendicular to the fibre axis. Figure 3 illustrates the situation. The internuclear vector r makes an angle β with H_0, and an angle ϕ with a chosen direction in the plane normal to the fibre axis. The angle between the fibre axis and H_0 is γ.

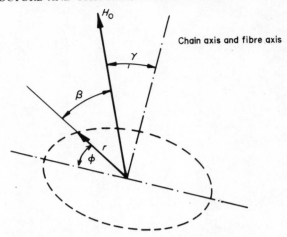

Fig. 3. The internuclear vector r *makes an angle* β *with* H_0 *and an angle* ϕ *with a chosen direction in the plane normal to the fibre axis.*

We have

$$\langle \Delta H^2 \rangle = \frac{G}{N} \frac{\overline{(3 \cos^2 \beta - 1)^2}}{r^6} = \frac{G}{N} \frac{\overline{(3 \cos^2 \phi \sin^2 \gamma - 1)^2}}{r^6}$$

$$= \frac{5}{4} \langle \Delta H^2 \rangle_{\text{iso}} \left(\frac{27}{8} \sin^4 \gamma - 3 \sin^2 \gamma + 1 \right)$$

where $\langle \Delta H^2 \rangle_{\text{iso}}$ is the second moment for an isotropic arrangement of similar CH_2 groups.

The experimental data for oriented polyethylene fibres and films (see for example Ref. 10) are consistent to a first approximation with this simple interpretation, provided that we add an additional isotropic second moment contribution to take into account the dipolar interactions external to the proton pairs (Fig. 4).

A more sophisticated treatment, however, allows for the fact that polymers do not possess complete orientation, but that there is always a distribution of molecular orientations. A suitable starting point for a theoretical discussion of this situation is the aggregate model, which has been used to describe optical and mechanical anisotropy of polymers[11] (see Chapter 8). On this model, the partially oriented polymer is considered

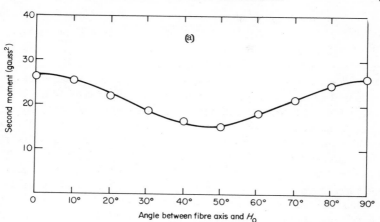

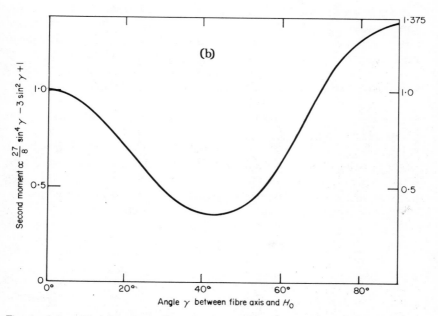

Fig. 4. *Anisotropy of the second moment in oriented polyethylene (a) experimental (after McColl and Slichter,[10] by permission of John Wiley & Sons Inc.); (b) theoretical, based on proton pair interaction (see text).*

to consist of an aggregate of anisotropic units of structure, each of which possesses the properties of the fully oriented polymer. The interpretation of the NMR anisotropy relies on the validity of a formal mathematical argument, which is presented below, and should not be taken to imply that there exist unique units of structure of the type described by the model. In the application of the model to polyethylene by McBrierty and Ward,[12] it was assumed that the magnetic anisotropy of the hypothetical unit of structure could be calculated on the basis of the crystalline unit cell. The model has also been applied by Ward and co-workers to amorphous polymers[13,14] and here the calculations were based on models for the molecular chains and their relationship with other chains. The reason for the success of such treatments, which may seem surprising, is that the NMR anisotropy depends very much on the local arrangement of magnetic nuclei, and in these polymers only proton interactions within a sphere of radius 5 Å need be considered.

In this chapter we will present the approach adopted by Ward and co-workers. Similar treatments have also been given independently by Yamagata and Hirota[15,16] and by Slonim and Urman.[17] Due to their appearance in the Japanese and Russian literature only, these latter previous treatments did not achieve prominence in the western literature. Furthermore, although it is perfectly possible to develop the theory in a very elementary manner, using Euler angle transformations, and this was the method of the earlier work, we choose to work here in terms of spherical harmonic analysis. The compactness of this representation has many advantages, particularly if the treatment is to be extended beyond transverse isotropy.

We consider first the situation of transverse isotropy. The transverse isotropy is assumed to arise as follows. First the units of structure are transversely isotropic. Secondly, there is no preferential orientation of the units of structure in a plane perpendicular to the draw direction. We rewrite the Van Vleck equation in terms of spherical harmonics:

$$\langle \Delta H^2 \rangle = \frac{4G}{N} \sum_{j>k} r_{jk}^{-6} \, P_2^2(\cos \beta_{jk})$$

where $P_2 (\cos \beta_{jk}) = \frac{1}{2}(3 \cos^2 \beta_{jk} - 1)$ is the second Legendre polynomial. Expressing $P_2^2(\cos \beta_{jk})$ as a series of Legendre polynomials,

$$P_2^2(\cos \beta_{jk}) = \sum_l a_l \, P_l(\cos \beta_{jk})$$

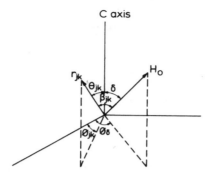

Fig. 5. This figure shows the angles θ_{jk}, ϕ_{jk} which define the direction of the internuclear vector r_{jk} with respect to the c-axis, and the angles δ, ϕ_δ which define the direction made by H_0 with the c-axis.

and

$$\langle \Delta H^2 \rangle = \frac{4G}{N} \sum_{j>k} r_{jk}^{-6} \sum_l a_l P_l(\cos \beta_{jk})$$

Now by a series of transformations in spherical harmonics we relate the second moment not to β_{jk} but to the angle between the symmetry axis of a unit of structure and the draw direction of the polymer together with some additional intermediate angles (Fig. 5).

Applying the Legendre addition theorem to $P_l(\cos \beta_{jk})$ we obtain:

$$P_l(\cos \beta_{jk}) = \frac{4\pi}{l+1} \sum_{m=-l}^{l} Y_{lm}^*(\theta_{jk}, \phi_{jk}) Y_{lm}(\delta, \phi_\delta)$$

where we have defined the intermediate angles δ, ϕ_δ between the applied field H_0 and the symmetry axis of the unit of structure. Our assumption of transverse isotropy for the unit enables us to average ϕ_{jk} over the range $0 \leq \phi_{jk} \leq 2\pi$ with a uniform probability function. This gives:

$$\langle P_2^2(\cos \beta_{jk}) \rangle_{\phi_{jk}} = \sum_l a_l P_l(\cos \theta_{jk}) P_l(\cos \delta)$$

The addition theorem is now applied to the $P_l(\cos \delta)$ term by introducing intermediate angles Δ, ϕ_Δ and γ, ϕ_γ, defined in Fig. 6. Thus:

$$\langle P_2^2(\cos \beta_{jk}) \rangle_{\phi_{jk}} = \sum_l a_l P_l(\cos \theta_{jk}) \sum_{m=-l}^{l} Y_{lm}^*(\Delta, \phi_\Delta) Y_{lm}(\gamma, \phi_\gamma) \frac{4\pi}{l+1}$$

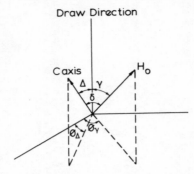

Fig. 6. This figure shows the angles Δ, ϕ_Δ *which define the direction of the* c-*axis with respect to the draw direction and* γ, ϕ_γ *which define the direction of* H_0 *with respect to the draw direction.*

For transverse isotropy of the sample, we may average ϕ_Δ over the range $0 \leqq \phi_\Delta \leqq 2\pi$ with a uniform probability function:

$$\langle P_2^2(\cos \beta_{jk}) \rangle_{\phi_{jk}, \phi_\Delta} = \sum_l a_l P_l(\cos \theta_{jk}) P_l(\cos \Delta) P_l(\cos \gamma)$$

Inserting this into the Van Vleck equation for the second moment gives:

$$\langle \Delta H^2 \rangle = 4G \sum_{l=0,2,4} a_l S_l P_l(\cos \gamma) \overline{P_l(\cos \Delta)}$$

where $a_0 = 1/5$, $a_2 = 2/7$ and $a_4 = 18/35$.

The S_l are defined by $S_l = \dfrac{1}{N} \sum_{j>k} P_l(\cos \theta_{jk}) r_{jk}^{-6}$. They are commonly called the lattice sums, and depend only on the positions of the magnetic nuclei in the structure of the material. The $\overline{P_l(\cos \Delta)}$ are the orientation distribution functions. For this case of transverse isotropy, l can only take the values 0, 2, 4, so that the second moment anisotropy relates to $\overline{P_2(\cos \Delta)}$ and $\overline{P_4(\cos \Delta)}$ only, or in simpler terms to $\overline{\cos^2 \Delta}$ and $\cos^4 \Delta$.

We have

$$\langle \Delta H^2 \rangle = \frac{9G}{64N} \sum_{j>k} A(35 \overline{\cos^4 \Delta} - 30 \overline{\cos^2 \Delta} + 3) \frac{\cos^4 \theta_{jk}}{r_{jk}^6}$$

$$-(30\,A\,\overline{\cos^4\Delta}-B\,\overline{\cos^2\Delta}+C)\frac{\cos^2\theta_{jk}}{r_{jk}^6}$$

$$+(3\,A\,\overline{\cos^4\Delta}-C\,\overline{\cos^2\Delta}+D)\frac{1}{r_{jk}^6}$$

where $A = (35\cos^4\gamma - 30\cos^2\gamma + 3)$
$B = (225\cos^4\gamma - 186\cos^2\gamma + 17)$
$C = 2/3\,(135\cos^4\gamma - 102\cos^2\gamma + 7)$
$D = \frac{1}{9}(81\cos^4\gamma - 42\cos^2\gamma + 49)$

Providing that the S_l can be computed either from the crystal structure or a prepared model of the structure, we can therefore obtain $\overline{P_2(\cos\Delta)}$ and $\overline{P_4(\cos\Delta)}$ from the experimental dependence of $\langle\Delta H^2\rangle$ on γ. One of the values of this technique stems from the fact that it is these moments of the orientation distribution which are required by the aggregate model to predict the mechanical anisotropy (see Chapter 8).

An interesting exercise is to calculate the $\overline{P_2(\cos\Delta)}$ and $\overline{P_4(\cos\Delta)}$ on

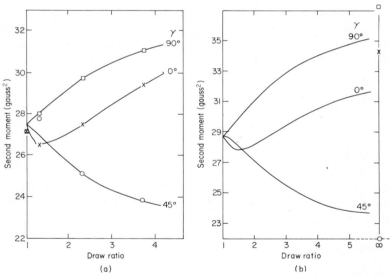

Fig. 7. Variation of second moment with draw ratio for oriented low density polyethylene (curves show results for angles γ between H_0 and draw direction). (a) experimental curve (b) theoretical curve based on pseudo-affine deformation scheme (after McBrierty and Ward).

the basis of a chosen deformation scheme and compare the predicted magnetic anisotropy with that observed. Figure 7 shows results obtained for polyethylene by McBrierty and Ward,[12] where the theoretical result was calculated on the basis of the Kratky or pseudo-affine deformation scheme[18,19] (see Chapter 8). There is reasonable qualitative agreement. Similar calculations were undertaken by Yamagata and Hirota[16] and by Chujo and Sudzuki.[20]

This method has also been applied to non-crystalline or very poorly crystalline polymers, *viz.* polymethylmethacrylate[13] and polyvinyl chloride.[14] In these cases, the lattice sums were calculated by assuming that the magnetic anisotropy arises entirely from the intramolecular interactions, and that the intermolecular interactions are isotropic. The lattice sums for the intramolecular interactions are then calculated on the basis of a model for the molecular chain, and the intermolecular interactions give a small additional term in the S_0 lattice sum which can be found from a least squares fit to the data. It was found that the adjusted value of S_0 was constant for samples of differing degrees of orientation thus justifying the initial assumptions.

The $\overline{P_2(\cos\Delta)}$ and $\overline{P_4(\cos\Delta)}$ were then used to compare with birefringence (for example see Fig. 8) to predict the mechanical anisotropy and to compare with theoretical values based on different deformation schemes. Figure 9 shows the $\overline{\cos^2\Delta}$ and $\overline{\cos^4\Delta}$ values for polymethylmethacrylate and polyvinylchloride, together with predicted curves for the

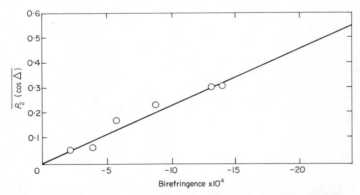

Fig. 8. *Variation of* P_2 *(cos* Δ*) for oriented PMMA, obtained from NMR measurements with birefringence. Circles are experimental results and the line shows the linear relationship expected on basis of the aggregate model.*

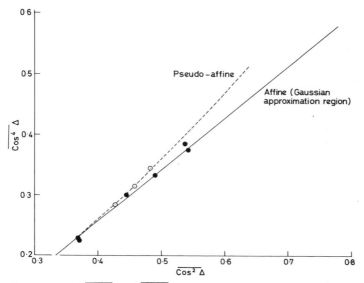

Fig. 9. Comparison of $\overline{\cos^4 \Delta}$ and $\overline{\cos^2 \Delta}$ obtained for oriented PVC ● and oriented PMMA ○. Lower and upper curves show predicted relationship according to the affine rubber elasticity model and the pseudo-affine deformation scheme respectively.

rubber-like affine deformation scheme and the Kratky or pseudo-affine deformation scheme. It can be seen that PMMA is closer to the affine deformation scheme, whereas PVC is more nearly pseudo-affine.

It is quite straightforward, in principle, to extend the second moment anisotropy calculations to cases of lower symmetry. Orthorhombic symmetry, which holds for film drawn at constant width, is of particular interest. This was first treated by Slonim and Urman[17] and more recently by Roe[21] and by Kashiwagi et al.[22] The latter workers examined the situation in considerable detail and pointed out that the complete description of molecular orientation requires a knowledge of the distribution in the three Euler angles θ, ϕ and ψ say, whereas the second moments are measured through only two parameters θ and ϕ say. We therefore cannot obtain any information about the distribution in ψ, in the absence of simplifying assumptions.

As remarked above, Van Vleck also showed how to calculate the NMR fourth moment. Whereas the second moment data relate to the values of $\overline{P_n(\cos \Delta)}$ with $n = 0, 2, 4$, the fourth moments relate to $n = 6, 8$ as

TABLE 1

RIGID LATTICE ANISOTROPY MEASUREMENTS

Polymer	Symmetry	Investigators
Polyethylene	Transverse isotropy	McCall and Slichter[10] Yamagata and Hirota[15,16] McBrierty and Ward[12] McBrierty et al.[23]
Polytetrafluoroethylene	Transverse isotropy	Yamagata and Hirota[16]
Polyformaldelyde	Transverse isotropy	Peterlin and Olf[24] Yamagata[25]
Polyethylene terephthalate	Orthorhombic	Slonim and Urman[17] Boye and Goodlet[9] Kashiwagi et al.[22]
Polymethylmethacrylate	Transverse isotropy	Kashiwagi et al.[13]
Polyvinyl chloride	Transverse isotropy	Kashiwagi and Ward[14]
Polyvinyl alcohol	Transverse isotropy	Yamagata and Hirota[16]
Polyoxymethylene	Transverse isotropy	Neiman et al.[26]

well. A theoretical treatment for a partially oriented transversely isotropic system has been presented by McBrierty et al.[23] It was shown that the fourth moment measurements for oriented polyethylenes provided values for $\overline{P_2(\cos \Delta)}$ and $\overline{P_4(\cos \Delta)}$ which were consistent with those obtained previously from second moment measurements. Because the higher moments depend critically on the shape of the 'wings' of the absorption curve, it was not possible to obtain precise values for the higher spherical harmonic functions. It is to be noted, however, that in principle the use of higher moments could be valuable in defining the orientation distribution more precisely.

Table 1 summarises the principal studies of rigid lattice anisotropy.

6.3 ANISOTROPY OF THE NMR RESPONSE WHEN MOLECULAR MOTIONS OCCUR

So far the discussion has been concerned with rigid lattice conditions such as pertain at low temperatures. At very high temperatures, such as at the melting point of the polymer, the NMR spectra are very narrow and are typical of a liquid-like structure. For intermediate temperatures, various types of molecular motion may arise, while the overall orientation distribution is maintained.

If motion is occurring it can be shown (see *e.g.* Ref. 3, p. 62) that to calculate the second moment the dipolar interaction term

$$\left\{ \frac{3\cos^2 \beta_{jk} - 1}{r_{jk}^3} \right\}$$

must be time averaged for the particular motion in question, and the result squared.

Thus

$$\langle \Delta H^2 \rangle_{\text{motion}} = \frac{G}{N} \left\langle \frac{3\cos^2 \beta_{jk} - 1}{r_{jk}^3} \right\rangle_{\text{motion}}^2$$

To fix our ideas consider again the simple model for the polethylene fibre with full orientation, but where there is rotation of the chain around its axis (at a frequency greater than the line width in Hz, *i.e.* 10^4–10^5 Hz for proton resonance with $H_0 \sim 10\,\text{kg}$). The length of the internuclear vector r is unchanged by the motion and we have for the proton pair interaction only:

$$\langle \Delta H^2 \rangle_{\text{motion}} = \frac{G}{N} \frac{\overline{(3\cos^2 \beta - 1)^2}}{r^6} = \frac{G}{N} \frac{\overline{(3\cos^2 \phi \sin^2 \gamma - 1)^2}}{r^6}$$

$$= \frac{5}{4} \langle \Delta H^2 \rangle_{\text{iso}} \frac{1}{4} (3\cos^2 \gamma - 1)^2$$

If this predicted angular dependence of the second moment is compared with that for the rigid chain, it can be seen that the major difference is the dramatic reduction in second moment for $\gamma > 30°$ (Fig. 10). In fact this type of anisotropy was first observed by Slichter for polytetra-fluoroethylene[27] (Fig. 11). Later, more detailed studies by Hyndman and Origlio[28] confirmed that at temperatures above a rotational disorder transition in this polymer at about 285°C, the NMR signal was consistent with coherent rotation of the molecules about the chain axis. It should be noted, however, that the latter workers observed a composite NMR signal at high temperatures, due to signals of distinctly different line width from the crystalline and amorphous regions, respectively. We choose to discuss the composite nature of NMR signals in the final section of this chapter.

The pioneering work of Slichter, and Hyndman and Origlio was followed by a systematic investigation of oriented linear polyethylene and oriented polyoxymethylene over a wide range of temperatures by Olf and

Peterlin.[29] In polyethylene, the reduction in second moment with temperature, although substantial, was much less than that expected for coherent rotation, and was qualitatively ascribed to coherent rotational oscillation about the chain axis (see also Ref. 28). This conclusion is consistent with the much steeper fall in second moment for $\gamma = 90°$ than for $\gamma = 0°$. We will find that this explanation was later rejected after more detailed studies. In polyoxymethylene the fall in second moment with temperature was not greatly dependent on the angle between the draw direction and H_0, and no particular mechanism was proposed for the molecular motion.

These earlier papers did not take into account the fact that the oriented polymers were partially oriented. It is therefore now necessary to extend the previous discussion of the aggregate model to the situation where molecular motion occurs. This has been done independently by Olf and Peterlin,[31] McBrierty and Douglass,[32] and Folkes and Ward.[33] McBrierty and Douglass' paper gives a formal theoretical treatment, as part of a general discussion of the influence of molecular motions on T_1, T_2 and $T_{1\rho}$. The other papers are particularly concerned with an analysis of the effects of molecular motion on the second moment anisotropy but nevertheless provide explicit results for special types of motion. The results are then applied to oriented mats of polyethylene single crystals[34] and to

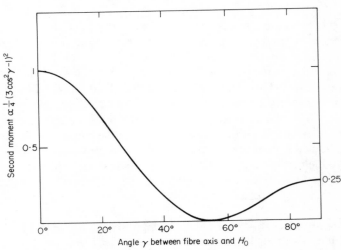

Fig. 10. Theoretical second moment anisotropy for a rotating pair of magnetic nuclei.

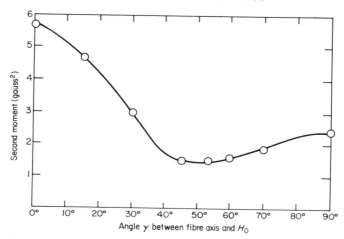

Fig. 11. Experimentally observed anisotropy of second moment in polytetrafluoro-ethylene at 20°C, where motional narrowing occurs (after Slichter,[27] by permission of John Wiley & Sons Inc.).

oriented long chain alkanes[35] notably hexatriacontane (n-$C_{32}H_{60}$) by Olf and Peterlin, and to oriented low density polyethylene[33] by Folkes and Ward.

The theoretical treatments are necessarily very involved, so only a brief summary of the Euler angle approach, adopted by Folkes and Ward,[33] will be given here. As discussed previously we wish to calculate

$$\left\langle \frac{3\cos^2 \beta_{jk} - 1}{r_{jk}^3} \right\rangle^2_{\text{motion}}$$

It is convenient to discuss the effect of motion on the intramolecular interactions (where r is constant) separately from the intermolecular interactions where both β and r vary with time. Similar transformations of $(3\cos^2 \beta_{jk} - 1)^2$ to those employed in the rigid lattice case, enable us to relate the second moment to the angle γ between the draw direction and H_0, and $\overline{\cos^2 \Delta}$ and $\overline{\cos^4 \Delta}$ which define the orientation functions for the transversely isotropic situation which has been analysed. We find

$$\langle \Delta H^2 \rangle_{\text{motion}} = \frac{G}{32N} \sum_{j>k} \frac{(3\cos^2 \theta_{jk} - 1)^2}{r_{jk}^6}$$

$$\{(11 - 30\cos^2 \gamma + 27\cos^4 \gamma)$$

$$+ \overline{\cos^2 \Delta}(252\cos^2 \gamma - 270\cos^2 \gamma - 30)$$
$$+ \overline{\cos^4 \Delta}(315\cos^4 \gamma - 270\cos^2 \gamma + 27)\}$$

for the case of coherent classical rotation around the chain axis. It is of interest to note that the motional second moment depends on the value of only one lattice sum, whereas the rigid lattice second moment is dependent on three.

The intermolecular second moment contribution is more complicated to calculate as we now require

$$\left\langle \frac{3\cos^2 \beta_{jk} - 1}{r_{jk}^3} \right\rangle_{\text{motion}}$$

with both β_{jk} and r_{jk} changing.

This may also be conveniently developed mathematically by the use of Euler angles. Thus we begin as before by relating β_{jk} to the angles $\theta_{jk}, \phi_{jk}, \delta$ and ϕ_δ as defined in Fig. 5:

$$\cos \beta_{jk} = \cos \theta_{jk} \cos \delta + \sin \theta_{jk} \sin \delta \cos(\phi_\delta - \phi_{jk})$$

Inserting this into the dipolar interaction term gives:

$$\left\langle \left\{ \frac{3\cos^2 \beta_{jk} - 1}{r_{jk}^3} \right\} \right\rangle = 3A\cos^2 \delta + \frac{3}{2}B \sin 2\delta \cos \phi_\delta + \frac{3}{2}C \sin 2\delta \sin \phi_\delta$$

$$+ \frac{3}{2}D \sin^2 \delta + \frac{3}{2}E \sin^2 \delta \cos 2\phi_\delta + \frac{3}{2}F \sin^2 \delta \sin 2\phi_\delta - G$$

where the following quantities have been defined, which must, in general, be averaged over θ_{jk}, ϕ_{jk} and r_{jk} for the particular motion:

$$A = \left\{ \frac{\cos^2 \theta_{jk}}{r_{jk}^3} \right\}_{\text{Av}} \qquad B = \left\{ \frac{\sin 2\theta_{jk} \cos \phi_{jk}}{r_{jk}^3} \right\}_{\text{Av}}$$

$$C = \left\{ \frac{\sin 2\theta_{jk} \sin \phi_{jk}}{r_{jk}^3} \right\}_{\text{Av}} \qquad D = \left\{ \frac{\sin^2 \theta_{jk}}{r_{jk}^3} \right\}_{\text{Av}}$$

$$E = \left\{ \frac{\sin^2 \theta_{jk} \cos 2\phi_{jk}}{r_{jk}^3} \right\}_{\text{Av}} \qquad F = \left\{ \frac{\sin^2 \theta_{jk} \sin \phi_{jk}}{r_{jk}^3} \right\}_{\text{Av}}$$

$$G = \left\{ \frac{1}{r_{jk}^3} \right\}_{\text{Av}}$$

The expression for the dipolar interaction term is now squared and averaged over ϕ_δ in the range $0 \leq \phi_\delta \leq 2\pi$ with a uniform probability function, to account for the transverse isotropy of the unit of structure. In the absence of any simplifying assumptions at this stage concerning the type of motion, this procedure leads to an extremely cumbersome expression and will not be shown in detail here. Further averaging is also required after δ is related to Δ, ϕ_Δ, γ and ϕ_γ (Fig. 6) to account for the transverse isotropy of the units of structure about the draw direction, as described in the previous section.

As an example, the result for the intermolecular second moment for a coplanar system of rotating protons will be quoted. This gives

$$
\begin{aligned}
\langle \Delta H^2 \rangle_{\text{motion}}^{\text{inter}} = \frac{G}{N} \sum_{j>k} [& \{3H - K + \cos^2 \gamma (2H - 3K) \\
& + 3H \cos^2 \gamma\} + \overline{\cos^2 \Delta} \{2H - 3K \\
& + \cos^2 \gamma (12H + 9K) - 30H \cos^4 \gamma\} \\
& + \overline{\cos^4 \Delta} \{3H - 30H \cos^2 \gamma + 35H \cos^4 \gamma\}]
\end{aligned}
$$

where we have defined two new lattice sums

$$
H = \frac{9}{64N} \{E^2 + F^2 + 2G^2\}; \qquad K = \frac{1}{2} \frac{G^2}{N}
$$

To fix our ideas, the results for oriented polyethylene of draw ratio 3·7 are shown in Fig. 12 following Andrew's[36] method for calculating the quantities E, F and G. It can be seen that the major contribution to the second moment comes from the intramolecular interactions, and that the intermolecular second moment is ~ 1–2 gauss^2 compared with its value of ~ 9 gauss^2 for the rigid lattice. Although the second moment anisotropy in polyethylene at temperatures close to the melting point is consistent with coherent classical rotation occurring, the second moment pattern from $-196°C$ to $60°C$ must be explained on a different basis. This is because the overall reduction in second moment, particularly in the $45°$ and $90°$ directions is small compared with that predicted from coherent classical rotation.

Olf and Peterlin[31] and Folkes and Ward[33] considered a number of different models in an attempt to explain the molecular motions in the range $-196°C$ to $20°C$, which are important in as much as they relate to the interpretation of the low temperature mechanical relaxation (the γ-relaxation[37]) in this polymer. In single crystal mats and hexatriacontane, Olf and Peterlin suggested that the best explanation for the low

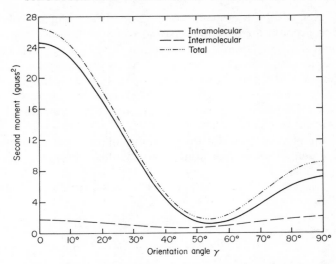

Fig. 12. Predicted variation of the second moment for low density polyethylene of draw ratio 3·7, based on the coherent classical rotation model (after Folkes and Ward).

temperature motion is a flip-flop mechanism, where each rotational jump of the chain through 180° is accompanied by a shift of the molecule along its axis by one CH_2 group. The results were not consistent, on the other hand, with either the kink model[38,39] or the crankshaft mechanism.[40] Folkes and Ward examined chain oscillation and chain twisting, and found these to be unsatisfactory. It was then shown that satisfactory general agreement between theory and experiment could be obtained by assuming that the NMR anisotropy relates to rigid lattice intramolecular interactions only with the intermolecular interactions averaged to a small value. This could occur by a chain sliding process, which was finally proposed as the mechanism. In fact, then, Olf and Peterlin and Folkes and Ward are in agreement as to the interpretation of the low temperature motions. It is to be noted, as both sets of authors point out, that their proposed mechanism is consistent with the earlier interpretation of the mechanical γ-relaxation by Hoffman et al.,[41] who envisage that an individual molecular chain rotates through 180° and is simultaneously translated along its axis. The introduction of rapid chain translational motion in order to reduce the intermolecular contribution has also been invoked by McCall and co-workers for polytetrafluoroethylene[42] and by McBrierty and McDonald[43] for oriented polyoxymethylene. The latter paper re-examines the previous data of Olf and Peterlin[29] and includes a discussion of the effects of molecular motion on both second and fourth moments.

6.4 COMPOSITE SPECTRA

In most crystalline polymers, a composite NMR signal is observed at intermediate temperatures. This effect was first described by Wilson and Pake[44] for isotropic polyethylene and polytetrafluoroethylene. It was concluded that the broad signal was associated with the rigid crystalline regions and the narrow signal with mobile non-crystalline or amorphous regions. Similar results have been obtained for polyethylene terephthalate,[45] nylon,[46] polypropylene,[47] and several other polymers. Careful examination of the structures over wide ranges of temperature and comparison with X-ray data by Fuschillo et al.,[48] and by Farrow and Ward,[49] have cast considerable doubt on this simple explanation. There is still much controversy regarding the exact way in which the proportion of mobile protons relates to the morphology of the sample, which really reflects an imperfect understanding of this morphology in some important respects.

The effect of orientation on the nature of the composite structure has also received some attention. In early work of this nature, Ward[45,50] observed that the proportion of the narrow component was much reduced by orientation of the polymer and moreover that the onset of molecular motion occurred at an appreciably higher temperature than in the isotropic polymer. It was considered reasonable that in the highly oriented specimens, even the non-crystalline regions will be oriented, and that such orientation would restrict the degree of molecular motion. Similar results were obtained by Statton for nylon,[51] even though the degree of crystallinity as determined by X-ray diffraction was reduced by drawing. Olf and Peterlin also obtained similar results in polyethylene.[29]

Finally, there is the very interesting result that in oriented linear polyethylene and polypropylene fibres, a three component NMR spectra has been observed, first by Hyndman and Origlio.[52] In addition to the broad crystalline component and the narrow 'amorphous' component, there was a third component of intermediate line width, which was attributed to strained amorphous material. This intermediate component was much reduced in intensity on annealing and consequent shrinkage of the fibre, which is consistent with the proposed interpretation. A three-component signal has also been observed in polyethylene crystals grown from dilute solution by Fischer and Peterlin,[53] and in conjunction with swelling experiments, used to investigate the morphology of these materials.

6.5 ANISOTROPY OF T_1 AND $T_{1\rho}$

The anisotropy of T_1 and $T_{1\rho}$ have been little studied. The only comprehensive treatment of this has been given by McCall and co-workers,[42] where results for tetrafluoroethylene–hexafluoropropylene copolymer fibres are presented, together with a theoretical treatment for the full orientation case. Any anisotropy in T_1 was found to be minimal; the anisotropy in $T_{1\rho}$ was comparable to that in T_2 or the second moment, as would be expected on theoretical grounds.

6.6 SUMMARY

The current application of NMR methods to the study of polymeric materials falls essentially into two categories. Of initial interest is its potential for characterising molecular orientation rather more precisely than has hitherto been possible in the past, using other methods. At present, experimental inaccuracies and mathematical complexities pose a limitation to its application to those polymer systems involving low symmetries. Set against this, is its well established success in characterising orientation in uniaxially drawn semi-crystalline polymers and perhaps even more impressively its application to non-crystalline polymers. Secondly, we have seen that the anisotropy of the second moment can also be quantitatively analysed in the case of various forms of molecular motion. Providing the frequency of molecular motion is comparable to the NMR line width it will have a predictable effect on the second moment anisotropy and this has already been studied in a number of oriented polymer systems. The implications for a molecular interpretation of mechanical relaxations are clear.

REFERENCES

1. Abragam, A. (1961). *Nuclear Magnetism,* Oxford University Press.
2. Andrew, E. R. (1955). *Nuclear Magnetic Resonance,* Cambridge University Press.
3. Slichter, W. P. (1963). *Principles of Magnetic Resonance,* Harper & Row, New York.

4. Pake, G. E. (1948). *J. Chem. Phys.*, **16**, 327.
5. Betsuyaku, H. (1970). *Phys. Rev. Letters*, **24**, 934.
6. Van Vleck, J. H. (1948). *Phys. Rev.*, **74**, 1168.
7. Slonim, I. Ya. and Lynbinov, A. N. (1970). *The NMR of Polymers*, Plenum Press, New York, p. 174.
8. Hyndman, D. and Origlio, G. F. (1960). *J. Polymer Sci.*, **46**, 259.
9. Boye, C. A. and Goodlet, V. W. (1963). *J. Appl. Phys.*, **34**, 59.
10. McCall, D. W. and Slichter, W. P. (1957). *J. Polymer Sci.*, **26**, 171.
11. Ward, I. M. (1962). *Proc. Phys. Soc.*, **80**, 1176.
12. McBrierty, V. J. and Ward, I. M. (1968). *Brit. J. Appl. Phys. (J. Phys. D)*, **1**, 1529.
13. Kashiwagi, M., Folkes, M. J. and Ward, I. M. (1971). *Polymer*, **12**, 697.
14. Kashiwagi, M. and Ward, I. M. (1972). *Polymer*, **13**, 145.
15. Yamagata, K. and Hirota, S. (1962). *Rep. Progr. Polymer Phys. Japan*, **5**, 261.
16. Yamagata, K. and Hirota, S. (1961). *J. Appl. Phys. Japan*, **30**, 261.
17. Slonim, I. Ya. and Urman, Y. G. (1963). *Zh. struht. Khin*, **4**, 216, quoted in Ref. 7, p. 180.
18. Kratky, O. (1933). *Kolloid-Z.*, **64**, 213.
19. Ward, I. M. (1971). *Mechanical Properties of Solid Polymers*, Wiley, London.
20. Chujo, R. and Sudzuki, T. (1961). *Busseiron Kenkyu*, **10**, 159, quoted in Ref. 7, p. 180.
21. Roe, R. J. (1970). *J. Polymer Sci.*, *A-2*, **8**, 1187.
22. Kashiwagi, M., Cunningham, A., Manuel, A. J. and Ward, I. M. (1973). *Polymer*, **14**, 11.
23. McBrierty, V. J., McDonald, I. R. and Ward, I. M. (1971). *J. Phys. D: Appl. Phys.*, **4**, 88.
24. Peterlin, A. and Olf, H. G. (1962). *J. Polymer Sci.*, **B2**, 769.
25. Yamagata, K. (1961). *J. Appl. Phys. Japan*, **30**, 940.
26. Neiman, M. B., Slonim, I. Ya. and Urman, Ya. G. (1964). *Nature*, **202**, 693.
27. Slichter, W. P. (1957). *J. Polymer Sci.*, **24**, 173.
28. Hyndman, D. and Origlio, G. F. (1960). *J. Appl. Phys.*, **31**, 1849.
29. Olf, H. G. and Peterlin, A. (1964). *J. Appl. Phys.*, **35**, 3108.
30. McCall, D. W. and Slichter, W. P. (1957). *J. Polymer Sci.*, **26**, 171.
31. Olf, H. G. and Peterlin, A. (1970). *J. Polymer Sci.*, *A-2*, **8**, 753.
32. McBrierty, V. J. and Douglass, D. C. (1970). *J. Magnetic Res.*, **2**, 352.
33. Folkes, M. J. and Ward, I. M. (1971). *J. Mater. Sci.*, **6**, 582.
34. Olf, H. G. and Peterlin, A. (1970). *J. Polymer Sci.*, *A-2*, **8**, 771.
35. Olf, H. G. and Peterlin, A. (1970). *J. Polymer Sci.*, *A-2*, **8**, 791.
36. Andrew, E. R. (1950). *J. Chem. Phys.*, **18**, 607
37. McCrum, N. G., Read, B. E. and Williams, G. (1967). *Anelastic and Dielectric Effects in Polymeric Solids*. Wiley, London, p. 371.
38. Pechhold, W. and Blasenbrey, S. (1967). *Kolloid-Z.*, **216/217**, 225.
39. Pechhold, W. (1968). *Kolloid-Z.*, **228**, 1.
40. Schatzki, T. (1962). *J. Polymer Sci.*, **57**, 496.
41. Hoffman, J. D., Williams, G. and Passaglia, E. (1966). *J. Polymer Sci.*, **C14**, 173.
42. McBrierty, V. J., McCall, D. W., Douglass, D. C. and Falcone, D. R. (1970). *J. Chem. Phys.*, **52**, 512.
43. McBrierty, V. J. and McDonald, I. R. (1973). *J. Phys. D: Appl. Phys.*, **6**, 131.
44. Wilson, C. W. and Pake, G. E. (1957). *J. Chem. Phys.*, **27**, 115.
45. Ward, I. M. (1960). *Trans. Faraday Soc.*, **56**, 648.
46. McCall, D. W. and Anderson, E. W. (1963). *Polymer*, **4**, 93.
47. McDonald, M. P. and Ward, I. M. (1962). *Proc. Phys. Soc.*, **80**, 1249.
48. Fuschillo, N., Rhian, E. and Sauer, J. A. (1959). *J. Polymer Sci.*, **25**, 381.
49. Farrow, G. and Ward, I. M. (1960). *Brit. J. Appl. Physics*, **11**, 543.
50. Ward, I. M. (1961). *Textile Res. J.*, **31**, 650.
51. Statton, W. O. (1963). *J. Polymer Sci.*, **C3**, 3.
52. Hyndman, D. and Origlio, G. F. (1959). *J. Polymer Sci.*, **36**, 556.
53. Fischer, E. W. and Peterlin, A. (1964). *Makromoles. Chem.*, **74**, 1.

CHAPTER 7

THE STIFFNESS OF POLYMERS IN RELATION TO THEIR STRUCTURE

L. HOLLIDAY

7.1 INTRODUCTION

In the majority of applications of polymers, we are interested in one or more of three basic mechanical properties—stiffness, strength and toughness. To these can be added creep, which becomes important in many engineering applications. Stiffness represents resistance to deformation, and is a much simpler property than strength and toughness, which relate to failure. Strength is the ultimate stress which a material can withstand before it fails, whether by fracture or by excessive deformation, whilst toughness represents the work required to fracture a material.

This chapter is concerned with stiffness, that is with elastic behaviour in the small strain region. There are a number of reasons for this. One is that stiffness is a very important property in engineering design, which is often limited by a permitted deflection. Furthermore, as Vincent[1] has pointed out one cannot expect to reach the theoretical strength before reaching the theoretical modulus. However, the most important reason is that it is much easier to relate stiffness to molecular structure than it is to relate the other more complicated mechanical properties. For that reason, stiffness is the most informative single mechanical measurement one can make on a material.

Values of elastic modulus will be given in SI units and to facilitate conversion to other units, some basic relationships between units are given in Table 1.

Engineering strain will be used in the discussion. This is defined as follows. If a bar of length l_0 is extended by an amount u to a new length l, then

$$l = l_0 + u$$

TABLE 1

CONVERSION TABLE FOR SI UNITS OF STRESS

$$1\,\mathrm{N\,m^{-2}} = 10\ \mathrm{dyne\ cm^{-2}}$$
$$1\ \mathrm{psi} = 6{\cdot}85 \times 10^4\ \mathrm{dyn\ cm^{-2}}$$
$$= 6{\cdot}85 \times 10^3\ \mathrm{N\,m^{-2}}$$
$$1\ \mathrm{kg\ cm^{-2}} = 10^5\ \mathrm{N\,m^{-2}}$$
$$1\ \mathrm{kg\ mm^{-2}} = 10^7\ \mathrm{N\,m^{-2}}$$
$$= 1{\cdot}45 \times 10^3\ \mathrm{psi}$$
$$1\ \mathrm{g\ denier^{-1}} = 10^8\ \mathrm{N\,m^{-2}}\ \mathrm{for}\ \rho = 1{\cdot}1$$
$$= 10\ \mathrm{kg\ mm^{-2}}$$
$$= 10^9\ \mathrm{dyn\ cm^{-2}}$$
$$= 1{\cdot}45 \times 10^4\ \mathrm{psi}$$

and the engineering strain is defined as

$$\varepsilon = \frac{u}{l_o}$$

In later chapters, when it is necessary to consider anisotropic elastic behaviour in detail, the symbol e is used for strain, with suitable suffixes depending on the notation such as e_{xx} or e_{ij}, but for general discussion the Greek symbol ε is used, in line with normal practice.

As a measurement of stiffness, Young's modulus is used, defined in the usual way.

$$E = \frac{\sigma}{\varepsilon}$$

where E is Young's modulus and σ is stress. The units of Young's modulus are thus stress units, e.g. $\mathrm{N\,m^{-2}}$. This equation is valid for small strains, say less than 1%. Since E is dependent on rate of strain, we define E as the limiting value at zero time as the strain rate tends to zero. For all practical purposes this is the value of E obtained with the low strain rates normally used in laboratory tensile testing machines.

Isotropic polymers have a single value of E, in which case the following relationships, between the four elastic constants E (Young's modulus), G (shear modulus), K (bulk modulus) and v (Poisson's ratio) apply:

$$E = 3G/(1 + G/3K)$$
$$G = E/2(1 + v)$$
$$K = E/3(1 - 2v)$$

For anisotropic materials, there are more elastic constants, depending on the direction of test. For example, for uniaxially aligned chains or fibres, with which this chapter is mostly concerned, there are two overall or macroscopic values of E: $E_\parallel$ and $E_\perp$ if the sample has transverse isotropy, these being the values of E parallel and perpendicular to the chain direction respectively.

The following account begins with an outline of the various methods used to determine the modulus in the chain direction and transverse to the chain for measurements at the molecular level, as well as the overall modulus in the direction of the fibre axis. It continues with a comparison of the experimental values at the molecular level, with a range of theoretical calculations based on the assessment of the intramolecular and intermolecular forces. It concludes with an attempt to reconcile the behaviour of bulk materials in their various forms—anisotropic and isotropic, semicrystalline and amorphous—with their behaviour at the molecular level.

7.2 EXPERIMENTAL METHODS OF OBTAINING ELASTIC MODULI

There are a number of alternative methods of measuring the elastic constants of polymers depending upon the purpose for which they will be used. The principal ones are (a) macroscopic stretching, as in a tensile testing machine, (b) dynamic mechanical methods, (c) sound velocity measurements, (d) from the shift in X-ray reflections under an applied stress, (e) Raman scattering spectra, (f) inelastic neutron scattering spectroscopy (INSS). Between them they cover the different levels of scrutiny, from the gross macroscopic to the molecular. In the very simplest of cases, such as with single crystals of argon, the various methods have been shown to give the same results, but this is not the case with polymers. These have a complex texture which leads to differences between macroscopic and molecular methods, and between static and dynamic. The various methods will now be briefly discussed in so far as they relate to estimates of theoretical stiffness. It should be remembered that there is *no one unique value for a particular elastic constant for a particular system* since the elastic constants depend on the test method.

7.2.1 Macroscopic stretching method
This requires little discussion since it is so well known. From the point of view of applications, it is by far the most important method. Strain

rates are low, within the range 0·01–10 min⁻¹ and over this range the effect of strain rate on stiffness is modest. As the strain rate is increased, Young's modulus may increase by a small amount—perhaps up to 10% —but the effect on strength (which generally goes up), and elongation (which tends to go down) is much greater.

7.2.2 Dynamic mechanical methods

These may be carried out in extension, torsion or flexure, and are concerned with alternating strains, normally over a range of frequencies and temperatures. Such methods are particularly suited to the study of molecular relaxation processes and have not been applied very extensively to oriented materials, where the studies to date have been more concerned with the static arrangement of structural elements in the materials. However, Takayanagi,[2] and Stachurski and Ward[3] have used dynamic tensile and torsion modulus to study the anisotropy of relaxations in oriented polyethylene.

7.2.3 Sound velocity measurements

The velocity of sound propagation in a material is simply related to the stiffness constants and density. It is a very convenient method in practice and gives simple and unambiguous results in the case of oriented glassy polymers, as in the work of Treloar[4] on polymethylmethacrylate. From a single experiment, two of the five elastic constants may be determined. The principle of the method is shown in Fig. 1.

A beam of ultrasonic waves generated by a transducer is directed on to the surface of the polymer which is immersed in a suitable liquid, for example water or silicone oil. Two refracted beams are produced, the higher velocity beam corresponding to a compressional or longitudinal wave, the other to a transverse or shear wave. There is also a reflected

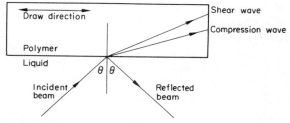

Fig. 1. Acoustic method for determining elastic constants.

wave at an angle equal to the angle of incidence θ. This is picked up by a transducer, and the apparatus is arranged so that the transmitter and receiver can be moved simultaneously, whilst keeping the angles of incidence and reflection equal. There are two angular positions θ_L and θ_T when the reflected intensity reaches 100%. These are the critical angles of internal reflection of the longitudinal and transverse waves respectively, when the angles of refraction of the waves reach 90°, and the waves are propagated along the solid–liquid boundary. At these points

$$V_L/V_0 = 1/\sin \theta_L \text{ and } V_T/V_0 = 1/\sin \theta_T$$

Since $V = (c/\rho)^{1/2}$ where c is the appropriate stiffness constant and ρ the density of the polymer, and since V_0, the velocity of sound propagation in the liquid can be separately determined, then c can be calculated.

7.2.4 X-ray determination

This method has largely been developed by Sakurada and his colleagues,[5] and depends on observing the shift under a static load of the X-ray diffraction spots associated with periodic structures in the polymers. It is fully described in Ref. 6. It was first used by Dulmage and Contois[7] to measure the elastic modulus of the crystalline phase in a range of polyesters and polyesteramides, and these results are discussed below.

This is a very powerful method when applied to oriented materials, since it yields information about the modulus of the crystalline region, *i.e.* information on a molecular scale. The only drawback is that, whilst the strain can be measured accurately, the stress can only be measured if a basic assumption is made—that the stress is homogeneous. This assumption amounts to the statement that the system can be represented by the series model, in which the strain is inhomogeneous and the stress homogeneous. Evidence for this has been set out in the paper of Holliday and White,[8] and can be regarded as satisfactory for the fibre direction.

7.2.5 Raman scattering measurements

This has been applied successfully by Mizushima and Shimanouchi[9] and by Shauffele and Shimanouchi[10] to measure the stiffness of polyethylene along the chain. The application of Raman spectroscopy to polymers has been greatly facilitated by the introduction of laser sources. To measure the elastic modulus of a hydrocarbon chain, one must identify the vibration frequencies of the 'accordion-like' motion of the planar zig-zag chain. In order to do this, it is necessary to examine spectra of a range of lower molecular weight alkanes, and from this it is possible to identify these vibrations in polyethylene unambiguously. If it is assumed that an extended zig-zag carbon backbone can be treated as an elastic rod, the following relationship for the Young's modulus in the chain direction is obtained:

$$E_{\parallel} = \{(2L/m)v_m\}^2 \rho$$

where L is the rod equilibrium length, m is the vibration order (a whole number 1, 2, 3, ...), v_m is the frequency of the longitudinal acoustic waves and ρ the density. The values obtained for $E_{\parallel}$ in this way agree very well with other measurements.

7.2.6 Inelastic neutron scattering spectroscopy (INSS)

The description of this method lies outside the scope of this chapter (see Ref. 8). It has only recently been applied to polymers, but has already yielded a figure for the modulus transverse to the chain direction $E_{\perp}$, as mentioned below.[11]

7.3 EXPERIMENTAL VALUES FOR THE MODULUS $E_{\parallel c}$ ALONG THE CHAIN—THE CRYSTALLINE REGION

In a highly oriented, semi-crystalline polymer, the chains in the crystalline regions are closely aligned to the draw direction. A large number of measurements of the elastic modulus of these regions along the chain have been carried out, using the X-ray method mentioned above. These have been based on the assumption that the stress on the crystalline region is the same as the overall stress on the fibre sample.

The results are shown on the left hand side of Table 2 (taken from Ref. 8), alongside the calculated values which will be discussed separately. Two of the measured values were obtained from low frequency Raman spectra, and one from INSS (in this case deuterated high density polyethylene was used). Otherwise all the results were obtained by the X-ray method. The same data is shown graphically on the left hand side of Fig. 2.

It will be seen that the measured elastic modulus parallel to the chain axis varies from 358×10^9 N m^{-2} for polyethylene, to $4 \cdot 1 \times 10^9$ N m^{-2} for polyvinyl *tert*. butyl ether—a factor of about 90 times. In contrast to this, as will be discussed below, the elastic modulus perpendicular to the chain is found to be about 3×10^9 N m^{-2}, with a range of only 3 times.

TABLE 2
EXPERIMENTAL AND CALCULATED VALUES FOR ELASTIC MODULI
ALONG THE CHAIN—$E_{\parallel}$
From L. Holliday and J. W. White[8]
(reproduced by kind permission of the International Union of Pure and Applied Chemistry.)

$E_{\parallel}$ measured

Ref.	Polymer	Method[c]	Value[d] of $E_{\parallel}$ ($N\,m^{-2}$)	Comments	Date	Authors
5	Cellulose I	X-ray[a]	130×10^9	Staggered ring	1964	Sakurada et al.
5	Cellulose II	X-ray	90×10^9	Staggered ring	1964	Sakurada et al.
5	Polyethylene	X-ray	240×10^9	Planar zig-zag	1964	Sakurada et al.
9		Raman	340×10^9		1949	Mizushima and Shimanouchi
10		Raman	358×10^9		1967	Shauffele and Shimanouchi
19		Neutron	329×10^9	$(CD_2)_n$	1968	Feldkamp et al.
5	Polypropylene (isotactic)	X-ray	42×10^9	3/1 helix	1964	Sakurada et al.
5	Polyoxymethylene	X-ray	54×10^9	9/5 helix	1964	Sakurada et al.
5	Polytetrafluoroethylene	X-ray	156×10^9	15/7 helix	1964	Sakurada et al.
21		Neutron	222×10^9		1972	Twisleton and White
7[b]	Polyethylene terephthalate	X-ray	140×10^9	Planar zig-zag	1958	Dulmage and Contois
5		X-ray	76×10^9		1964	Sakurada et al.
5	Polyvinyl alcohol	X-ray	255×10^9	Planar zig-zag	1964	Sakurada et al.
5	Polystyrene	X-ray	12×10^9	Helix	1964	Sakurada et al.
5	Nylon 6	X-ray	25×10^9 α form 21×10^9 γ form	Planar zig-zag	1964	Sakurada et al.
	Nylon 66					
5	Polyethylene oxide	X-ray	10×10^9	Helix	1964	Sakurada, Ito and Nakamae
	Polyvinyl chloride			Helix		
	Polyisobutylene			Helix		
5	Polybutene I	X-ray	25×10^9	Helix	1964	Sakurada et al.
5	Polyvinyl tert. butyl ether	X-ray	4.1×10^9	Helix	1964	Sakurada et al.
5	Polytetrahydrofuran	X-ray	55×10^9	Planar zig-zag	1964	Sakurada et al.
5	Polyvinylidene chloride	X-ray	41.5×10^9	?	1964	Sakurada et al.
	Poly-3,3-bis-halomethyl oxycyclobutanes			Planar zig-zag		
20	Diamond		800×10^9			
20	Graphite		450×10^9 1000×10^9	Carbon fibre Graphite (basal plane)		

[a] For further information on the lattice planes used for measurements see Refs. 5 and 7.
[b] This important paper contains modulus data on seven other condensation polymers, with $E_{\parallel}$ ranging from 3·7 to $130 \times 10^9\,N.m^{-2}$.
[c] The temperature at which the modulus was measured in the various experiments falls within the range 20–28°C. See Refs. 5 and 7.

TABLE 2—*contd.*

$E_\parallel$ calculated

Ref.	Method	Value of $E_\parallel$ ($N\,m^{-2}$)	Comments	Date	Authors
12	2 constant valence force field	77 or 121 × 10⁹		1936	Meyer and Lotmar
13	2 constant valence force field	(180 × 10⁹)		1958	Lyons[e]
14	2 constant valence force field	56 × 10⁹		1960	Treloar
14	2 constant valence force field	182 × 10⁹		1960	Treloar
15	4 constant Urey–Bradley force field	340 × 10⁹		1962	Shimanouchi *et al.*
16	8 constant Urey–Bradley force field	256 × 10⁹		1966	Odajima and Maeda
17	4 constant Urey–Bradley force field	49 × 10⁹		1962	Asahina and Enomoto
17	4 constant Urey–Bradley force field	(220 or) 150 × 10⁹		1962	Asahina and Enomoto
15	4 constant Urey–Bradley force field	160 × 10⁹		1962	Shimanouchi *et al.*
13	2 constant valence force field	(146 × 10⁹)		1958	Lyons[e]
14	2 constant valence force field	121 × 10⁹		1960	Treloar
13	2 constant valence force field	(157 × 10⁹)		1958	Lyons[e]
14	2 constant valence force field	196 × 10⁹		1960	Treloar
18	4 constant Urey–Bradley force field	13 × 10⁹		1969	Matsuura and Miyazawa
17	4 constant Urey–Bradley force field	200, 230, 160 × 10⁹	syndiotactic	1962	Asahina and Enomoto
17	4 constant Urey–Bradley force field	8, 4, 7 × 10⁹		1962	Asahina and Enomoto
17	4 constant Urey–Bradley force field	77–110 × 10⁹	(passing from iodine to fluorine as halogen)	1962	Asahina and Enomoto

[d] The specimen moduli (as opposed to the modulus E of the chain in the crystalline region) fall within the range 2–23 × 10⁹ N m⁻² (see Refs. 5 and 7).
[e] The accuracy of these calculations has been criticised by Treloar;[14] corrected values would be much lower.

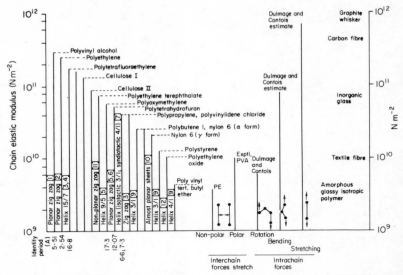

Fig. 2. *Range of measured polymer elastic constants in the chain direction (left-hand side) compared with moduli based on fundamental deformation mechanisms (right-hand side). (Reproduced by kind permission of the International Union of Pure and Applied Chemistry.)*

Table 2 shows that, apart from diamond and graphite, the highest moduli are encountered with the planar zig-zag conformation. The examples are polyethylene and polyvinyl alcohol. There is a marked drop in going to the very lazy helix (15/7) of polytetrafluoroethylene, which departs only slightly from the planar zig-zag.

A number of polymers studied by Dulmage and Contois[7] in their pioneering paper have not been included in Table 2, owing to lack of space. Their results support the foregoing conclusions, since they showed that a low modulus was invariably associated with a contracted fibre identity period.

There are two inconsistencies which show up in Table 2. The first is the significant difference in results obtained by X-ray measurements from those obtained by Raman and neutron spectra, although some difference is to be expected. The X-ray results are much lower, 240×10^9 N m^{-2} compared with 340×10^9 N m^{-2} for polyethylene. One possible reason for this difference may be the assumption of uniform stress[8] which is involved in the X-ray method. The other inconsistency is found in the results from

X-ray measurements for polyethylene terephthalate, these being 140 and $70 \times 10^9 \, \text{N m}^{-2}$. It is suggestive that the former is closer to the theoretical figure.

7.4 THEORETICAL VALUES FOR THE MODULUS $E_{\|c}$ ALONG THE CHAIN

The first theoretical estimates of the modulus of elasticity of a polymer was made by Meyer and Lotmar[12] in 1936. They showed that the modulus in the chain direction of a polymer could be calculated from spectroscopic data, using the force constants of the chemical bonds of the chain derived from the vibration frequencies of these bonds in other molecules. They were interested in cellulose and, for the calculation, considered two modes of deformation—bond stretching and bond angle opening. The method has since been extended and improved by Lyons[13] and Treloar,[14] and the calculations of the latter represent the latest refinement of this type of calculation. The simplicity of the method is very attractive, as can be seen from the following derivation.[14] It should be noted that only the chain atoms are considered, and that interchain forces are neglected.

Consider a planar zig-zag molecule like polyethylene (Fig. 3), with a force F acting along the chain. The angle of inclination of each carbon–carbon bond to the chain axis is θ. The change of length δL caused by the application of the force is:

$$\delta L = nF(\cos^2 \theta/k_1 + \sin^2 \theta/4k_p)$$

where n is the number of bonds, k_1 is the force constant for bond stretching and k_p is the force constant for valence angle opening (for definitions see Ref. 14; for the force constants used by the different workers see Ref. 8).

Since $E = (F/A)/(\delta L/L)$ and knowing A, the cross-sectional area of the chain, the elastic modulus can be calculated. The method tends to give low values when compared with experiment. Despite the somewhat unrefined nature of this type of calculation, it has given useful results and it is the only method available for complicated molecules.

In connection with this valence force field type of calculation, the

Polyethylene chain

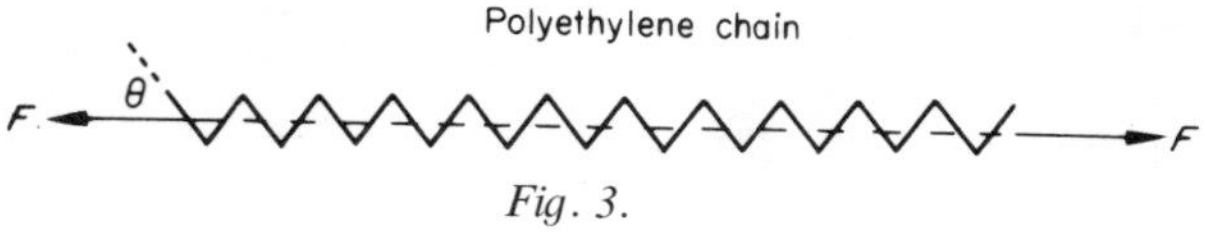

Fig. 3.

estimates of Dulmage and Contois[7] are very illuminating. These are shown diagrammatically on the right hand side of Fig. 2, and show the values of modulus which would be expected based on three different modes of intrachain deformation. To these have been added[8] typical values of modulus based on stretching non-covalent *interchain* bonds, for comparison. Dulmage and Contois considered extension in the chain direction based on either (a) bond extension, (b) valence angle opening, or (c) rotation around bonds. The resistance to deformation in the three cases is roughly in the ratio $100:10:1$.

Although these calculations are simple, they give an insight into what happens when a polymer crystal chain is extended, as can be seen on comparing the right and left hand sides of Fig. 2. It is obvious that in the known types of stiff chain, such as polyethylene and polyvinyl alcohol, valence angle opening plays an important part in the chain extension since the chain modulus falls between the Dulmage and Contois estimates for bond extension and valence angle opening. For the lower modulus materials shown, such as polystyrene and polyethylene oxide, the internal rotation of chains becomes the controlling process.

The next point to consider is the improvement of the two-constant valence force field type of calculation, upon which the preceding estimates were based. A number of Japanese workers have made significant contributions to this field[15-18] and their estimates are shown on the right hand side of Table 2, where they can be compared with the experimentally determined values. More recently, there has been a decline of interest in this type of calculation, which becomes increasingly laborious as the molecules under consideration become more complicated.

The calculations have included a more complete range of interatomic interchain interactions,[15,17] and finally intermolecular forces have been taken into account as well.[16] These are referred to in Table 2 as the 4 constant and 8 constant Urey–Bradley force field calculations. Generally speaking, the 4 constant calculations give good agreement with the X-ray method of measurement, which in fact gives a static modulus. The exception to this is the case of polyethylene, where the picture is further confused by the Raman and neutron data.

Generally speaking, the agreement between the experimental and calculated values is satisfactory. Where there are discrepancies, it should be remembered that both are based on assumptions and simplifications, but the major assumption in the experimental method—of homogeneous stress—is probably more reasonable than the assumptions made in the calculations.

7.5 EXPERIMENTAL AND THEORETICAL VALUES FOR ELASTIC MODULUS TRANSVERSE TO THE CHAIN IN THE CRYSTALLINE REGION, $E_{\perp c}$

Compared with the amount of experimental and theoretical work which has been done on the longitudinal modulus $E_{\parallel c}$, relatively little has been done in the transverse direction. This is summarised in Table 3,[8] from which it is seen that the data are quite consistent with our understanding of intermolecular forces. In the first place, the moduli are much lower than in the chain direction, and secondly the range covered varies by a factor of only three based on the X-ray data. Later information[21,22] is that these transverse moduli in the crystalline regions may be higher, based on measurement by neutron spectroscopy. These points are clear from the two vertical bars denoting interchain forces on the right hand side of Fig. 2. It will be seen that the interchain forces offer approximately the same resistance to a deforming force as does the rotation mode along the chain.

In Fig. 4 these experimental values based on X-ray measurement, are plotted against the cohesive energy density of the material.[8] It might be expected that the more polar the polymer, the higher the transverse modulus. This hypothesis seems to be supported by the data. Two further values for polyethylene have not been included since they were obtained by other methods. Samuels,[23] using the sound wave method obtained a

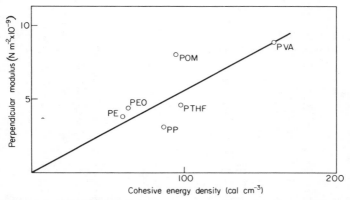

Fig. 4. Elastic modulus perpendicular to the chain axis direction as a function of cohesive energy density. (Reproduced by kind permission of the International Union of Pure and Applied Chemistry.)

value of 4×10^9 N m^{-2} for $E_{\perp c}$, which is in good agreement with the foregoing. On the other hand, Twisleton and White[11] using INSS found 6×10^9 N m^{-2} which is much higher. These differences will remain unresolved until further experimental work clarifies the situation. Nevertheless, it is worth repeating that there is no unique value for a particular type of modulus, since this depends on the experimental conditions such as rate of strain or frequency of vibration, etc. Accepting this, it remains to be seen what are the upper and lower bounds to be expected for such a mechanical property as $E_{\perp c}$ or $E_{\parallel c}$, for a given polymer crystal.

Turning to the theoretical values for $E_{\perp c}$, the calculations have been restricted to polyethylene, and these show considerable variations. The results, which are given in Table 3 are complicated by the fact that small differences are found for the three crystallographic directions. These individual values are averaged when we come to consider oriented materials in practice. For example, with an oriented fibre, it is reasonable to suppose that it is isotropic around the fibre, that is at right angles to the fibre direction, in which case the values normal to the lattice planes, 110, 200 and 020 emerge as an average figure. Nevertheless, for a given crystallographic direction, the values vary widely depending upon the model used in the calculation. This problem has only been mentioned here for interest, and is discussed further in Ref. 8. For the time being, the experimental values must be regarded as the important figures, whilst the calculated values represent a tentative step in tackling a complex problem. With this in mind, we shall assume a value of 3.5×10^9 N m^{-2} for $E_{\perp c}$ for polyethylene, based on the X-ray data.

7.6 EXPERIMENTAL VALUES FOR THE ELASTIC MODULUS OF THE AMORPHOUS REGIONS $E_{\parallel a}$ ALONG A FIBRE

The previous discussion has been confined to the crystalline regions. It is now necessary to consider the amorphous regions in an oriented material, since they are of overriding importance in determining the modulus in the draw direction. Although there is an extensive literature on the texture and fine structure of oriented polymers, we shall assume that—to a first approximation—a highly oriented specimen can be regarded as a series arrangement of crystalline and amorphous regions, as shown diagrammatically in Fig. 5. This is equivalent to the assumption made earlier in connection with the calculation of $E_{\parallel c}$ by X-ray measurement of the strain.

TABLE 3
EXPERIMENTAL AND CALCULATED VALUES FOR ELASTIC MODULI TRANSVERSE TO THE CHAIN—$E_{\perp c}$
From L. Holliday and J. W. White[8]
(reproduced by kind permission of the International Union of Pure and Applied Chemistry.)

Polymer	110	$E \times 10^{-9}$ 200(E_a)	$N\,m^{-2}$ 020(E_b)	Other	Method	Authors (Ref.)
				$E_\perp$ measured Modulus normal to lattice planes		
Polyethylene	4·3	3·2	3·9	—	X-ray	Sakurada et al.[5]
	—	3·2	3·9	—	X-ray	Quoted by Odajima and Maeda[16]
		2·5	1·9			
		5·0				
	6·0^a	6·0^a	—	—	Neutron	Twisleton and
PTFE				18·2 (1010)	Neutron	White[11,21]
Polypropylene		2·9		3·2 (040)	X-ray	Sakurada et al.[5]
				4·0	Sonic	Samuels[23]
Polyvinyl alcohol		8·9		9·0 (10Ī)	X-ray	Sakurada et al.[5]
				4·7 (002)	X-ray	
Polyoxymethylene				8·0 (10Ī0)	X-ray	Sakurada et al.[5]
Polytetrahydrofuran		4·6	4·6		X-ray	Sakurada et al.[5]
Polyethylene oxide				4·4 (120)	X-ray	
Graphite ⊥ to layer planes				39	Neutron	
Hexadecane crystal	1·5				Neutron	Twisleton
Argon				3·0	Sonic	

Polymer	110	$E \times 10^{9}$ 200(E_a)	$N\,m^{-2}$ 020(E_b)	Comments	Authors (Ref.)
				$E_\perp$ calculated Modulus	
Polyethylene		5·7	2·1	Urey–Bradley molecular field Lennard–Jones intermolecular	Miyazawa and Kitagawa[24]
		6·7	10·8	Quoted by Sakurada et al.[5]	Miyazawa and Kitagawa[24]
		2·1	2·1		Enomoto and Asahina[25]
	4·91	4·76	8·33	Set 1 6 exponential potential with two sets of interatomic constants	Odajima and Maeda[16]
	6·84			Set 3 ditto	

^{a} The same number is quoted for each direction here. Latest information is that this figure may be as high as $12 \times 10^9\,N\,m^{-2}$ (Ref. 22).

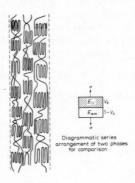

Diagrammatic series
arrangement of two phases
for comparison

Fig. 5. Diagrammatic arrangement of crystalline and amorphous regions in an
oriented polymer fibre.

If the moduli are represented by $E_\parallel$, $E_{\parallel c}$, and $E_{\parallel a}$ for the overall
oriented specimen in the draw direction, the crystalline regions and the
amorphous regions respectively, then

$$I/E_\parallel = (1 - V_c)/E_{\parallel a} + V_c/E_{\parallel c}$$

where V_c is the volume fraction of the crystalline phase. One problem in
using this equation is that $E_{\parallel a}$—the modulus for the amorphous region—
cannot be regarded as invariant, since the amorphous region will itself be
oriented to a greater or lesser extent by the drawing operation. In their
classical work, Dulmage and Contois[7] found that $E_{\parallel a}$ could change by
a factor of three on drawing, depending on the nature of the polymer.
Where a large increase in modulus of the amorphous phase was obtained
by orientation, the overall stiffness of the oriented fibre was also very high,
suggesting that the cause of the increase was the availability of a
considerable driving force for orientation of the amorphous phase.

That the influence of the amorphous region is overwhelming in a series
arrangement is shown by the following calculation (the figures are not too
far from the example of polyethylene).

If $V_c = 0.7$, $E_{\parallel a} = 2 \times 10^9 \, \mathrm{N\,m}^{-2}$ and $E_{\parallel c} = 300 \times 10^9 \, \mathrm{N\,m}^{-2}$, then
$I/E_\parallel = 0.3/2 \times 10^9 + 0.7/300 \times 10^9 \, \mathrm{m}^2 \, \mathrm{N}^{-1}$ from which $E_\parallel = 10 \times 10^9$
$\mathrm{N\,m}^{-2}$.

Despite the importance of the amorphous region in determining the
overall modulus in the fibre direction, there is very little information to be
gleaned from the literature on this subject. In order to fill this gap, an

attempt has been made to calculate $E_{\|a}$ from the data in Refs. 5 and 7, and using the series formula for modulus; this is shown in Table 4.

Despite the arbitrary nature of some of these calculations, the figures are very revealing. They show (1) that the stiffness of the amorphous region depends upon the degree of orientation (it can vary by a factor of six), (2) the amorphous region itself can have a relatively high value of $E_{\|a}$. Presumably the amorphous region has become both oriented and therefore anisotropic.

A comparison of these values with the values on the right hand side of Fig. 2 shows that—in most cases—they depend upon the mechanisms of chain separation or bond rotation. These mechanisms will suffice to explain the data. In the case of polyoxymethylene, however, with $E_{\|a}$ of $14 \cdot 0 \times 10^9 \, N \, m^{-2}$, it appears that there may also be a significant contribution from valence angle opening. Some very recent work by Ward and his collaborators[26] has shown that ultra-high modulus polyethylene can be produced by cold drawing linear polyethylene of the appropriate molecular weight and molecular weight distribution to a very high draw ratio. In this way, a Young's modulus of $\sim 7 \times 10^{10} \, N \, m^{-2}$ has been reached—a figure which is comparable to inorganic glass. Examination of Fig. 2 shows that in this material, the deformation modes must be

TABLE 4

VALUES OF $E_{\|a}$ CALCULATED FROM DATA IN REFS. 5 AND 7

Polymer	E	$\times 10^{-9} \, N \, m^{-2}$ $E_{\|c}$	$E_{\|a}$	Ref.
Polyethylene terephthalate[a]	11·1	140	6·0	7
Poly(pentamethylene 4,4′ sulphonyldibenzoate)[b]	2·8	4·2	2·5	7
Polyethylene x	2·4	240	1·2	5
Polyethylene x	15·0	240	7·6	5
Polyethylene x	3·1	230	1·6	5
Polyvinyl alcohol x	11·0	255	5·5	5
Polypropylene x	6·4	42	6·0	5
Polypropylene x	7·0	48	3·8	5
Polyoxymethylene x	23·0	54	14·0	5
Polyoxymethylene x	6·0	58	3·1	5

[a] 47% crystallised.   [b] 40% crystallised.

x In order to arrive at a rough estimate of $E_{\|a}$, a crystallinity of 50% has been assumed since it is not quoted in the original reference.

largely valence-angle opening and carbon–carbon bond extension (these are equivalent very approximately to moduli of 8×10^{10} and 7×10^{11} $N\,m^{-2}$ respectively). This means that the rotation mode (equivalent to a modulus of around $6 \times 10^9\,N\,m^{-2}$) has almost disappeared. If this is so, then the overall modulus of $7 \times 10^{10}\,N\,m^{-2}$ found by Ward can be explained in terms of a structure in which the drawn ultra-high modulus fibre is made up of extended chain crystallites and folded chain crystallites. There can be little 'amorphous' material left in the fibre.

7.7 THE PROBLEM OF ISOTROPIC POLYMERS

In any discussion on orientation, the isotropic material cannot be ignored, since it forms the starting point for the operation, which means that the properties of the oriented material must be latent or inherent in the isotropic polymer. However, it may seem illogical to end this chapter with a discussion of the isotropic polymer, rather than to begin at this point. The reason why the present pattern has been adopted is quite simple—the isotropic material represents a much more complex system, and can best be regarded as made up of the structural elements which have already been discussed.

Although the isotropic polymer may be amorphous or semi-crystalline, it will suffice if we confine the discussion to the latter type. Since semi-crystalline polymers include an amorphous component, they may be regarded as subsuming the class of amorphous polymers as a consequence.

For convenience, we can regard an isotropic semicrystalline polymer as being made up of an isotropic polycrystalline phase and an isotropic amorphous phase, as shown in Fig. 6, which is purely diagrammatic (it shows the classical fringed micelle model for simplicity). From a geometrical standpoint, we cannot discriminate *a priori* between two continuous interpenetrating phases, a dispersion of a crystalline phase in an amorphous phase, or of an amorphous phase in a crystalline phase.[27] The distinction may depend upon the volume fractions.

The modulus of an isotropic semi-crystalline polymer can then be described by the following general expression :

$$E_o = f(E_a, E_{pc}, V_{pc}, S_{pc})$$

E_o is the Young's modulus of the isotropic material, E_a is the modulus of the isotropic amorphous phase, E_{pc} is the modulus of the polycrystalline phase, V_{pc} is the volume fraction and S_{pc} an omnibus geometrical term

Fig. 6. Very diagrammatic representation of isotropic semi-crystalline polymer.

allowing for factors such as spherulite size which affect E_o. It will be recalled that the expression takes the following simple forms for series and parallel arrangement of the two phases respectively.

$$E_o = E_a(1 - V_{pc}) + E_{pc} V_{pc}$$

$$1/E_o = (1 - V_{pc})/E_a + V_{pc}/E_{pc}$$

The actual modulus of a material made up of two randomly arranged phases will fall somewhere between these two limits or bounds, and it becomes necessary to estimate values for E_{pc} and E_a.

The value of E_{pc} is the average elastic modulus of the crystallites assumed randomly arranged, in a wholly crystalline aggregate. However, clearly

$$E_{\|c} \gg E_{pc} > E_{\perp c}$$

that is, E_{pc} is bounded by $E_{\|c}$ and $E_{\perp c}$ and will be closer to the lower figure.

A more sophisticated analysis admits the contribution of the shear moduli and Poisson's ratios and enables the aggregate elastic constants to be calculated from the elastic constants of the anisotropic elements of which the aggregate is composed. Such considerations are the basis of the Ward aggregate model[28] which is discussed extensively in Chapters 8 and 9. The calculation can be made in two ways, either assuming uniform strain (which gives the Voigt bound) or uniform stress (the Reuss bound). Odajima and Maeda[16] have made such calculations for polyethylene, based on calculated elastic constants for polyethylene crystals. In this instance, the bounds are relatively close together and the experimental value is found to be closer to the lower or Reuss average. The following are the actual results.

Calculated and experimental values of Young's modulus of polycrystalline polyethylene of 100% crystallinity E_{pc}[16] are as follows.

	Reuss average		Voigt average		Experimental
	Set 1	Set 3	Set 1	Set 3	
$E_{pc} \times 10^{-9} \, \text{N m}^{-2}$	4·90	5·82	15·6	15·8	5·05

The distinction between Set 1 and Set 3 represents the alternative force constants used in the theoretical calculations of the individual elastic constants. Since for polyethylene the experimental value for $E_{\parallel c}$ is of the order of $300 \times 10^9 \, \text{N m}^{-2}$ and for $E_{\perp c}$ of the order of $3 \times 10^9 \, \text{N m}^{-2}$, the overriding importance of $E_{\perp c}$ in determining E_{pc} can clearly be seen.

The experimental value for E_{pc} quoted above ($5\cdot05 \times 10^9 \, \text{N m}^{-2}$) comes from the work of Davidse et al.[29] The method was based on extrapolating the modulus of semi-crystalline polyethylene to 100% crystallinity. A more recent estimate based on a dynamic mechanical method at $-190°C$[30] gave a value of $13 \times 10^9 \, \text{N m}^{-2}$.

The uncertainty over the modulus of the amorphous phase E_a, represents another difficulty in estimating E_0. One approach to the problem of calculating this figure is that of Bowden,[31] but at best this gives a very approximate answer. He considers an amorphous glassy polymer, which typically has a Young's modulus of the order of $3-4 \times 10^9 \, \text{N m}^{-2}$. If this modulus is controlled by secondary intermolecular forces, a reasonable estimate of the modulus can be made from the interatomic potential.

$$\phi = Aa^{-n} - Bb^{-m}$$

$$E = (1/a)(\text{d}^2 \phi/\text{d}a^2)$$

Using an $m = 6$, $n = 15$ potential, it can be calculated that the modulus will be $\ll 4 \times 10^9\,\mathrm{N\,m^{-2}}$. This is low compared with experiment, and it appears that some bond rotation must be involved to account for the observed value of Young's modulus which is of the order of $3\text{--}4 \times 10^9\,\mathrm{N\,m^{-2}}$ for a rigid amorphous glassy polymer, as already mentioned (see Fig. 2).

This argument applies to a rigid glassy polymer, and it might be anticipated that the amorphous phase in a semi-crystalline polymer will have a much lower Young's modulus, since it is most commonly in the rubbery state. This means that if the chains were relaxed, the modulus of the amorphous region would be very low, of the order of $10^6\text{--}10^7\,\mathrm{N\,m^{-2}}$. There are good grounds for believing, however, that this is not so, *i.e.* that the amorphous chains are under tension and this effectively stiffens them.

This point is dealt with specifically by Krigbaum *et al.*[32] They use a model in which the small crystallites are interconnected by amorphous chains, for which they act as cross-links. This is tantamount to assuming that the system can be treated as a dispersion or suspension of a discontinuous crystalline phase, in a continuous amorphous phase, to use the terminology of composite materials. On the assumption that the amorphous chains are under tension as the result of the process of crystallisation, they are able to account convincingly for the modulus of the two different polyethylenes they discuss. These have moduli E_0 in the range $0.5\text{--}1.0 \times 10^9\,\mathrm{N\,m^{-2}}$.

We may use the same data to obtain a very approximate figure for E_a of the isotropic amorphous phase. Assuming that the Krigbaum model approximates to a series arrangement of the phases (this is not unreasonable, based on general experience with composites), and assuming $E_{pc} = 5 \times 10^9\,\mathrm{N\,m^{-2}}$, we obtain the following order of magnitude figures for their materials.

$$\text{Marlex} \quad E_a = 0.1 \times 10^9\,\mathrm{N\,m^{-2}}$$

$$\text{Lupolen} \quad E_a = 0.4 \times 10^9\,\mathrm{N\,m^{-2}}$$

These figures compare favourably with the determination of E_a by Gray and McCrum[30] at $-80^\circ\mathrm{C}$, using linear polyethylene, for which they obtained a value of $0.5 \times 10^9\,\mathrm{N\,m^{-2}}$.

Based on the foregoing arguments, we may now set out a plausible picture of how the overall modulus E_0 of isotropic polyethylene (which can be regarded as a representative semi-crystalline polymer) is built up. The figures are very approximate.

Isotropic
semi-crystalline
material

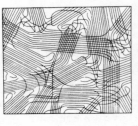

$E_{\|c} = 300 \times 10^9 \, \mathrm{N\,m^{-2}}$ $E_{\perp c} = 3 \times 10^9 \, \mathrm{N\,m^{-2}}$

$E_{pc} \approx 5 \times 10^9 \, \mathrm{N\,m^{-2}}$ $E_a = 0.5 \times 10^9 \, N\,m^{-2}$

$E_o = 1 \times 10^9 \, \mathrm{N\,m^{-2}}$

Oriented material
(the figures are very
    approximate)

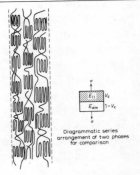

Drawing process
$E_{\|} = 10 \times 10^9 \, \mathrm{N\,m^{-2}}$
made up of
$E_{\|a} = 2 \times 10^9 \, \mathrm{N\,m^{-2}}$
$E_{\|c} = 300 \times 10^9 \, \mathrm{N\,m^{-2}}$

Diagrammatic series
arrangement of two phases
for comparison

7.8 CONCLUSION

The foregoing account has been deliberately limited to the problem of stiffness in relation to orientation. Other mechanical properties such as strength, toughness, creep, fatigue resistance, etc., have not been mentioned, since they are dealt with in subsequent chapters. So far as stiffness is concerned, it has been seen that it is possible to measure the stiffness of oriented polymers on a molecular scale in certain cases, and to reconcile the results obtained for the component parts with calculated figures based on spectroscopic data. The agreement is good, and leads us to believe that we have a reasonable insight into the molecular processes involved in the deformation. On a more qualitative level, it is also possible to visualise the macroscopic phase arrangements in oriented polymers, and even to understand the more complex situation with isotropic semi-crystalline polymers. The overall picture is sufficiently illuminating to form a satisfactory introduction to the more complex phenomena which are discussed in later chapters.

REFERENCES

1. Vincent, P. I. (1964). *Proc. Roy. Soc.*, **A282**, 113.
2. Takayanagi, M., Imada, K. and Kajiyama, T. (1966). *J. Polymer Sci.*, **C15**, 236.
3. Stachurski, Z. H. and Ward, I. M. (1968). *J. Polymer Sci.*, **A2**, 6, 1083 and 1817.
4. Treloar, L. R. G. (1970). *Plastics and Polymers*, **39**, 29.
5. Sakurada, I., Ito, T. and Nakamae, K. (1966). *J. Polymer Sci.*, **C15**, 75.
6. Sakurada, I., Nukushima, Y. and Ito, T. (1962). *J. Polymer Sci.*, **57**, 651.
7. Dulmage, W. J. and Contois, L. E. (1958). *J. Polymer Sci.*, **28**, 275.
8. Holliday, L. and White, J. W. (1971). *Pure and Applied Chemistry*, **26**, 545.
9. Mizushima, R. and Shimanouchi, T. (1949). *J.A.C.S.*, **71**, 1320.
10. Shauffele, R. F. and Shimanouchi, T. (1967). *J. Chem. Phys.*, **47**, 3605.
11. Twisleton, J. F. and White, J. W. (1972). *Inelastic Scattering of Neutrons*, I.A.E.A., Grenoble.
12. Meyer, K. H. and Lotmar, W. (1936). *Helv. Chim. Acta*, **19**, 68.
13. Lyons, W. J. (1958). *J. Appl. Phys.*, **29**, 1429; (1959). 30, 796.
14. Treloar, L. R. G. (1960). *Polymer*, **1**, 95/279/290.
15. Shimanouchi, T., Asahina, M. and Enomoto, S. (1962). *J. Polymer Sci.*, **59**, 93.
16. Odajima, A. and Maeda, T. (1966). *J. Polymer Sci.*, **C15**, 55.
17. Asahina, M. and Enomoto, S. (1962). *J. Polymer Sci.*, **59**, 101/113.
18. Matsuura, H. and Miyazawa, T. (1969). *J. Polymer Sci.*, **B7**, 65.
19. Feldkamp, L. A., Venkataraman, G. and King, J. S. (1968). *Neutron Inelastic Scattering*, Vol. 2, I.A.E.A., Vienna, p. 159.
20. Cottrell, A. H. (1964). *Proc. Roy. Soc.*, **A282**, 2.
21. Twisleton, J. F. and White, J. W. (1972). *Polymer*, **13**, 40.
22. White, J. W. Private Communication.
23. Samuels, R. J. (1965). *J. Polymer Sci.*, **A3**, 1741.
24. Miyazawa, T. and Kitagawa, T. (1964). *J. Polymer Sci.*, **B2**, 395.
25. Enomoto S. and Asahina, M. (1964). *J. Polymer Sci.*, **A-2**, 3523.
26. Capaccio, G. and Ward, I. M. (1973). *Nature Phys. Sci.*, **243**, 143; (1974). *Polymer*, **15**, 223.
27. Holliday, L. (Ed.) (1966). *Composite Materials*, Ch. 1, Elsevier, Amsterdam.
28. Ward, I. M. (1962). *Proc. Phys. Soc.*, **80**, 1176.
29. Davidse, R. H., Waterman, H. I. and Wasterdijk, J. B. (1962). *J. Polymer Sci.*, **59**, 389.
30. Gray, R. W. and McCrum, N. G. (1969). *J. Polymer Sci.*, **A-2**, 7, 1329.
31. Bowden, P. B. (1969). *Polymer*, **9**, 449.
32. Krigbaum, W. R., Roe, R. J. and Smith, K. J. (1964). *Polymer*, **5**, 533.

CHAPTER 8

THE MACROSCOPIC MODEL APPROACH TO LOW STRAIN PROPERTIES

D. W. HADLEY and I. M. WARD

8.1 INTRODUCTION

Two distinct types of macroscopic theoretical model for the low strain mechanical behaviour of oriented solid polymers will be considered in this chapter. First, models which predict the changes in elastic constants with the development of orientation; these will be referred to as 'orienting element' models. Secondly, models which seek to explain the mechanical behaviour of both isotropic and oriented polymers in terms of a two phase material with separate components representing crystalline and amorphous fractions; these we shall call 'composite structure' models.

Both types of model are phenomenological, based on experimental measurements, and the model elements do not generally correspond directly with fine structure at the molecular level. They should not, however, be inconsistent with structural observations, and possible similarities between the models and molecular structure will be discussed.

The models have been developed mainly for semi-crystalline polymers, which in general show the largest mechanical anisotropy, but some of the discussion is equally relevant to oriented non-crystalline polymers. Although an oriented polymer is strictly a non-linear viscoelastic solid (see Chapters 10 and 11) the present discussion is restricted to theoretical models which represent linear elastic or linear viscoelastic behaviour.

Predictions for the values of the moduli of idealised fully oriented crystalline polymers will not be discussed, since this topic is dealt with extensively in Chapter 7. Neither shall we consider one-dimensional rheological models of the spring and dashpot type, reviews of which have been given by Kennedy[1] and Reiner[2]; Sobotka[3] has generalised such models in two dimensions.

264

8.2 THE GENERALISED HOOKE'S LAW

The mechanical properties of an anisotropic elastic solid can be described by the generalised Hooke's law

$$e_{ij} = s_{ijkl}\,\sigma_{kl}$$

or

$$\sigma_{ij} = c_{ijkl}\,e_{kl}$$

relating strains e_{ij} to stresses σ_{kl}, where the s_{ijkl} are compliance constants and the c_{ijkl} are the stiffness constants.

It is customary to use an abbreviated matrix rotation (*e.g.* Ref. 4, p. 25) which relates the six independent components of the engineering strains to the six independent components of stress

$$e_p = s_{pq}\,\sigma_q$$

$$\sigma_p = c_{pq}\,e_q$$

where e_p is $e_{xx}, e_{yy}, e_{zz}, e_{xz}$, etc., and σ_q is $\sigma_{xx}, \sigma_{yy}, \sigma_{zz}, \sigma_{xz}$, etc.

Studies of mechanical anisotropy in polymers have been made on specimens of two distinct types. Uniaxially drawn filaments or films have fibre symmetry, with isotropy in the plane perpendicular to the draw direction. Films drawn at constant width or films drawn uniaxially and subsequently rolled and annealed under closely controlled conditions, show orthorhombic symmetry. For fibre symmetry (also called transverse isotropy) the number of independent elastic constants reduces to five and the compliance matrix is

$$\begin{bmatrix} S_{11} & S_{12} & S_{13} & 0 & 0 & 0 \\ S_{12} & S_{11} & S_{13} & 0 & 0 & 0 \\ S_{13} & S_{13} & S_{33} & 0 & 0 & 0 \\ 0 & 0 & 0 & S_{44} & 0 & 0 \\ 0 & 0 & 0 & 0 & S_{44} & 0 \\ 0 & 0 & 0 & 0 & 0 & 2(S_{11}-S_{12}) \end{bmatrix} \tag{1}$$

where the $z(3)$ axis has been chosen as the axis of symmetry.

For orthorhombic symmetry there are nine independent elastic constants and the compliance matrix is

$$
\begin{bmatrix}
S_{11} & S_{12} & S_{13} & 0 & 0 & 0 \\
S_{12} & S_{22} & S_{23} & 0 & 0 & 0 \\
S_{13} & S_{23} & S_{33} & 0 & 0 & 0 \\
0 & 0 & 0 & S_{44} & 0 & 0 \\
0 & 0 & 0 & 0 & S_{55} & 0 \\
0 & 0 & 0 & 0 & 0 & S_{66}
\end{bmatrix}
\tag{2}
$$

Equivalent matrices exist for the stiffness constants, but are less frequently used, since experimentally it is most convenient to apply a stress and measure the corresponding strains.

8.3 ORIENTING ELEMENT MODELS

In the course of a study of the properties of polyethylene terephthalate fibres, Ward observed[5,6] that the low strain mechanical anisotropy was governed primarily by the degree of molecular orientation as determined from birefringence, with features such as crystallinity having only a small effect. He proposed that the unoriented polymer might be regarded as a random aggregate of anisotropic elastic units, whose elastic properties were those of the fully oriented material, i.e. for the uniaxially oriented fibres which were being studied the elastic properties were defined by a compliance matrix of the form of (1) above.

When the polymer was stretched the units of the aggregate would be aligned, without the units themselves being altered. Thus, in principle, the mechanical properties of partially oriented polymers might be derived in terms of the degree of molecular orientation and the properties of specimens with maximum orientation. We shall consider only uniaxial stretching, i.e. fibre orientation, and the treatment will follow that given by Ward,[7,8] and by Hadley et al.[9]

The calculation of the elastic constants of the partially oriented polymer can be made in two ways. One can assume either uniform strain throughout the aggregate, which involves a summation of stiffness constants, or uniform stress, which implies a summation of compliance constants. In the former case the tractions across the boundaries of the unit do not satisfy the stress equilibrium conditions; in the latter case there is discontinuity of strain at the boundaries of the units so that they cannot remain bonded throughout the orientation process.

For an *isotropic* aggregate, the stiffness averaging procedure had been proposed by Voigt,[10] and the compliance averaging procedure by Reuss,[11] many years previously. Each had been used to compare the elastic constants of single crystals with those of an isotropic aggregate of single crystals (see for example Ref. 12).

For an isotropic aggregate, it was shown by Bishop and Hill,[13] that the Voigt and Reuss averaging procedures give upper and lower bounds for the moduli of the aggregate, the correct value lying between these extremes; closer bounds have subsequently been derived by Hill,[14] Hashin and Shtrikman,[15] and Peselnick and Meister.[16] Comparisons between the predicted Voigt and Reuss bounds for an isotropic aggregate, and experimental data for five unoriented polymers are shown in Table 1.[9] For polyethylene terephthalate and low density polyethylene the experimental data lie within the calculated bounds, suggesting that the mechanical anisotropy is primarily determined by molecular orientation. The experimental values lie outside the predicted limits for high density polyethylene, nylon and polypropylene, which implies that other factors in addition to molecular orientation determine the mechanical anisotropy. In polypropylene, Pinnock and Ward[17] have suggested that simultaneous changes occur in both morphology and molecular mobility.

The equations which predict the elastic constants of a partially oriented polymer involve orientation functions to define the orientation of the aggregate units. For example, the average extensional compliance S'_{33} for a transversely isotropic aggregate of transversely isotropic structural units is given by $S'_{33} = S_{11} \overline{\sin^4 \theta} + S_{33} \overline{\cos^4 \theta} + (2S_{13} + S_{44}) \overline{\sin^2 \theta \cos^2 \theta}$.

TABLE 1

MEASURED EXTENSIONAL AND TORSIONAL COMPLIANCES OF
UNORIENTED POLYMERS COMPARED WITH REUSS AND VOIGT
AVERAGE PREDICTIONS FOR AN AGGREGATE MODEL.
(ALL VALUES IN UNITS OF 10^{-10} m^2 N^{-1}.)
(Adapted from Ref. 9 with errors corrected.)

Polymer	Extensional ($S'_{33} = S'_{11}$)			Torsional (S'_{44})		
	Measured	Reuss	Voigt	Measured	Reuss	Voigt
Low density polyethylene	81	139	26	238	416	80
High density polyethylene	17	10	2·1	26	30	6
Polypropylene	14	7·7	3·8	27	23	11
Polyethylene terephthalate	4·4	10	3·0	11	25	7·6
Nylon 6·6	4·8	6·6	5·2	12	17	13

In this equation S_{11}, S_{33}, S_{13} and S_{44} are the elastic compliance constants for the completely oriented polymer; the quantities $\overline{\sin^4 \theta}$, $\overline{\cos^4 \theta}$ and $\overline{\sin^2 \theta \cos^2 \theta}$ define averages of trigonometrical functions involving the angle θ between the symmetry axis of the unit and the draw direction, which is the symmetry axis of the aggregate.

It is possible in principle to test the aggregate model by comparing the observed and predicted mechanical anisotropy, provided that the units of structure can be identified and their orientation functions determined using structural techniques, for example, X-ray diffraction or nuclear magnetic resonance, which permit measurement of quantities such as $\overline{\sin^4 \theta}$, etc. The first attempts to test the aggregate model by-passed these considerations by noting the analogy with models for predicting the optical anisotropy of polymers (Ref. 7, see also Ref. 4, p. 257 *et seq.*). These optical anisotropy models derive from the work of Kratky[18] and Kuhn and Grün.[19] In the Kuhn and Grün theory for the optical anisotropy of rubbers, it is assumed that the deformation is affine; that is the vector length of each molecular chain segment between cross-links changes in the same ratio as the corresponding dimensions of the bulk material.

Crawford and Kolsky[20] first applied these ideas to thermoplastic polymers in an attempt to account for the general form of the increase in birefringence with strain in low density polyethylene. They observed that although initially the birefringence increased rapidly with strain it asymptotically approached a maximum value at high draw ratios. Moreover, since the birefringence was a univalued function of strain and not stress, it was concluded that its origin lay in the orientation of the crystallites rather than in distortion due to stress. The observed behaviour was therefore modelled by the orientation at constant volume of rod-like transversely isotropic units, the symmetry axes of which rotated on drawing in an identical manner to lines joining pairs of points in the uniaxially drawn specimen as in the Kratky scheme. The units did not change in length as required in the true 'affine' deformation process of Kuhn and Grün; for this reason it has been termed 'pseudo-affine'.[4]

The orientation of a unit is defined by the angle θ between its symmetry axis and the draw direction, and by the angle ϕ between the projection of the symmetry axis on the plane perpendicular to the draw direction and a selected direction in that plane. Uniaxial stretching to a draw ratio λ leaves ϕ unchanged, but changes θ to θ', where $\tan \theta' = \tan \theta / \lambda^{3/2}$. The orientation distribution function for the aggregate of units can then be derived in terms of λ.

The birefringence Δn for any degree of orientation is given by

$\Delta n = \Delta n_{max}(1 - 3/2 \overline{\sin^2 \theta})$ where $\overline{\sin^2 \theta}$ is the average value of $\sin^2 \theta$ for the aggregate of units and Δn_{max} is the birefringence for complete orientation.

A qualitative fit to experimental data was found for polyethylene,[20] and by later workers for nylon,[21] polyethylene terephthalate,[22] and polypropylene,[17] but the predicted birefringence at small draw ratios tended to be low, with the approach to full orientation being more gradual than found experimentally (Fig. 1).

Corresponding predictions for the variation of the mechanical anisotropy with draw ratio are shown in Fig. 2, where the experimental data of Hadley et al.[9] are given for comparison. These results will be discussed more fully in Chapter 9.

Considerable efforts have been devoted to the study of mechanical anisotropy in low density polyethylene, particularly because the aggregate model for compliance averaging with the pseudo-affine deformation scheme was successful in predicting qualitatively the anomalous behaviour of the extensional moduli with draw ratio already indicated in Fig. 2. At first sight it would appear that a complicated explanation in terms of changes in morphology with draw might be required, yet this phenomenological model reproduced the main features of the behaviour very well.[7] It did appear that there was a slower predicted change in the compliances with orientation than observed experimentally, and Gupta and Ward[23] noted that an improved fit could be obtained by replacing the

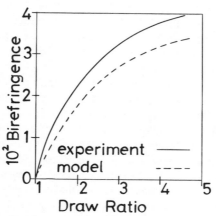

Fig. 1. Low density polyethylene. Variation of birefringence with draw ratio compared with aggregate model predictions. (Adapted from Ref. 4.)

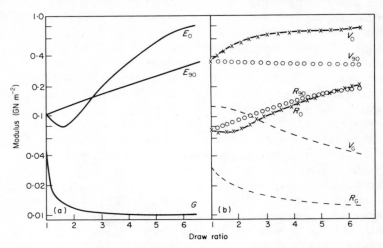

Fig. 2. Low density polyethylene. Variation of extensional (E_o) transverse (E_{90}) and torsional (G) moduli with draw ratio; (a) experimental, (b) aggregate model predictions. V = Voigt average, R = Reuss average. (Adapted from Ref. 9.)

pseudo-affine deformation scheme by an arbitrary distribution function proposed by Raraty,[24] the effect of which was equivalent to the more rapid orientation of material close to the equatorial plane. It was then realised that the orientation of the crystalline regions in low density polyethylene as shown by X-ray scattering was of the requisite form,[25] and when X-ray pole figure data were used to calculate the orientation functions,[26,27] the predictions were much improved. Concurrent measurements of the orientation functions by broad line nuclear magnetic resonance gave similar improvements.[28,29]

During the course of these and related studies, notably those concerned with the temperature dependence of the mechanical anisotropy and the identification of relaxation processes in structural terms,[30-33] it became apparent that the aggregate model was successful in low density polyethylene because it described effectively the influence of the very anisotropic α-relaxation process on the mechanical behaviour. Stachurski and Ward were even able to extend the aggregate model to deal with the anisotropy of dynamic loss factor.[32] (See Chapter 9 for further discussion.) It was, however, more in the spirit of the original conception of the aggregate model[5] that it would deal with mechanical anisotropy in glassy polymers, where morphology was of secondary importance.

Two examples where the model has been applied successfully to glassy polymers are shown by polyethylene terephthalate and polymethylmethacrylate. Allison and Ward[34] obtained a series of polyethylene terephthalate fibres by drawing at a temperature below the glass transition temperature. When cold drawn, a fibre of given initial orientation has a single natural draw ratio, which is a physical characteristic of the polymer system. The natural draw ratio is, however, strongly influenced by the orientation before stretching, and to obtain a range of draw ratios, melt extruded specimens are required over a wide range of birefringence, i.e. pre-oriented fibres.

A highly pre-oriented fibre cannot be regarded as isotropic, and Ward[35] modified the aggregate theory to take into account initial orientation. In such circumstances the *increase* of birefringence due to cold drawing may be related to the natural draw ratio in a manner analogous with that for the total birefringence of an initially isotropic specimen. Agreement between theory and experiment is good (Fig. 3); the characteristic curve is unique, and independent of temperature within the range for which cold drawing is observed. Measured values of the extensional modulus of these fibres always fell between the Voigt and

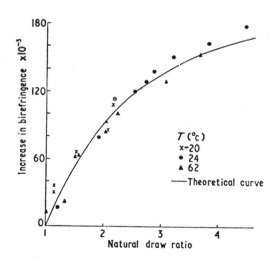

Fig. 3. Polyethylene terephthalate. Variation of increase in birefringence on drawing with draw ratio for filaments cold drawn at different temperatures. (From Ref. 34.)

TABLE 2

MEASURED VALUES OF EXTENSIONAL MODULI OF POLYETHYLENE
TEREPHTHALATE FILAMENTS (IN GN m^{-2}) FOR A RANGE OF NATURAL
DRAW RATIOS (NDR) AND INITIAL BIREFRINGENCE (Δn) COMPARED
WITH REUSS AND VOIGT PREDICTIONS FOR AN AGGREGATE MODEL.
(Adapted from Ref. 34.)

$10^3 \Delta n$	NDR	Measured E_{33}	Reuss E_R	Voigt E_v	$\dfrac{2E_{33}}{E_R + E_v}$
67·8	1·25	4·94	1·81	6·76	1·15
46·0	1·59	5·35	1·93	7·58	1·12
28·0	1·92	5·58	1·98	7·94	1·12
13·1	2·36	6·28	2·14	8·48	1·18
9·52	2·52	6·09	2·22	8·62	1·12
5·43	2·64	6·27	2·24	8·70	1·15
3·52	3·19	7·02	2·68	9·62	1·14
1·37	3·99	7·85	3·31	10·6	1·13

Reuss averaging bounds (Table 2), exceeding the mean predicted value by
a ratio ranging only between 1·12 and 1·18.

The aggregate model also worked well for a series of drawn polymethylmethacrylate samples.[36] The amorphous polymer was regarded as
an aggregate of transversely isotropic units, whose structure was based on
considerations derived from the crystal structure of crystallite poly-
methylmethacrylates. Orientation functions obtained from broad line
nuclear magnetic resonance measurements were shown to give excellent
correlation with birefringence values. The elastic constants of the samples
were then predicted using the NMR orientation functions and the
stiffness constants for the aggregate unit estimated by extrapolating
measured values to full orientation. The predicted elastic constants for
the series of samples were found to lie approximately mid-way between
the bounds for the averaging of compliances and stiffnesses.

8.4 APPLICATION OF THE AGGREGATE MODEL

In a somewhat later study of mechanical anisotropy, Hennig and
Kausch–Blecken von Schmeling have both independently considered the
application and possible development of the aggregate model. Kausch[37]
reviewed the applicability of compliance and stiffness averaging predictions
for several polymers. He noted that the compliance averaging predictions with the pseudo-affine deformation scheme were close to the
experimentally observed behaviour for nylon 66, Dacron and regenerated

cellulose, but that the fit was poor for the polyethylene and nylon results of Hillier and Kolsky.[38] In a later paper along similar lines,[39] (see also Ref. 40) Kausch obtained excellent predicted fits for the variation of extensional modulus with draw ratio in polystyrene, polymethylmethacrylate, polyvinyl chloride, polyacrylonitrile and nylon 66. It was also noted that the increase in the extensional modulus in some other polymers is greater than can be explained on this model. Kausch suggested that this could be attributed to additional orientation of segments within the molecular zones (*i.e.* the aggregate units) as they are being aligned. A more sophisticated proposal would be that this increase in stiffness arises in crystalline polymers from pulling out of tie molecules at high draw ratios (see Chapter 2). Later still[41] Kausch emphasised that it is valuable to reformulate the reorienting element models in terms of a molecular network rather than orienting rods, partly to allow for this modification in domain properties with draw, and partly to permit the representation of high strain properties, *i.e.* yield and fracture.

In a series of related publications, Hennig has reported the measurements of elastic constants for oriented polymers which are either amorphous or of low crystallinity. In his earliest work,[42] Hennig showed that in polyvinyl chloride and polymethylmethacrylate the relationship $3/E_o = S_{33} + 2S_{11}$, where E_o is the modulus of the isotropic polymer, holds to a good approximation. Results for the anisotropy of the linear compressibility γ in polyvinyl chloride, polymethylmethacrylate, polystyrene and polycarbonate were also reported.[43] In this experiment Hennig measured the linear compressibility parallel to the draw direction $\gamma_\parallel$, and that in the plane perpendicular to the draw direction $\gamma_\perp$. For uniaxially oriented polymers $\gamma_\parallel = 2S_{13} + S_{33}$; $\gamma_\perp = S_{11} + S_{12} + S_{13}$. It was found that $\gamma_\parallel + 2\gamma_\perp$ was constant to a very good approximation, and that the differences in the anisotropy of the linear compressibility for different materials were due to different degrees of molecular orientation rather than to intrinsic differences in their ultimate linear compressibility.

On the aggregate model, it can readily be shown that, for the compliance averaging

$$\gamma_\parallel - \gamma_\perp \propto \tfrac{1}{2}\overline{(3\cos^2\theta - 1)}$$

In a further paper Hennig showed that the linear compressibility results for polyvinyl chloride stretched in the temperature range 90–120°C fitted this relationship very well, with values of $\overline{\cos^2\theta}$ calculated from the draw ratio using the pseudo-affine deformation scheme. It was also found that the anisotropy of the thermal conductivity and the thermal expansion coefficient fitted a similar theoretical model for this and other polymers (see also Refs. 45–47 and related work by F. H. Müller and his collaborators reported in Refs. 48 and 49). However, in the other polymers (hot-stretched polystyrene and polymethylmethacrylate, and cold-drawn polycarbonate) the molecular orientation function did not fit that obtained from the pseudo-affine deformation scheme. In the detailed study of polyvinyl chloride,[44] Hennig combined measurements of linear compressibility with measurements of Young's moduli and torsional modulus to obtain a more exacting test of the aggregate model. He showed that the data fitted the compliance averaging predictions, with the pseudo-affine deformation scheme, by making an exact fit to the measured compliance at a draw ratio of 1·5. This result is, of course, exactly that reviewed by Kausch,[39] and it must be pointed out that the fit to the compliance averaging procedure for this polymer is at variance with the conclusions of Ward and co-workers for other polymers where the measured values appeared to lie between the two bounds of the compliance and stiffness averaging procedures. In the latter case, the averaging calculations were performed using the elastic constants for high draw ratio, which accounts for some of the discrepancy between the two sets of results.

8.5 ALTERNATIVE THEORIES OF MECHANICAL ANISOTROPY

Linear viscoelastic behaviour can be represented in a simplified form by spring and dashpot models, which offer a pictorial representation of the behaviour with the loss of some generality. In a somewhat analogous fashion, Hsiao[50,51] has considered the behaviour of oriented polymers in terms of oriented linear spring elements, and attempted to deal with fracture and time dependence as well as low strain elasticity. The polymer was regarded as homogeneous on a macroscopic scale, but within a small sub-volume the presence of molecular chains and their orientation was taken into account. Load carrying chain molecules were represented schematically by linear elastic chains, and, assuming continuity of strain between the sub-volumes, the mechanical anisotropy could be expressed as a function of the extensional modulus of the springs, deduced from measurements on the unoriented polymer. It was collaborative studies of this type of model by Kausch and Hsiao,[52] which led Kausch to his proposals for oriented networks discussed above, but in the present case the relationship between the model and molecular structure is less direct. Although these theories are of interest, they predict inter-dependent elastic constants, implying similar relative variations in moduli for all polymers, a prediction refuted by experimental evidence.

More recently[53] this model has been modified so that it resembles a system of oriented elements, which are subjected to elongation as well as rotation from their initial position. The orientation distribution function is obtained by optimising the total energy of the system.

8.6 THE SONIC VELOCITY

It has been suggested[54,55] that the extensional modulus measured by wave propagation methods might be used to obtain a direct measure of molecular orientation, analogous to the derivation of the optical orientation function from birefringence measurements. Ward[56,4] has shown that by making approximations appropriate for low degrees of orientation the extensional compliance obtained from the compliance averaging procedure is related directly to birefringence through $\overline{\sin^2\theta}$ giving

$$S_{33} = \tfrac{2}{3}S_{11}(\Delta n_{max} - \Delta n)$$

This prediction of a linear relation between S_{33} and birefringence, extrapolating to zero compliance at maximum birefringence was tested

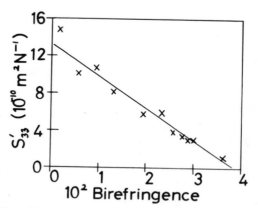

Fig. 4. Polypropylene. Linear relation between extensional compliance (S'_{33}) and birefringence for drawn filaments. (Adapted from Ref. 4.)

using the data (not obtained by wave propagation) of Hadley *et al.*[9] As Fig. 4 shows, the relation was linear to relatively high degrees of orientation, but the values of S_{11} deduced did not agree with directly measured experimental values, throwing doubt on the validity of the method.

8.7 LIMITATIONS OF THE AGGREGATE MODEL

The aggregate model was proposed in the knowledge that it was likely to provide only a first approximation to actual behaviour, because the assumptions of perfectly elastic behaviour and no changes in morphology during the orientation process are known to be untrue for crystalline polymers. The model works quite well for 'low density polyethylene primarily because it expresses the variation in the contribution of the c-shear relaxation (see Chapter 9) to the deformation to a reasonable degree of approximation. Even here, however, it has been proposed that there is an additional contribution to the deformation at low draw ratios,[57] due to reversible mechanical twinning, *i.e.* the boundaries of the aggregate units are not fixed under stress, which violates the principle of the Reuss–Voigt–Hill type averaging procedures. Similar effects may also occur in nylon 6 where recent work[58] has shown the aggregate model to be inapplicable, at least to the room temperature behaviour. In polypropylene,[17] the failure of the aggregate model was attributed to the changes in morphology and molecular mobility which occur at high draw ratios.

The aggregate model appears most successful for glassy polymers, or perhaps crystalline polymers in the glassy state, provided that changes in morphology may be neglected in the first instance. Read and Dean[59] have suggested that the aggregate model should be extended to consider contributions to the modulus anisotropy from both crystalline and amorphous regions, coupled together in a manner dependent on the morphological structure. For highly drawn and annealed polyethylene specimens it is postulated, following Hosemann[60] and Fischer,[61] that a series coupling would be appropriate, *i.e.* continuity of stress and the compliance averaging procedure. Read and Stein suggested that the amorphous contribution to the modulus might be obtained by optical and spectroscopic methods[62,63] (see also Chapters 3, 4, 5 and 6) and showed how the orientation functions of the crystalline component could be obtained from X-ray pole figure data. They considered that the symmetry of the crystal must be taken into account even for overall macroscopic transverse isotropy, which implies that for polyethylene, which is orthorhombic, nine, rather than five, independent elastic constants are required.

The considerations of Read and Dean are, however, moving away from the spirit of the aggregate model, which is essentially a single phase model, and it is appropriate to consider them as a link with composite structure models, which will now be discussed in some detail.

8.8 COMPOSITE STRUCTURE MODELS: THE TAKAYANAGI MODEL

The Takayanagi model was developed to account for the viscoelastic relaxation behaviour of two phase polymers, as recorded by dynamic mechanical testing.[64] It was then extended to treat both isotropic[65,66] and oriented[67] semi-crystalline polymers. The model does not deal with the development of mechanical anisotropy on drawing, but attempts to account for the viscoelastic behaviour of either an isotropic or a highly oriented polymer in terms of the response of components representing the crystalline and amorphous phases. Hopefully, comparisons between the predictions of the model and experimental results may throw light on the molecular processes occurring.

In its original form the model sought to derive the temperature dependence of the relaxation behaviour of a composite amorphous polymer having two distinct phases in terms of the properties of the individual components. The resultant response would depend on whether the components were in parallel or series (Fig. 5). For the parallel model the complex modulus is given by:

$$E^* = \lambda E_A^* + (1 - \lambda) E_B^*$$

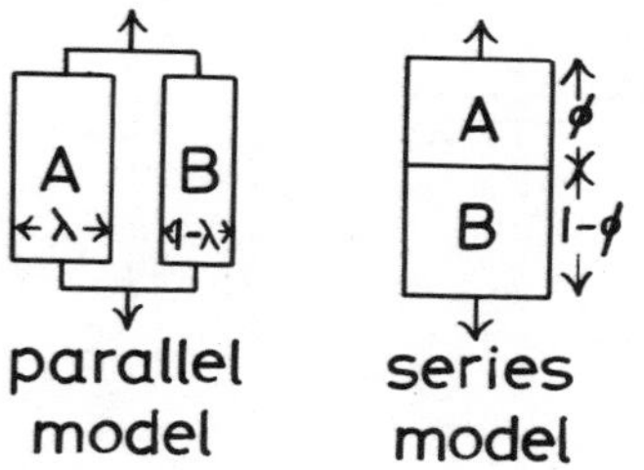

parallel model   series model

Fig. 5. Parallel and series Takayanagi models. (Adapted from Ref. 64.)

where subscripts refer to the two components, and λ is the volume fraction of A. This situation gives a Voigt average modulus. By contrast the series model yields a Reuss average modulus:

$$\frac{1}{E^*} = \frac{\phi}{E_A^*} + \frac{(1-\phi)}{E_B^*}$$

where ϕ now represents the volume fraction of A.

Trials with composite films of polyvinyl chloride intimately bonded to nitrile butadiene rubber, but with distinctly separate phases, confirmed the predictions over a temperature range covering a major relaxation for each component. If, however, the components were dispersed so that they were separated by a layer containing a mixture of each polymer the experimentally determined relaxations were broader than predicted.

When the two amorphous phases are dispersed two alternative models arise (Fig. 6). The complex modulus of model (i) could be expressed as

$$\frac{1}{E^*} = \frac{\phi}{\lambda E_A^* + (1-\lambda)E_B^*} + \frac{1-\phi}{E_B^*}$$

where ϕ is the volume fraction of A.

For model (ii)

$$E^* = \lambda \left(\frac{\phi}{E_A^*} + \frac{1-\phi}{E_B^*} \right)^{-1} + (1-\lambda)E_B^*$$

With a fixed volume ration of A the relative sizes of λ and ϕ are adjustable to give the best experimental fit. Figure 7 indicates that the

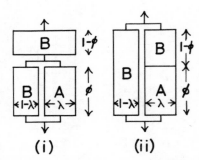

(i)   (ii)

Fig. 6. Takayanagi models for A-phase dispersed in B-phase. (Adapted from Ref. 64.)

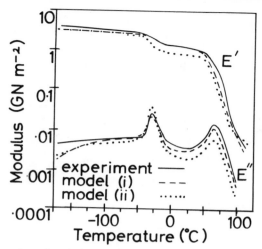

Fig. 7. Temperature dependence of storage and loss moduli for PVC–NBR film bonded in parallel to a PVC film. Takayanagi model type I gives better fit to experiment. (Adapted from Ref. 64.)

performance of a composite of polyvinyl chloride and nitrile butadiene rubber having the form shown, can be better fitted by model (i).

The performance of polymer blends could be well represented by a Takayanagi model in which the relative values of λ and ϕ were related to the shape of the dispersed phase, *e.g.* $\lambda = \phi$ for homogeneous dispersions, and $\lambda \gg \phi$ for dispersions in the form of elongated molecular aggregates. But for semi-crystalline polymers, with A and B representing amorphous and crystalline components, the dispersions were generally broader than predicted, suggesting that the unordered material was not identical with that in a completely amorphous state.

Takayanagi's first comparison[65] between the predictions of his model and the observed mechanical behaviour covered a wide range of crystalline polymers, including polyethylene, polyvinyl alcohol, polytetrafluoroethylene, polyamide, polyethylene oxide, polyoxymethylene and polypropylene. Attempts were made to define relaxation processes as associated with either the crystalline regions or the non-crystalline regions, and in the former case specific molecular mechanisms were proposed, *e.g.* a local twisting mode of molecular chains around their axes and a translational mode of molecular chains along their axes.

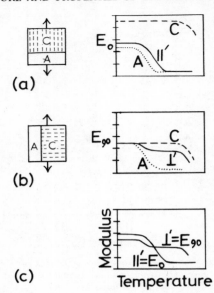

Fig. 8. High density polyethylene. Schematic representation of variation of extensional (E_o) and transverse (E_{90}) moduli of oriented specimens with temperature, in terms of a Takayanagi model. (Adapted from Ref. 67.)

A particularly interesting application of the theory was to oriented high density polyethylene which had been annealed after drawing.[67] It was observed that although the real part of the extensional modulus was greater than the real part of the transverse modulus at low temperatures, the reverse situation held at temperatures above the region of the α relaxation for the unordered material ($\sim 30°C$). A simplified schematic representation of the behaviour is shown in Fig. 8. Application of stress when A and B are in series (a) would give a large fall of modulus at temperatures above the relaxation transition for A. In the perpendicular direction (b) (A and B in parallel) the fall in modulus is less pronounced, because the crystalline region still supports the applied stress. The over-all picture thus predicts the crossover behaviour shown at (c). Such behaviour was not observed with as-drawn material, because annealing was necessary to remove the stress in the amorphous regions.

More detailed study of the relaxations produced a model in which there was a small parallel component of crystalline material, additional to the series component indicated in Fig. 8. This representation was then interpreted on a molecular basis in terms of the ultrafibrils suggested by Hosemann.[60]

8.9 THE LOGARITHMIC MIXING RULE

A critique of the Takayanagi model, as applied to semi-crystalline polymers with overall isotropy, has been given by Gray and McCrum.[68] Evidence is produced to refute Takayanagi's assertion that if mechanical relaxation occurs in the amorphous phase the peak value of the out of phase modulus is proportional to the volume fraction of the amorphous phase. It is suggested that the stress and strain fields must differ between the crystalline and amorphous components, and that the Takayanagi model yields only a Voigt average solution[69] representing upper bounds to moduli. A Reuss type average is equally inadmissible, and the correct picture must lie between these limits. An acceptable law of mixing must be found, and a purely empirical logarithmic mixing hypothesis is advanced:

$$\log G^* = v_1 \log G_1^* + v_2 \log G_2^*$$

where G^* represents the complex shear modulus and v_1 and v_2 denote the volume fractions of components 1 and 2.

It can then readily be shown that the logarithmic decrement of the polymer $\Lambda = \pi G''/G = \pi \tan \delta$ is given in terms of the logarithmic decrements of the two components Λ_1 and Λ_2 by the relationship

$$\Lambda = v_1 \Lambda_1 + v_2 \Lambda_2$$

This assertion that the logarithmic decrement for the composite is a linear combination of the weighted logarithmic decrement of the two phases is a mathematical statement of the assumptions of some earlier workers (*e.g.* Ref. 70).

The mixing law has, however, only empirical justification, and is possibly a working approximation valid for many practical combinations of crystalline and amorphous moduli. The initial Takayanagi model gave an adequate description for the composites with separate phases reported earlier,[64] and the energy dissipation in a sandwich containing a thin polymeric layer between metal plates ('damping tape') is greater than would be anticipated on the grounds of proportion of polymer alone, presumably because most of the shear occurs across the polymer, due to its lower modulus, and because of dissipative processes associated with the boundary regions.[71]

8.10 MORPHOLOGICAL DEVELOPMENTS OF THE COMPOSITE MODEL

In parallel research to that of Takayanagi, McCrum and Morris[72] studied the α and α' relaxations in high density polyethylene. They proposed that the α' relaxation should be attributed to slip at the boundaries of the lamellae, and put forward a model similar to that of Iwayanagi[73] in which elastic lamellae are separated by a viscous liquid. To ensure recoverability, the lamellae are pinned at points along their length. The composite solid then shows linear viscoelastic behaviour, with a characteristic relaxation which depends on structural parameters.

More detailed information on the relaxation processes in both low density and high density polyethylene has come from a series of investigations by Ward and co-workers, which provide the basis for more sophisticated composite models in which morphological considerations play a major role. In addition to drawn and annealed sheets of low density and high density polyethylene, drawn rolled and annealed sheets of low density polyethylene were prepared with well defined textures (following the structural studies of Hay and Keller[74] and Point[75]). Three types of special structure sheet can be produced. In the first type, the c-axes of the crystallites lie along the initial draw direction, the b-axes lie in the plane of the sheet and the a-axes normal to the plane of the sheet. This has been termed b–c sheet. In the second type the a-axes lie along the draw direction, b-axes again in the plane of the sheet and c-axes normal to the plane of the sheet. This has been termed a–b sheet. In both these sheets, structural studies[76,77] (principally low-angle X-ray diffraction) show that there is a lamellar texture with the lamellar planes inclined at about $45°$ to the c-axis direction (or the a-axis direction). A similar texture, but with transverse isotropy, is observed in the cold drawn and annealed specimens. The third type of special structure sheet is the parallel lamellae sheet, where the lamellar plane normals lie along the initial draw direction, and the c-axes now lie out of the plane of the sheet making an angle of about $45°$ with the initial draw direction.

Gupta and Ward[78] found modulus crossover points in drawn and annealed sheets of low density polyethylene and also in the b–c and a–b sheets similar to that observed by Takayanagi in drawn and annealed high density polyethylene. The crossover points were attributed to interlamellar shear. When the tensile stress is applied along the draw direction (E_o measurements in a drawn and annealed sheet) or along either the c or a direction in the special structure sheets (E_c and E_a measurements respectively) there is a maximum shear stress on the lamellar planes. A tensile stress in the b direction (E_b measurement), on the other hand, will produce a small shear stress on the lamellar planes because these are approximately parallel to the b-axis. The argument is that above the β-relaxation in low density polyethylene inter-lamellar shear can occur, which gives $E_b > E_a \sim E_c$ as observed. The transversely isotropic drawn and annealed sheet shows $E_o < E_{90}$ which is also consistent with this hypothesis.

The cross-over point in high density polyethylene occurs above the α-relaxation, which it is proposed is an inter-lamellar shear relaxation. This highlights a major difference between high density polyethylene and low density polyethylene, where the α-relaxation is attributed to the c-shear process which gives rise to the anomalous mechanical anisotropy.

A comparative study of the viscoelastic behaviour of these sheets by Stachurski and Ward[31,33] which showed that the anisotropy of the loss factor was consistent with the above conclusions, led to an attempt to quantify this anisotropy in terms of a mathematical model for the inter-lamellar shear relaxation process.[79] The model represents the semi-crystalline polyethylene as rigid lamellae set in a deformable matrix. It is assumed: (a) there is homogeneous stress; the lateral dimensions of the lamellae are large compared with their thickness; (c) the relaxation occurs in the intercrystalline material only; (d) the relaxation is activated by inter-lamellar shear. The calculation of the loss anisotropy can be performed either by considering the contribution of the inter-lamellar shear mechanism to the total compliance and calculating the change which occurs when the inter-lamellar material relaxes or by calculating the energy losses. The result gives the loss anisotropy as proportional to $\sin^2 \gamma \cos^2 \gamma$ where γ is the angle between the applied stress direction and a lamellar plane normal. The measured loss anisotropy in low density and high density polyethylene was shown to be consistent with this model, when the required distribution of orientations of the lamellar plane normals was estimated from low angle X-ray diffraction data.

The success of the model for the loss anisotropy led Owen and Ward[80,81] to use equivalent assumptions to calculate the modulus anisotropy. The loss anisotropy calculations assume *simple* shear between the lamellae only, which for parallel lamellae sheet would imply that inter-lamellar shear is not activated when the tensile stress is applied along the initial draw direction (*i.e.* parallel to the lamellar plane normals). A very appreciable fall in tensile modulus was, however, observed in this case, although as expected by comparison with the corresponding loss factor in the previous experiments[33] the fall in tensile modulus for a strip cut at 45° to this direction was greater. To account for this behaviour, Owen and Ward assumed that the lamellae were effectively of infinite extent only in the direction of the crystallographic *b*-axis, so that under a normal stress component the interlamellar material would undergo *pure* shear. It was shown that the observed anisotropy of the parallel lamellae and *a–b* sheets was in reasonable agreement with this model, in which the inter-lamellar regions of the polymer deform under load by both simple shear and pure shear (see Table 3).

The prediction of the moduli of *isotropic* crystalline polymers has been considered by Halpin and Kardos[82] using theories previously developed for fibrous composites.[83] They point out that the specification of the two components in terms of their elastic properties and volume concentration is not sufficient to define the resultant properties of the composite solid, because geometrical factors will be crucial in determining where the moduli lie between the Voigt and Reuss bounds. Evidence is presented from various sources to show that the amount by which crystallinity increases the shear modulus over its value for the amorphous polymer is greatest where the crystalline phase has large length to width ratios. The analogy is then made with a fibre reinforced composite, based on equations for a matrix reinforced with discontinuous fibres.[84]

The extensional modulus E_{33} of the uniaxially oriented system is given by

$$E_{33} = E_m(1 + \xi \eta v_f)/(1 - \eta v_f)$$

where

$$\eta = \left(\frac{E_f}{E_m} - 1 \right) \left(\frac{E_f}{E_m} + \xi \right)^{-1}, \xi = l/d$$

TABLE 3

PREDICTED AND EXPERIMENTAL MODULUS ANISOTROPY

| | Parallel lamellae | | | *a–b* sheet | |
	Predicted	Experimental		Predicted[a]	Experimental
E_o/E_{45}	4/3·2	1·50/1·33	E_a/E_{45}	4/5·33	1·24/1·71

[a] These results differ from those quoted in Ref. 80, where there is a small error due to neglect of one strain contribution (see Ref. 81).

v_f is the volume fraction of fibres of length l and diameter d, and the subscripts f and m refer to fibres and matrix. The transverse modulus E_{11} is given by this equation with $\xi = 2l/d$, and the inplane shear modulus G_{23} is equal to $G_m(1 + \xi\eta v_f)/(1 - \eta v_f)$.

This theory was tested by making quasi-isotropic nylon–rubber laminates (*i.e.* laminates where there is isotropy in one plane, all the fibres lying in this plane) and showing that the elastic properties of these laminates were close to average values predicted by Tsai and Pagano,[85] varying in the expected manner with the fibre aspect ratio l/d. These calculations were then applied to isotropic butadiene acrylonitrile copolymers[86] in which acetanilide had been polymerised *in situ* to provide crystalline filler particles. Kardos and his colleagues studied variations of dynamic moduli with crystalline morphology, as observed by electron microscopy. It was shown that the results for low filler concentrations were consistent with measured aspect ratios for the crystalline component, and the theory described above.

This dependence on the aspect ratio of the reinforcing component is at variance with what was effectively a complementary experiment reported by Brody and Ward,[87] in which the Ward aggregate theory was applied to short carbon and glass fibre-reinforced composites. The length–diameter ratio is not a parameter of this model, and yet for all fibre lengths the moduli lay close to the lower Reuss bounds. This condition meant that where fibres lay at appreciable angles to the stressed direction they contributed only as a filler, with the high modulus of the fibre not being utilised. The size of fibres in these composites was much larger than the microscopic fillers investigated by Halpin and Kardos, with the lengths being always adequate for stress transfer, and this fact may account for the apparently contradictory results.

It is clear that the more sophisticated composite model theories emphasise that this is a powerful method of examining the structure/ property relationships in crystalline polymers and that the success or failure of particular models can be fruitfully used as a guide to the underlying morphology.

8.11 CONCLUSIONS: MACROSCOPIC MODELS AND MOLECULAR STRUCTURE

Other chapters in this book are devoted to the texture of oriented polymers, and to a detailed discussion of experimental studies of low strain anisotropy (Chapters 3 and 9 respectively).

Morphological investigations have tended quite naturally to concentrate on polyethylene. Of the many published studies, three in particular may be singled out as of particular relevance. First, there is the work of Hay and Keller[88] on cold drawing polyethylene, which emphasises the non-affine deformation of the spherulitic structure. Even if the aggregate model has some relevance to crystalline polymers, the orientation functions cannot be predicted by simple theoretical schemes. Secondly, Peterlin[89,90] has shown that in drawing high density polyethylene whole blocks of folded chains are broken out of the original spherulites; the deformation proceeds by phase transformation, twinning, chain tilting and slip, with the new fibre structure containing many tie molecules. The aggregate model cannot deal with such a complex transformation. However, annealing causes relaxation of these tie molecules, so that a simpler composite structure is produced. Here, the Takayanagi model has relevance, although it must be borne in mind that more sophisticated composite models which incorporate both the orientation and shape of the lamellae are desirable objectives. Finally, Hosemann and co-workers[91,92] have proposed that oriented high density polyethylene retains a memory of its shape before stretching, and that drawing produces an affine deformation of a superstructure of colloidal dimensions. Each crystalline domain is postulated as transforming or stretching into about twenty smaller domains, which lie together like a string of pearls and contribute to the ultrafibrils of the stretched material. The quantification of such ideas would form the link with the simpler reorienting aggregate model.

However, at present it has not proved possible to construct a quantitative model for such complex structures, and generalised models which do not take account of fine structure at the molecular level are needed as working approximations. Of such models the Ward aggregate model and the Takayanagi two phase model have been most widely used.

The aggregate model takes no account of the way in which crystalline and amorphous material are interrelated, but assumes that this relation does not change substantially as a result of stretching. Where the theory gives a reasonable fit to experimental data it is strongly suggestive that orientation does occur principally in the simple manner proposed. If the theory is inappropriate then light may be shed on the alternative mechanisms that may occur.

Investigation of cold drawn specimens of low density polyethylene by low angle X-ray scattering indicates that there is little clearly defined lamellar structure in this material, in contrast with the strongly lamellar structure deduced from the distinct four point pattern observed in samples which were annealed after drawing. For cold drawn polymers the assumption of a single phase structure implicit in the aggregate theory may therefore be valid. Some of the materials to which the aggregate theory has been applied were hot stretched (*i.e.* drawn above the glass transition temperature), but the annealing in these cases would be far from complete, since the temperature was maintained for a relatively short time, and the filaments were always under tension. Thus again a distinct two phase structure may not have been well developed.

Despite its two-dimensional nature, the Takayanagi model is successful in explaining the main features of low strain mechanical anisotropy in oriented annealed crystalline polymers. Perhaps even more important it points the way to more sophisticated composite structure models which are at present being developed for crystalline polymers.

It is in terms of these two complementary models, one assuming a single phase, and the other two essentially independent phases, that much of the experimental evidence presented in Chapter 9 will be discussed.

REFERENCES

1. Kennedy, A. J. (1962). *Processes of Creep and Fatigue in Metals,* Oliver and Boyd, Edinburgh, p. 7.
2. Reiner, M. (1971). *Advanced Rheology,* Lewis, London.
3. Sobotka, Z. (1971). *J. Macromol. Sci., Phys.,* **B5,** 393.
4. Ward, I. M. (1971). *Mechanical Properties of Solid Polymers,* Wiley, London.
5. Ward, I. M. (1961). *Text. Res. J.,* **31,** 650.
6. Pinnock, P. R. and Ward, I. M. (1963). *Proc. Phys. Soc.,* **81,** 260.
7. Ward, I. M. (1962). *Proc. Phys. Soc.,* **80,** 1176.
8. Ward, I. M. (1966). *Appl. Mater. Res.,* **5,** 224 and 228.
9. Hadley, D. W., Pinnock, P. R. and Ward, I. M. (1969). *J. Mater. Sci.,* **4,** 152.
10. Voigt, W. (1928). *Lehrbuch der Kristallphysik Teubner,* Leipzig, p. 410.
11. Reuss, A. (1929). *Z. Angew. Math. Mech.,* **9,** 49.
12. Hearmon, R. F. S. (1956). *Adv. Phys.,* **5,** 323.
13. Bishop, J. and Hill, R. (1951). *Phil. Mag.,* **42,** 414 and 1298.
14. Hill, R. (1963). *J. Mech. Phys. Solids,* **11,** 357.
15. Hashin, Z. and Shtrikman, S. (1962). *J. Mech. Phys. Solids,* **10,** 343.
16. Peselnick, L. and Meister, R. (1965). *J. Appl. Phys.,* **36,** 2870.
17. Pinnock, P. R. and Ward, I. M. (1966). *Brit. J. Appl. Phys.,* **17,** 575.
18. Kratky, O. (1933). *Kolloid-Z.,* **64,** 213.
19. Kuhn, W. and Grün, F. (1942). *Kolloid-Z.,* **101,** 248.
20. Crawford, S. M. and Kolsky, H. (1951). *Proc. Phys. Soc.,* **B64,** 119.
21. Cannon, C. G. and Chappel, F. P. (1959). *Brit. J. Appl. Phys.,* **10,** 68.
22. Pinnock, P. R. and Ward, I. M. (1964). *Brit. J. Appl. Phys.,* **15,** 1559.
23. Gupta, V. B. and Ward, I. M. (1967). *J. Macromol. Sci. Phys.,* **B1,** 373.

24. Raraty, L. E. (1966). *Appl. Mater. Res.,* **5,** 104.
25. Gupta, V. B., Keller, A. and Ward, I. M. (1968). *J. Macromol. Sci. Phys.,* **B2,** 139.
26. Gupta, V. B. and Ward, I. M. (1970). *J. Macromol. Sci. Phys.,* **B4,** 453.
27. Gupta, V. B. and Ward, I. M. (1971). *J. Macromol. Sci. Phys.,* **B5,** 629.
28. McBrierty, V. J. and Ward, I. M. (1968). *Brit. J. Appl. Phys. (J. Phys. D.),* **1,** 1529.
29. McBrierty, V. J., McDonald, I. R. and Ward, I. M. (1971). *Brit. J. Appl. Phys. (J. Phys. D.),* **4,** 88.
30. Stachurski, Z. H. and Ward, I. M. (1968). *J. Polymer Sci.,* **A2(6),** 1083.
31. Stachurski, Z. H. and Ward, I. M. (1968). *J. Polymer Sci.,* **A2(6),** 1817.
32. Stachurski, Z. H. and Ward, I. M. (1969). *J. Macromol. Sci. Phys.,* **B3,** 427.
33. Stachurski, Z. H. and Ward, I. M. (1969). *J. Macromol. Sci. Phys.,* **B3,** 445.
34. Allison, S. W. and Ward, I. M. (1967). *Brit. J. Appl. Phys.,* **18,** 1151.
35. Ward, I. M. (1967). *Brit. J. Appl. Phys.,* **18,** 1165.
36. Kashiwagi, M., Folkes, M. J. and Ward, I. M. (1971). *Polymer,* **12,** 697.
37. Kausch–Blecken von Schmeling, H. H. (1967). *J. Appl. Phys.,* **38,** 4213.
38. Hillier, K. W. and Kolsky, H. (1949). *Proc. Phys. Soc.,* **B62,** 111.
39. Kausch–Blecken von Schmeling, H. H. (1970). *Kolloid-Z.,* **237,** 251.
40. Kausch–Blecken von Schmeling, H. H. (1969). *Kolloid-Z.,* **234,** 1148.
41. Kausch–Blecken von Schmeling, H. H. (1971). *J. Macromol. Sci. Phys.,* **B5,** 269.
42. Hennig, J. (1964). *Kolloid-Z.,* **200,** 46.
43. Hennig, J. (1965). *Kolloid-Z.,* **202,** 127.
44. Hennig, J. (1967). *J. Polymer Sci.,* **C16,** 2751.
45. Hellwege, K.-H., Hennig, J. and Knappe, W. (1963). *Kolloid-Z.,* **188,** 121.
46. Hennig, J. (1964). *Kolloid-Z.,* **196,** 136.
47. Hennig, J. (1967). *Kunststoffe,* **57,** 385.
48. Müller, F. H. (1967). *J. Polymer Sci.,* **C20,** 61.
49. Hellmuth, W., Kilian, H. G. and Müller, F. H. (1967). *Kolloid-Z.,* **218,** 10.
50. Hsiao, C. C. (1959). *J. Appl. Phys.,* **30,** 1492.
51. Kao, S. R. Hsiao, C. C. (1964). *J. Appl. Phys.,* **35,** 3127.
52. Kausch–Blecken von Schmeling, H. H. and Hsiao, C. C. (1968). *J. Appl. Phys.,* **39,** 4915.
53. Hsiao, C. C. and Moghe, S. R. (1971). *J. Macromol. Sci. Phys.,* **B5,** 263.
54. Charch, W. H. and Moseley, W. W. (1959). *Text. Res. J.,* **29,** 525.
55. Morgan, H. M. (1962). *Text. Res. J.,* **32,** 866.
56. Ward, I. M. (1964). *Text. Res. J.,* **34,** 806.
57. Frank, F. C., Gupta, V. B. and Ward, I. M. (1970). *Phil. Mag.,* **21,** 1127.
58. Owen, A. J. and Ward, I. M. (1973). *J. Macromol. Sci. Phys.,* **B7,** 279.
59. Read, B. E. and Dean, G. (1970). *Polymer,* **11,** 597.
60. Hosemann, R. (1963). *J. Appl. Phys.,* **34,** 25.
61. Fischer, E. W. (1969). *Kolloid-Z.,* **231,** 458.
62. Read, B. E. and Stein, R. S. (1968). *Macromolecules,* **1,** 116.
63. Stein, R. S. and Read, B. E. (1969). *J. Appl. Polym. Sci., Appl. Polym. Symp.,* **8,** 255.
64. Takayanagi, M., Harima, H. and Iwata, Y. (1963). *Mems. Fac. Eng., Kyushu Univ.,* **23,** 1.
65. Takayanagi, M. (1963). *Mems. Fac. Eng., Kyushu Univ.,* **23,** 41.
66. Uemura, S. and Takayanagi, M. (1966). *J. Appl. Polym. Sci.,* **10,** 113.
67. Takayanagi, M., Imada, K. and Kajiyama, T. (1966). *J. Polymer Sci.,* **C15,** 263.
68. Gray, R. W. and McCrum, N. G. (1969). *J. Polymer Sci., A-2,* **7,** 1329.
69. Bondi, A. (1967). *J. Polymer Sci., A-2,* **5,** 83.
70. Willbourn, A. H. (1958). *Trans. Farad. Soc.,* **54,** 717.
71. Hadley, D. W. and Pretlove, A. J. (Unpublished work).
72. McCrum, N. G. and Morris, E. L. (1966). *Proc. Roy. Soc.,* **A292,** 506.
73. Iwayanagi, S. (1962). *Rep. Prog. Polym. Phys. Japan,* **5,** 135.
74. Hay, I. L. and Keller, A. (1966). *J. Mater. Sci.,* **1,** 41.

75. Point, J. J. (1953). *J. Chim. Phys.,* **50,** 76.
76. Seto, T. and Hara, T. (1963). *Rep. Prog. Polym. Phys., Japan,* **7,** 63.
77. Hay, I. L. and Keller, A. (1967). *J. Mater. Sci.,* **2,** 538.
78. Gupta, V. B. and Ward, I. M. (1968). *J. Macromol. Sci. (Phys.),* **B2,** 89.
79. Davies, G. R., Owen, A. J., Ward, I. M. and Gupta, V. B. (1972). *J. Macromol. Sci. (Phys.),* **B6,** 215.
80. Owen, A. J. and Ward, I. M. (1971). *J. Mater. Sci.,* **6,** 485.
81. Owen, A. J. (1972). *Ph.D. Thesis,* Bristol University.
82. Halpin, J. C. and Kardos, J. L. (1972). *J. Appl. Phys.,* **43,** 2235.
83. Ashton, J. E., Halpin, J. C. and Petit, P. H. (1969). *Primer on Composite Materials,* Technomic, Stamford, Conn., U.S.A.
84. Halpin, J. C. and Tsai, S. W. (1967). Air Force Materials Lab., Report TR67-423.
85. Tsai, S. W. and Pagano, N. J. (1968). *Composite Materials Workshop* (Ed. S. W. Tsai, J. C. Halpin and N. J. Pagano), Technomic, Stamford, Conn., U.S.A., p. 233.
86. Kardos, J. L., McDonnell, W. L. and Raisoni, J. (1972). *J. Macromol. Sci. (Phys.),* **B6,** 397.
87. Brody, H. and Ward, I. M. (1971). *Polym. Engng. Sci.,* **11,** 139.
88. Hay, I. L. and Keller, A. (1965). *Kolloid-Z.,* **204,** 43.
89. Peterlin, A. (1966). *J. Polymer Sci.,* **C15,** 427.
90. Peterlin, A. (1971). *J. Mater. Sci.,* **6,** 440.
91. Hosemann, R., Loboda-Cackovic, J. and Cackovic, H. (1972). *Z. Naturforsch.,* **27a,** 478.
92. Hosemann, R., Loboda-Cackovic, J. and Cackovic, H. (1972). *J. Mater. Sci.,* **7,** 963.

CHAPTER 9

SMALL STRAIN ELASTIC PROPERTIES

D. W. HADLEY

9.1 INTRODUCTION

Young's modulus, and to some extent torsional rigidity, have been characteristic properties of polymeric materials quoted in both scientific and advertising literature for several decades, but, with few exceptions, only within the past 15 years has much attention been given to the less directly accessible elastic constants.

Interpretation of mechanical measurements in terms of molecular structure was until fairly recently confined essentially to identification of the temperatures of the major viscoelastic relaxations through extensional or torsional dynamic mechanical studies.[1] Now, however, investigations of the elastic constants and their temperature dependence—allied with dynamic mechanical, creep and both wide and small angle X-ray diffraction—are yielding fairly detailed pictures of the interrelation of the crystalline and less well ordered regions of some oriented solid polymers.

The major emphasis has been on the more highly crystalline materials—particularly polyethylene, because of its simple chemical composition and high crystallinity—but investigations of some amorphous polymers, which show considerably reduced molecular orientation, have been reported. In these latter cases additional structural information from X-ray scattering is not available.

The emphasis in this chapter will be on elastic properties at strains small enough for departures from Hookean behaviour to be neglected. Dynamic mechanical data will be quoted to supplement elastic measurements, but the principles of this type of experiment will not be discussed, since a very detailed review of this field (not confined to anisotropic specimens) has been given by McCrum et al.[1]

9.2 ELASTIC CONSTANTS AND THEIR DETERMINATION

Oriented polymers are most readily obtainable as uniaxially drawn sheets or as fibres, with the molecular chain (c) axis aligned more or less closely along the stretch direction, to an extent dependent on the draw ratio (see Chapter 2). It is now possible to produce biaxially oriented sheets, typical production techniques being uniaxial stretching, followed by rolling and a carefully controlled anneal,[2,3] or stretching lengthwise at constant width.[4,5] Material formed by these latter processes has overall orthorhombic symmetry as indicated by wide angle X-rays (this symmetry must not be confused with the symmetry of the unit cell of the normal polymer, which may be of a much lower class), and needs 9 independent elastic constants to describe its low strain properties.

The compliance matrix is

$$\begin{bmatrix} S_{11} & S_{12} & S_{13} & 0 & 0 & 0 \\ S_{12} & S_{22} & S_{23} & 0 & 0 & 0 \\ S_{13} & S_{23} & S_{33} & 0 & 0 & 0 \\ 0 & 0 & 0 & S_{44} & 0 & 0 \\ 0 & 0 & 0 & 0 & S_{55} & 0 \\ 0 & 0 & 0 & 0 & 0 & S_{66} \end{bmatrix}$$

where S_{11}, S_{22} and S_{33} represent extensional compliances; S_{12}, S_{13} and S_{23} occur in expressions for the contraction ratios; and S_{44}, S_{55} and S_{66} represent torsional deformations.

For fibre type orientation there is transverse symmetry, and the compliance matrix becomes

$$\begin{bmatrix} S_{11} & S_{12} & S_{13} & 0 & 0 & 0 \\ S_{12} & S_{11} & S_{13} & 0 & 0 & 0 \\ S_{13} & S_{13} & S_{33} & 0 & 0 & 0 \\ 0 & 0 & 0 & S_{44} & 0 & 0 \\ 0 & 0 & 0 & 0 & S_{44} & 0 \\ 0 & 0 & 0 & 0 & 0 & 2(S_{11}-S_{12}) \end{bmatrix}$$

Because polymers are essentially viscoelastic it is usual to measure deformations at a fixed time after application of load; to obtain, *e.g.* the

10 second compliance. This treatment follows a suggestion of Biot[6] of direct equivalence between elastic and linear viscoelastic behaviour. Recently the time dependence of the deformation has been considered by Ladizesky and Ward.[7-9] (Creep studies are discussed fully in Chapter 10.)

Stresses should be small enough for linear behaviour to hold to a good approximation, a situation which is rarely in doubt in torsion experiments, but which may not hold for tensile strains greater than about 0·001 in particular cases. The linear range of each experimental material must therefore be investigated.

First cycle strain is invariably uncharacteristic, enforcing mechanical conditioning until behaviour is reproducible. There are suggestions that in some materials complete conditioning is unattainable.[10]

Extensional measurements involve a simple loading process, but with compression and torsion measurements simplifying assumptions and/or corrections need to be applied to experimental data. In compression it is necessary to assume plane strain,[11] and Poisson's ratio measurements may be performed either at constant strain[11] or constant stress.[7] For a non-linear viscoelastic polymer these two methods are not equivalent.

Torsional work is normally based on St. Venant's theory,[12] with a correction being made for the axial load.[13] Ladizesky and Ward[8,9] have shown that the St. Venant theory for small strain elasticity is not adequate in the situation where two shear moduli contribute to the elastic behaviour and have demonstrated that experimental extrapolation to zero axial stress is also required in this case.

With fibres the possible modes of deformation are longitudinal extension, transverse compression and axial torsion, but sheet materials permit more varied experiments; extensional measurements on samples cut at several angles to the stretch direction allow combinations of compliances to be obtained. For example, Young's moduli of strips cut at 0°, 45° and 90° to this direction in an orthorhombic sheet are related to compliances by[14]

$$E_0 = \frac{1}{S_{33}}, E_{90} = \frac{1}{S_{11}} \text{ and } E_{45} = \tfrac{1}{4}[S_{11} + S_{33} + (2S_{13} + S_{55})]$$

Torsional samples may be cut with the width of the specimen rather than the breadth in the thickness of the sheet when sufficiently thick sheets can be produced.

9.3 EXPERIMENTAL TECHNIQUES

Equipment for measuring mechanical properties has been reviewed extensively. General descriptions are given in Nielsen,[15] Ferry[16] and Ward,[14] and dynamic modulus techniques are discussed in the above, and also more particularly by Hillier[17] and McCrum et al.[1] Apparatus is described in most of the experimental investigations cited in this chapter, so that only a brief review need be given here.

9.3.1 Static (elastic) measurements

Any sensitive form of creep machine is adequate for extensional measurements with strips or filament; counterbalancing is not needed unless recovery after unloading is to be investigated. Extensions are most accurately recorded using a linear differential transformer (l.d.t.). The device is most sensitive for a null reading, and Gupta and Ward[18] have described apparatus in which the sample is raised using a calibrated micrometer screw head until the l.d.t. attached to its lower end has reached the null position. Extension is then obtained from the micrometer reading.

Ladizesky[7] has measured longitudinal extension, transverse contraction, and hence Poisson's ratio under constant stress, of sheets on which an accurate grid had been evaporated. Photographs of the grid, taken on glass plates at various times after loading, were measured using a microscope.

Similar measurements were made by Darlington and Saunders using sensitive extensometers, with displacements measured by a capacitative transducer.[53] The lateral contraction extensometer, in which rigid feelers rested lightly against the specimen sides, showed some irregularities at strains below 0·01.

When a filament is compressed between two parallel glass flats a contact zone is formed along its length, and the diameter parallel to the glass plates is increased slightly. The width of the contact zone ($2b$) and the increase in diameter (U_1) have been measured using a device which fits onto a microscope stage.[11,19] The load acting on the filament may be varied by attaching small weights to a lever arm incorporating the upper glass flat, and a mechanism is attached which can extend the filament when the loaded flat is out of contact; the change in diameter for a given extension gives the Poisson's ratio v_{13}, and hence S_{13}. To remove irregularities, particularly those in the region of the end clamps, changes in length and diameter are measured between closely spaced reference ink marks.

It can be shown that

$$b^2 = \frac{4FR}{\pi}(S_{11} - v_{13}^2 S_{33})$$

and

$$U_1 = F\left\{\left(\frac{4}{\pi} - 1\right)\left(S_{11} - \frac{S_{13}^2}{S_{33}}\right) - \left(S_{12} - \frac{S_{13}^2}{S_{33}}\right)\right\}$$

where F is the load per unit length applied to a filament of unloaded radius R. From these expressions S_{11} and S_{12} can be derived.

Two earlier methods of obtaining Poisson's ratio v_{13} of filaments have been reported: Davis[20] measured the changes due to extension in the diffraction pattern formed by a single filament; and Frank and Ruoff[21] deduced the diameter change of a filament confined in a capillary from the displacement of mercury occupying the annulus between the filament and the capillary walls. Both methods were relatively insensitive, and their use was confined to nylon, since this material shows negligible permanent set for extensional strains up to 5%.

Typical apparatus for torsional measurements is that described by Raumann[22] and in a modified form by Ladizesky and Ward.[8] The sample is fastened between two clamps, the lower of which may be fixed in any required orientation, and the upper of which is attached rigidly to a small mirror and then to the lower end of a long phosphor-bronze strip. When the top of this strip is rotated both the suspension and the sample will twist, the former through an angle ϕ and the latter through θ. At equilibrium torques in suspension and sample are equal, giving

$$\frac{C\phi}{L} = \frac{M\theta}{l}$$

where C is the torque for unit twist per unit length of suspension of length L, and M is the torque for unit twist per unit length of specimen of length l. As mentioned in Section 9.2 M must be corrected for finite aspect ratio and finite axial load.

9.3.2 Dynamic measurements

Dynamic extensional moduli are most conveniently measured over a wide frequency range using a forced vibration technique: a sinusoidally varying strain is applied and the resultant stress and phase angle are measured. The specimen must not go into compression, and thus the varying strain must be superposed on a slightly greater static extensional strain. Typical devices are the Vibron Viscoelastometer due to Takayanagi[23] and the Transfer Function Analyser equipment described by Pinnock and Ward.[24] With suitable attachments such equipment may cover the frequency range 10^{-3}–10^3 Hz over a wide range of temperature.

Forced vibration torsion pendula have been described, but have found little use, probably because of the relatively low sensitivity of those built. In a typical design[25] a sample is clamped at the lower end, and to its upper end is fixed a coil, which is suspended between the poles of a magnet. A sinusoidally varying input to the coils will cause the sample to rotate, the angle of rotation being measured by changes in capacity between a pair of fixed plates and small vanes fastened to the specimen immediately below the energising coil.

Dynamic measurements in torsion are most frequently made using a free vibration torsion pendulum. A particularly simple device for single filaments is described by Meredith;[26] more typical is the design of Heijboer et al.[27] in which an inertia bar with sliding weights is fixed to the upper end of the sample, the lower end of which is clamped. The inertia bar is suspended from a wire supported by a counterweight to give a negligible extensional load. Measurements are usually limited to logarithmic decrement ($\cong \pi \tan \delta$), but Inoue and Kobatake[28] have derived a relation for the shear modulus.

The specimen is enclosed in a thermosetted jacket to enable the temperature variation of logarithmic decrement to be obtained. As modulus changes with temperature it is necessary to move the masses on the inertia bar to maintain the frequency constant. The most frequent experimental frequency is about 1 Hz, but frequencies up to about 50 Hz may be attained before the wavelengths of stress waves within the specimen become comparable with its dimensions.

For particular resonant values of higher frequencies, usually in the range 100–2000 Hz, systems of standing waves are set up in vibrating samples, and relaxation processes may be studied through examination of the variation of the amplitude of vibration with frequency and temperature.

In the vibrating reed technique one end of a small sample is clamped in a gramophone record cutter head driven by a variable frequency oscillator.[29,30] The amplitude of vibration is measured opto-electrically, enabling tan δ to be calculated from the width of the resonance, since tan $\delta = \Delta f / f_r$ where f_r is the resonant frequency and Δf the bandwidth.

A disadvantage of the technique is that only resonant frequencies are accessible, and these frequencies, which are proportional to the square root of the extensional modulus, vary with temperature.

9.4 SOME EARLY INVESTIGATIONS, AND PROBLEMS RAISED

Investigations of some of the elastic constants of uniaxially stretched polymers were made on nylon 66 and polyethylene terephthalate by

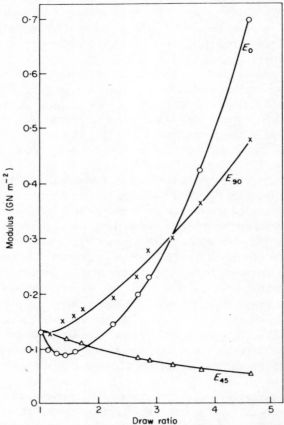

Fig. 1. Low density polyethylene. Variation of room temperature modulus with draw ratio for samples cut at various directions from cold drawn sheet. (Adapted from Raumann and Saunders.[35])

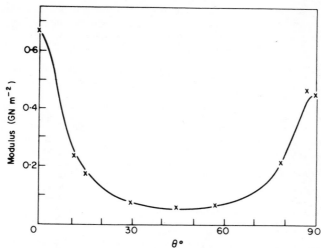

Fig. 2. Low density polyethylene. Room temperature modulus of highly oriented strips cut at various angles to direction of cold drawing. (Adapted from Raumann and Saunders.[35])

Wakelin *et al.*[31] in 1955, on polyethylene terephthalate by Kawaguchi[32] in 1957, and on regenerated cellulose by Shinohara and Tanzawa[33] in 1957, and Wilson in 1960;[34] but the first determination of all 5 constants was that of Raumann and Saunders on low density polyethylene in 1961 and 1962.[35,22]

The polyethylene was alkathene 7 of only 47% crystallinity, as measured by X-rays, and as can be seen from Figs. 1 and 2 there were several unexpected features: sheets cut at 45° to the stretch direction showed the lowest stiffness (E_{45}); and shallow minima in extensional moduli (E_0) and (E_{90}) occurred at low draw ratios for strips cut parallel with and perpendicular to the stretch direction; also the torsional stiffness ($1/S_{44}$) was very low by comparison with the other 4 stiffnesses. It was this low value which accounted for the observed behaviour of E_{45}. The compliance at an angle θ to the stretch direction is given by $S_\theta = S_{11} \sin^4 \theta + S_{33} \cos^4 \theta + (2S_{13} + S_{44}) \sin^2 \theta \cos^2 \theta$. For $\theta = 45°$ this reduces to $S_{45} = \frac{1}{4}[S_{11} + S_{33} + 2S_{13} + S_{44}]$, and will be large if $S_{44} \gg$ other compliances.

A complete X-ray investigation of the samples was not performed, but the minima in moduli could not be attributed to a fall in crystallinity.

TABLE 1

MODULI OF TYPICAL ISOTROPIC AND HIGHLY DRAWN FILAMENTS, IN
UNITS OF GN m^{-2}; AND POISSON'S RATIOS OF DRAWN FILAMENTS.

(Adapted from Hadley et al.,[38] with misprints corrected.)

State	Highly drawn					Isotropic	
Modulus, etc. Material	E_{90}	E_0	$1/S_{44}$	v_{13}	v_{12}	$E_0 = E_{90}$	$1/S_{44}$
Low density polyethylene	0·33	0·82	0·012	0·55 ±0·20	0·61 ±0·20	0·10	0·04
High density polyethylene	0·60	5·2	0·7	0·46 ±0·15	0·52 ±0·08	0·60	0·30
Polypropylene	0·80	5·4	0·75	0·42 ±0·16	0·68 ±0·18	0·80	0·35
Polyethylene terephthalate	0·80	18	0·75	0·43 ±0·06	0·44 ±0·09	2·2	0·90
Nylon	1·3	4·6	0·50	0·48 ±0·05	0·26 ±0·08	2·1	0·90

Birefringence increased steadily with draw ratio, and so no simple explanation of the phenomena in terms of overall morphology was advanced. These experimental data were used by Ward[36] as a test for a model of orientation in terms of alignment of anisotropic structural elements, and qualitative agreement was obtained with the general form of the stiffness v. draw ratio relations, including the minimum in E_0 (see Chapter 8).

Most notable among slightly later measurements were those of Raumann on polyethylene terephthalate[37] and Ward and his colleagues on filaments of low and high density polyethylene, polypropylene, nylon 6·6 and polyethylene terephthalate, summarised in Ref. 38. From these measurements, all made at room temperature, it was obvious that the development of mechanical anisotropy was not similar in all semi-crystalline polymers (see Table 1). For instance, with increasing orientation S_{11} rose slightly in nylon 6·6 and polyethylene terephthalate, remained approximately constant in polypropylene, and decreased significantly in low density polyethylene. It was only in this last material that S_{44} was very much greater than S_{11}, and changed rapidly with draw ratio.

Attempts to fit the orienting element model to the data met with mixed success, indicating that in some cases, such as polypropylene, more complex structural changes were probably occurring. It was thus advisable that mechanical measurements should be combined with as full as possible a study of the detailed morphology and structure of the specimens, using techniques such as wide and small angle X-ray scattering (for these techniques see Chapter 3). Further, it would be necessary to investigate the effect of changes in temperature on the elastic constants, since it was possible that their relative magnitudes would alter after passing through a transition region.

9.5 LOW DENSITY POLYETHYLENE

Because of its ready availability and simple chemical composition this material has been a favoured polymer for structural and mechanical studies. The relative ease with which doubly oriented specimens of overall orthorhombic symmetry can be obtained has intensified investigations of the relation between physical properties and molecular and supermolecular structure. But these investigations have demonstrated that the structure of this material is in fact exceedingly complex (see Appendix).

Gupa and Ward[18,39] have shown that the anomalous behaviour at room temperature, namely the minimum in E_0, which had first been observed in dynamic experiments by Hillier and Kolsky,[40] the minimum in E_{90}, and the very low value of E_{45} disappeared as the temperature was reduced (Figs. 3 and 4). Also E_{90}, which at room temperature increased

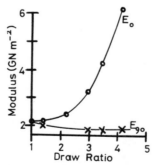

Fig. 3. Low density polyethylene. Variation of modulus with draw ratio for samples cut from cold drawn sheet examined at $-125°C$. Compare with Fig. 1. (Adapted from Gupta and Ward.[18])

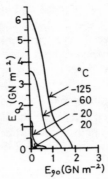

Fig. 4. Low density polyethylene. Angular variation of modulus at several tempera-tures for samples cut from cold drawn sheet. (Adapted from Gupta and Ward.[18])

with draw ratio, after the initial small minimum, showed a steady decrease with draw ratio at temperatures below $-35°C$.

They investigated, over the temperature range $-125°C$ to $50°C$, the extensional moduli E_0, E_{45} and E_{90} of strips of Alkathene WNF15: (i) uniaxially stretched, both in the cold drawn state, and after annealing; (ii) sheets of orthorhombic symmetry, having either the c-axis or the a-axis along the original stretch direction, in both cases the b-axis being in the plane of the sheet perpendicular to the original stretch (these are referred to as b–c and a–b sheets respectively).

Wide angle X-ray scattering was used to examine crystallite orientation, small angle scattering to determine the orientation of any lamellar structure, infra-red dichroism to study the orientation of non-crystalline regions, and nuclear magnetic resonance to record molecular mobility.

A major part of this study was to examine the applicability of the Ward rotating element aggregate model to low density polyethylene. As discussed in Chapter 8 satisfactory agreement was obtained when account was taken of the non-affine processes which occur in the early stages of drawing.[41-44] In addition, however, these experiments formed the basis of our present knowledge of the mechanisms responsible for deformation at a molecular level.

The most important features in the variation of modulus with tempera-ture for highly anisotropic specimens, clearly seen in Figs. 5 and 6, are that as the temperature is increased the ranking of moduli may change, indicating the onset of additional deformation mechanisms. There was,

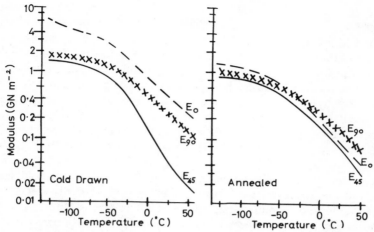

Fig. 5. Low density polyethylene. Variation of modulus with temperature for highly oriented samples cut from uniaxially drawn sheet. Note crossover in annealed samples. (Adapted from Gupta and Ward.[39])

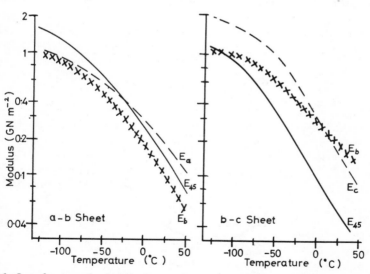

Fig. 6. Low density polyethylene. Variation of modulus with temperature for sheets of orthorhombic symmetry. (Adapted from Gupta and Ward.[39])

however, no change in ranking for cold drawn sheet, the only type of sample in which small angle scattering failed to indicate a clearly defined lamellar structure. Neither annealed uniaxially drawn sheet, nor orthorhombic sheets (which had undergone annealing during the final stage of production), showed any non-crystalline orientation, indicating the relaxation of any previously strained tie molecules between crystallites.

The onset of deformation mechanisms at particular temperatures indicated that it was appropriate to investigate viscoelastic relaxation processes by dynamic mechanical methods. Some preliminary studies were made in 1956 and 1957 by Hellwege et al.,[45,46] who demonstrated that the magnitude of the damping peak in torsion differed in samples cut parallel with and perpendicular to the stretch direction of a cold drawn film, but the principal contribution has been that of Stachurski and Ward,[47-50] who examined four grades of Alkathene, in similar states to those used in the static experiments of Gupta and Ward. Forced vibration methods at 50 Hz and 1–10^{-2} Hz were used for the study of extensional moduli, except where small sample size dictated the introduction of vibrating reed equipment. A free vibration torsion pendulum, operating at approximately 1 Hz was used for torsional studies.

The mechanical relaxations occurring in unoriented low density polyethylene are indicated schematically in Fig. 7(a). Studies on unoriented

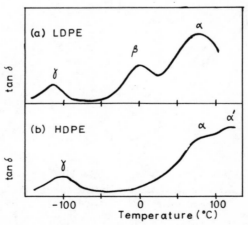

Fig. 7. Schematic representation of relaxations in (a) low density polyethylene; (b) high density polyethylene.

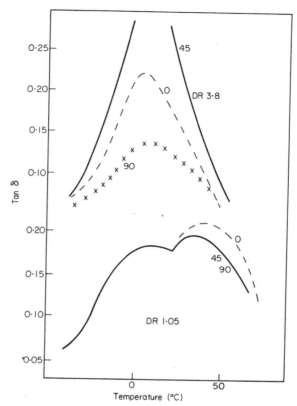

Fig. 8. Low density polyethylene. Effect of draw ratio on temperature variation of tan δ for samples from cold drawn sheet. (Adapted from Stachurski and Ward.[49])

material, summarised by McCrum *et al.*[1] had suggested that the β peak was associated with side chain activity, being enhanced in highly branched specimens, and that the α peak was a crystalline relaxation.

Cold drawn samples showed both α and β peaks at low draw ratios, with tan $\delta_{45} \cong$ tan $\delta_{90} <$ tan δ_0 (see Fig. 8), but as draw ratio increased the higher temperature peak moved to lower temperatures and swamped the β peak.[49] The dominant peak was highly anisotropic, with tan $\delta_{45} \gg$ tan δ_{90}. It was demonstrated that tan δ_{45} was a direct measure of losses associated with the S_{44} shear compliance, which static measurements had shown to be large,[35,18] and concluded that the relaxation involved shear deformation in the direction of drawing in planes containing the draw direction (or perpendicular to it). This mechanism was called c/c shear.

Analysis of the anisotropy in terms of a Ward aggregate model[36] permitted close reproduction of experimental data provided that orientation of the crystalline units was taken to determine the anisotropy, which suggested that the relaxation process was related to the crystalline regions. Experiments with a range of alkathenes showed the temperature of the relaxation to depend on percentage crystallinity, further evidence in favour of it being the α rather than the β relaxation.

Specimens from uniaxially stretched and annealed sheets, and from orthorhombic sheets showed two relaxations at high draw ratios[49] (see Fig. 9), which represents data obtained with vibrating reed equipment; the resonant frequency was about 250 Hz for the lower temperature peak and between 150 and 50 Hz for the upper relaxation. The anisotropy of the peak at about 70°C was similar to that of the 0° peak in cold drawn

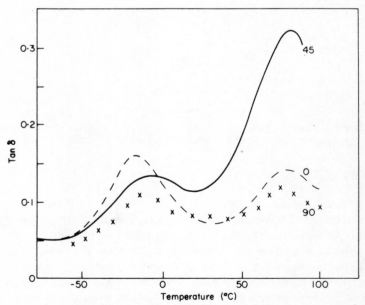

Fig. 9. Low density polyethylene. Temperature variation of tan δ for samples from annealed sheet. (Adapted from Stachurski and Ward.[49])

specimens, with $\tan \delta_{45} > \tan \delta_{90}$. This peak represented the α relaxation and was attributed to c/c shear. Specimens from ideal a–b orthorhombic sheet should not exhibit the relaxation since the c-axis is then perpendicular to the sheet surface. Any relaxation observed in the appropriate region might be attributed to imperfect molecular orientation.

For the low temperature (β) relaxation $\tan \delta$ is largest in the $0°$ direction for the annealed sheet, in the c direction for the b–c sheet, and in the a direction for the a–b sheet. It was in similar directions at the appropriate temperatures that specimens rapidly became more compliant in static loading experiments (Figs. 5 and 6). Reference to the schematic representation of morphology (Fig. 17, Appendix) indicates that blocks of crystalline lamellae are oriented at approximately $45°$ to the crystallographic axes. The blocks are built of folded molecular chains oriented axially, and thus aligned at approximately $45°$ to the basal planes of the lamellae. They are separated by relaxed amorphous material and tie molecules.

The most compliant directions, and thus those in which $\tan \delta$ is a maximum, occur when the maximum resolved shear stress is parallel with the lamellar planes. An inter-lamellar shear mechanism is suggested, the mechanical anisotropy of the β relaxation being determined by the configuration of the lamellae, and not by the molecular orientation within the lamellae, which differs between annealed, a–c and b–c sheets. Differences in the magnitudes of the relaxation peaks for different types of specimen will depend on the spread of deviations from the mean angle of the lamellae, and on the lamellae in the annealed sheet being arranged in cones around the stretch direction rather than in flat layers.

Support for the inter-lamellar shear hypothesis is provided by the isotropy of the dielectric β process,[51] which shows that the mechanical anisotropy cannot be a consequence of an ordered inter-lamellar material.

The original analysis of the relaxation mechanism by Stachurski and Ward was qualitative, and based on simplifying assumptions such as all lamellae at $45°$ to the draw direction. More recently Davies et al.[52] have presented a quantitative theory for mechanical isotropy of cylindrically symmetric specimens based on an inter-lamellar shear mechanism. As will be discussed later the α relaxation in high density polyethylene shows a similar anisotropy to the β relaxation in low density polyethylene, and is therefore proposed as being a consequence of inter-lamellar shear.

A thorough study, at 20°C, of the deformation of cold drawn Alkathene WNF15 has been published recently by Ladizesky and Ward.[7,8] Young's moduli and Poisson's ratios were determined from the changes in dimensions under load, and shear moduli using extrapolations from the

St. Venant theory. Time dependence, and the anisotropy of the strain dependence of the compliances were discussed.

The significance of these experiments is that the Poisson's ratio values, which yield S_{12} and S_{13}, were obtained directly, with an accuracy of about 5%, on the same small area of material over which extensional compliances were measured. With all compliances measured accurately over a range of strains it was possible to test under what circumstances the deformation proceeded at constant volume. By comparison the pioneering work of Raumann[22] had been so limited in its accuracy that the data were not inconsistent with the deformation occurring at constant volume—a situation in which the behaviour may be described in terms of only 3, rather than 5, elastic constants. In the earlier measurements of Ward and his colleagues[11,19,38] v_{13} was measured directly, but with an accuracy no better than 10%, while v_{12} was determined indirectly, applying an expression containing other compliances obtained with different test samples.

Ladizesky and Ward found S_{33} and v_{13} ($= 0.50$) independent of strain up to an extension of 1.5%, and the torsional rigidity was independent of twist up to 0.35 radians. In contrast S_{11} and v_{21} were strain dependent (Table 2).

When extrapolated to zero strain $v_{21} = v_{13} = 0.50$, implying that the total deformation takes place at constant volume only for infinitesimally small strains.

These results were in agreement with the somewhat less accurate data of Darlington and Saunders,[54,55] in which changes in thickness of thin sheets were measured with a lateral contraction extensometer.[53] Their measurements on strips cut at various angles (θ) to the stretch direction indicated that departures from linearity of the stress–strain relation increased as θ was made larger.

TABLE 2

S_{11} AND v_{21} FOR COLD DRAWN POLYETHYLENE (D.R. 3.8)

Extension %	$10^{10} \times S_{11}$ $(m^2 N^{-1})$	v_{21}
0.4	28.92	0.477
0.7	30.23	0.468
1.0	32.19	0.461
1.3	34.20	0.451
1.6	36.11	0.441

The c/c shear process already discussed was responsible for the high value of S_{44}, and dominated tensile behaviour for specimens cut at $\theta = 45°$. Such a shear process does not involve contraction in the thickness direction, and so gave v_{21} (45°) close to zero.

For a highly oriented sample at $\theta = 0°$ c/c shear makes a very small contribution, because only a few crystallites experience a sufficiently large shear stress. The observed linear recoverable and incompressible behaviour, with $v_{13} = 0.50$, corresponded to that of a linear elastic solid with an extremely high bulk modulus, and was thought to be associated with the non-crystalline regions.

On the basis of plastic deformation visible at 2% extension, the original draw direction being contracted while the thickness of the sheet remained unchanged, the non-linearity at $\theta = 90°$ was attributed to a slip mechanism involving rigid body rotation. This process occurred in conjunction with the linear mechanism associated with unordered regions. Again c/c shear should be negligible.

Studies of possible deformation mechanisms in drawn, rolled and annealed sheets have been made by Owen and Ward,[56] as an extension of the model of Davies et al. for specimens having transverse isotropy.[52] Because of the anneal involved in the production process c/c shear is a less important deformation mechanism in this form of material. If the lamellae exist as platelets of very large extent compared with the thickness of inter-lamellar material this latter could undergo only simple shear, with the shear direction parallel with the surface of the lamellae; but if the platelets are effectively of infinite extent only in one direction then inter-lamellar material could deform when the platelets are subjected to a normal stress. The inter-lamellar material then undergoes pure shear. (Some consequences of this model have been discussed in Chapter 4.)

A very full investigation of the deformation mechanisms in material with parallel lamellar morphology has recently been published by Ladizesky and Ward.[57] (The authors remark that the tensile data of Owen and Ward[56] are not in accord with their own measurements, almost certainly because of errors of temperature measurement in the earlier experiment.) Data have been interpreted in terms of a model of perfect parallel lamellae, with no spread of orientation, separated by isotropic amorphous regions. It was concluded that the main deformation mechanisms were an inter-lamellar pure shear, and a highly time-dependent intra-lamellar c-shear, which significantly affected the values of S_{22} and S_{55}.

9.6 HIGH DENSITY POLYETHYLENE

The initial published reports on high density polyethylene were dynamic mechanical studies, but before considering them it is necessary to compare the mechanical relaxations in isotropic material with those observed in unoriented low density polyethylene. From the schematic curve of tan δ v. temperature [Fig. 7(b)] it can be seen that the β relaxation, which was ascribed to branch point mobility, is not present, and that the high temperature relaxation is frequently resolvable into α and α' peaks.

In studies of single crystal mats, which had all c-axes approximately aligned, but were randomised in a and b directions, Sinnott[58] measured the dynamic rigidity modulus and the corresponding logarithmic decrement; only a single α peak was found. At annealing temperatures above 100°C the peak moved to higher temperatures and its magnitude diminished. Single crystal mats were examined also by Takayanagi and Matsuo,[59] who measured mechanical losses and tensile modulus in the plane of the mats, obtaining an average Young's modulus for the a–b plane. Again the α relaxation appeared as a single peak.

By 1966 it was thought that lamellae and their orientations might be involved in the α relaxation, and in an attempt to investigate this possibility Crissman and Passaglia[60] compared torsional measurements on oriented and isotropic samples. The most highly oriented sample, prepared by zone crystallisation, had the b-axis parallel with its length, with a- and c-axes distributed randomly in the transverse plane. Because of the large number of variables involved an unambiguous interpretation was not possible. Similar measurements in torsion were made by Eby and Colson[61] using samples with a highly oriented surface layer in which the b-axis was normal to the surface of the sheet. The surface layers were then removed, and repeat measurements on the now unoriented specimen showed the α peak to be reduced in magnitude. It was concluded that the average of shearing over all directions in the a–c plane enhanced the α relaxation.

McCrum and Morris[62] also compared torsion data for samples originally with an oriented surface layer and then with the layer removed. They did resolve the α' peak, which they found diminished by removal of the oriented layer. Interpretation was made in terms of a two phase model of Iwayanagi[63] involving lamellar boundary slip. It was concluded that the α' relaxation was caused by an inter-lamellar shear process.

In the same year Takayanagi et al.[64] reported measurements of complex Young's moduli in the 0° and 90° directions in both cold drawn and

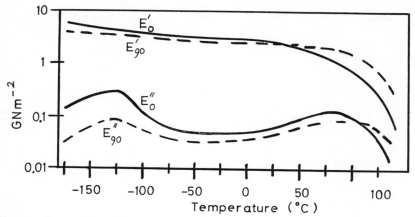

Fig. 10. High density polyethylene. Dynamic moduli as a function of temperature for samples from annealed sheets. (Adapted from Takayanagi et al.[64])

annealed sheets. For the former both in-phase and out-of-phase components of E_0 were greater than those for E_{90} over the complete temperature range, but in annealed material $E'_{90} > E'_0$ above 40°C (at 110 Hz) and $E''_{90} > E''_0$ above about 100°C (Fig. 10). Explanation was made in terms of the two phase model of separate crystalline and amorphous phases, essentially in series, discussed in Chapter 8. Relaxation of the amorphous phase, which was assumed to occur at a lower temperature than that of the crystalline phase, would have a much greater effect in the 0° direction. Although the model itself did not represent the morphology of the structure an interpretation was given in terms of the Hosemann paracrystalline hypothesis.[65]

The absence of a crossover in cold drawn material was considered as caused by anisotropy of modulus in the amorphous regions, due to tension in intercrystalline tie molecules.

The static measurements of Hadley *et al.*[38] using a grade of Rigidex polymer showed considerable differences compared with low density material. Apart from shallow minima in E_0 and G at low draw ratios the behaviour at room temperature appeared rather straightforward (Fig. 11). With increase of draw ratio E_0 increased steeply, the torsion modulus increased slightly and E_{90} varied only a little; v_{12} and v_{13} were generally not inconsistent with a value of 0·50, and seemed insensitive to the

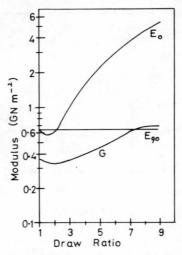

Fig. 11. High density polyethylene. Room temperature measurements of the variation with draw ratio of extensional, transverse and torsion moduli of hot drawn monofilaments. (Adapted from Hadley et al.[38])

amount of draw. Deformation thus occurred without much change of volume.

Poor agreement was found with the Ward aggregate model (not surprisingly, since the model assumes a single phase, while subsequent studies have shown high density polyethylene to exhibit a lamellar structure even in the cold drawn state). This suggested, by analogy with polypropylene, which showed a superficially very similar behaviour, that orientation was not the only significant mechanism operating at room temperature.

Gupta et al.[41] investigated Rigidex 9 in parallel with the studies of low density polyethylene already reviewed. X-ray observations were interpreted as showing a structure after annealing of crystalline lamellae at about 45° to the stretch direction. The lamellae are poorly defined in cold-drawn specimens, which showed also some orientation of non-crystalline material.

At low temperatures in the annealed samples $E_0 > E_{45} > E_{90}$, but above 20°C $E_{90} > E_0$.

Cold drawn samples showed no crossover, but at the highest

temperatures attained E_0 was dropping rapidly towards the value of E_{90} (Fig. 12). Data were interpreted in terms of an inter-lamellar shear mechanism; the enhanced value of E_0 compared with E_{90} at low temperatures, when the inter-lamellar layers are stiff, was attributed to the influence of covalent forces in the chain direction; the reverse behaviour above the transition temperature was accounted for by the conical distribution of lamellae about the stretch direction, which caused only some of the lamellae to be in the most favoured position for a force applied in the 90° direction.

The relaxation associated with interlamellar shear has been further investigated by Stachurski and Ward,[47,50] who confirmed and extended Takayanagi's earlier measurements. The anisotropy of the α peak in annealed samples, with $\tan \delta_0 > \tan \delta_{45} > \tan \delta_{90}$, was similar to that of the β relaxation in low density polyethylene, being attributed to an inter-lamellar shear mechanism. Evidence of this process was present also in cold drawn samples, but was less clearly defined.

Although the mechanical mechanisms of the β relaxation in low density polyethylene and the α relaxation in high density polyethylene are similar, this does not imply that both relaxations are associated with the same molecular processes. There are considerable structural differences between the materials; not only does Rigidex have a higher crystallinity than Alkathene, but the unordered fraction might be rather a defect region than an amorphous component. These factors, plus the

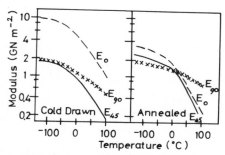

Fig. 12. *High density polyethylene. Variation of modulus with temperature for highly oriented samples from uniaxially drawn sheet. Note crossover in annealed samples. (Adapted from Gupta and Ward.[39])*

difference of almost 100°C between the two relaxation peaks, imply that two different molecular processes are operative. Ward[14] suggests that the α relaxation in high density polymer requires mobility of the fold surfaces, while the β relaxation in low density material requires mobility of chains close to branch points.

Neither in cold drawn nor annealed high density samples did Stachurski and Ward see any evidence for the c/c shear mechanism observed in low density material. It was noticed that for a range of low density polyethylenes the activation energy of this process increased with density. Provided that the molecular processes were similar in Rigidex an activation energy of about 400 kJ mol^{-1} was predicted, which implied that the relaxation would not be observed below the melting point at the lowest accessible frequency of 10^{-2} Hz.

Young et al.[92] recently reported a study of plastic deformation in samples of single crystal texture obtained by compression at constant width under ambient temperature conditions, followed by annealing at high pressure and temperature. Specimens cut at various angles to the chain axis were compressed at room temperature in a manner favouring simple shear deformations parallel with and perpendicular to the chain direction. From a combination of strain measurements, small angle and wide angle X-ray scattering, the latter both under load and after the sample had relaxed, it was deduced that the reversible elastic component of the deformation was due essentially to an inter-lamellar shear process. This conclusion was in agreement with the work on drawn and annealed material by Ward and his colleagues discussed earlier.

The authors commented that their results contrasted with those for low density polyethylene, in which both inter-lamellar shear and slip between folded chains within lamellae were found to be reversible,[93] but Owen and Ward[56] have shown that in low density polyethylene which has been annealed the inter-lamellar slip contribution is enhanced.

The low temperature γ relaxation has been examined by Buckley, Gray and McCrum in Rigidex 9 samples both uniaxially drawn and annealed[66] and drawn at constant width to give biaxial orientation.[67] In each case the modulus parallel to the stretch relaxed more than that in the perpendicular direction. This was interpreted in terms of a series Takayanagi model as evidence for deformation of amorphous material in an inclined lamellar structure. The work by Stachurski and Ward already discussed did not throw any light on the nature of the γ transition.

Some preliminary measurements of extensional and compression moduli of extended chain polyethylene, produced by annealing conventional drawn material under pressures of several kilobars at temperatures in the region of 240°C, have been reported by Bassett and Carder.[68,69] All measurements were made at room temperature, using a Cambridge extensometer. The starting polymer was usually Rigidex 2, but other experiments with high molecular weight samples of Hifax showed comparable results.

High pressure annealing increased fold length without destroying orientation, which was assumed to be identical in all cases. The higher the annealing temperature the greater the fold length (and the higher the density), and at 241°C E_0 was more than 3.5 GN m^{-2}, about 6 times the pre-annealed value for Rigidex and 20 times that for Hifax—for which only a low draw ratio was possible.

A further increase in annealing temperature resulted in a first order transition, with which was associated a change in the nature of the inter-crystalline surface, resulting in a dramatic fall in room temperature modulus. Specimens containing low molecular weight polymer became extremely brittle and tensile testing for higher anneal temperatures was not practicable. Compression tests were, however, carried out for anneal temperatures up to 248°C. The compression modulus in the chain direction increased with anneal temperature up to a value of 4 GN m^{-2} for 241°C. Softening of the interface then led to a fall to 0.3 GN m^{-2} for 246°C, followed by a rapid increase to 3 GN m^{-2} after annealing at 248°C. This form of variation in stiffness was accounted for in terms of the large increase in chain length between folds permitted due to the change from folded chain to extended chain texture above 241°C. Completely extended chains would be expected to exhibit an extremely high modulus, and the modulus increase associated with the approach to full chain extension more than compensated for the intercrystalline softening mechanism at anneal temperatures above 246°C.

Morphology suggested a Takayanagi type model in terms of crystalline lamellae perpendicular to the stretch direction, separated by inter-lamellar material, and modulus, density and melting data were consistent with this model up to the point where a textural change was observed. It was concluded that the mechanical properties in this range were a consequence of the properties of the relatively small less well ordered intercrystalline regions rather than of the extended chains.

For material annealed at 248°C the compression modulus in the chain direction was 3.1 GN m^{-2}, compared with a transverse modulus of 5.4 GN m^{-2}. In terms of a Takayanagi model this implied that the disordered material had passed through a softening transition—but at this high anneal temperature E_0 was no longer explicable by the simple model used at lower temperatures.

9.7 POLYPROPYLENE

The variation of compliances with draw ratio for cold drawn polypropylene filaments examined at 20°C appeared very similar to that of high density polyethylene,[70,38] with an increase in all compliances but S_{11}, which was insensitive to draw ratio. Ward aggregate theory was not applicable except for low draw ratios, implying that other processes intervened in addition to an orientation of pre-existing units. It was probable that even above the glass transition temperature increasing orientation led to a reduction in molecular mobility, as was known to occur in polyethylene terephthalate.[71]

Except at high draw ratios the compliances were sensitive to the orientation existing in the filaments before drawing, in addition to the final level of overall orientation. This feature was a consequence of filaments with a high degree of orientation at the spinning stage having some crystallites with their c-axis aligned perpendicular to the fibre axis.

Takayanagi et al.[64] reported some data on the dynamic testing of polypropylene, but the most comprehensive treatment is that of Owen and Ward[72] who performed static and dynamic measurements on sheets in the cold drawn (20°C), cold drawn and annealed, and hot drawn (120°C) states, over the temperature range $-130°C$ to $+140°C$. Low angle X-ray scattering suggested a structure of stacks of lamellae, with the orientation of lamellae within stacks varying; lamellae normals lay between 0° and 45° to the draw direction, implying the lamellar thickness to be dependent on the angle of tilt.

For both cold and hot drawn specimens $E_0 > E_{45} > E_{90}$ over the complete temperature range. This suggested that an inter-lamellar shear process was not taking place, presumably because inter-lamellar ties were restraining the non-crystalline regions.

Samples annealed at 145°C showed an increase of E_{90} relative to E_0 and E_{45}, giving $E_0 > E_{90} > E_{45}$ above 20°C; and for specimens annealed at 158°C $E_{90} > E_0 > E_{45}$ above 40°C. The low value of E_{45} implied that a shearing process was taking place, and evidence favoured an inter-lamellar shear mechanism, which could occur presumably because of relaxation of intercrystalline tie molecules.

Dynamic studies at 50 Hz showed a β peak at about 20°C, whose anisotropy was not consistent with a simple model for inter-lamellar shear, but for material annealed at 158°C the α peak (above 120°C) showed $\tan \delta_{45} > \tan \delta_0$, as expected for such a process (see Davies *et al.*[52]). This peak occurred, however, in a region where static studies had suggested that inter-lamellar shear was already the dominating mechanism, and presumably it related to a further degree of molecular mobility giving increased inter-lamellar shear.

9.8 POLYETHYLENE TEREPHTHALATE

In 1956 Thompson and Woods[73] reported that dynamic experiments in extension indicated that orientation increased the temperature of the β transition, about 80°C, for oriented crystalline fibres, and reduced the drop in modulus occurring at higher temperatures. Subsequently nuclear magnetic resonance was used to demonstrate that orientation reduced molecular mobility above the glass transition temperature.[71] Measurements of dynamic extensional and torsional moduli of hot stretched filaments and films were reported in 1963 by Pinnock and Ward,[74] who found that the relations between measured compliances below the glass transition temperature were consistent with the deformation of an incompressible elastic solid.

Also in 1963 Raumann[37] reported measurements on specimens prepared from amorphous sheets stretched below the glass transition temperature. The experimental method was identical with that reported earlier for polyethylene. Major features were that as the result of orientation E_0 increased to over five times the isotropic value, and the shear modulus at 90° dropped to about half the isotropic shear modulus; other moduli showed only small variations from the isotropic values (Ladizesky and Ward[75] have commented that Raumann's torsional moduli are probably in error, owing to her assumptions of the validity of the St. Venant theory, and an inadequate correction for tensile stress).

Room temperature data on hot stretched films and fibres[76,38] (Fig. 13) followed the general form of change of modulus with draw ratio predicted by an aggregate model, suggesting that an orientation mechanism plays an important role in the drawing process, but the model predicted a more gradual approach to the fully oriented values than was found in practice. In the course of these investigations Pinnock and Ward discovered that detailed changes in the drawing process could affect the attainable tensile modulus, and that certain partially drawn filaments

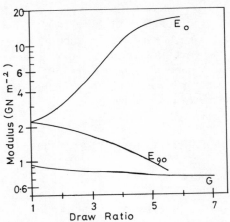

Fig. 13. Polyethylene terephthalate. Room temperature moduli of hot drawn mono-filaments. (Adapted from Hadley et al.[38])

showed a skin having different tensile, but not torsional, properties from the bulk of the filament.

Allison and Ward[77] have shown that for cold drawn polyethylene terephthalate the aggregate model provides a good representation of the variation of moduli with drawing, the fit to theory lying almost midway between the constant stress and constant strain bounds over the complete range of natural draw ratios investigated. As with other polymers cold drawing appears to produce the closest approach to a single phase material; the non-crystalline regions are certainly oriented, and Farrow and Ward[78] have shown, by comparing X-ray and density estimates of percentage crystallinity, that they have a density higher than that of the amorphous polymer.

Room temperature measurements of the extensional and shear compliances of specimens of orthorhombic symmetry, produced by stretching sheets at constant width, have been made by Ladizesky and Ward[75] in a similar manner to their measurements on polyethylene discussed earlier. Low angle X-ray scattering showed no lamellar morphology. The z and x directions lay in the plane of the sheet, which contained the (100) plane, and lay close to the plane containing the benzene ring. Compliances, in units of $10^{-10}\,\mathrm{m^2\,N^{-1}}$ were: S_{33} (draw direction) $= 0.58$; $S_{11} = 4.45$; $S_{44} = 26.2$; $S_{55} = 5.88$; $S_{66} = 27.7$. The easiest deformation was a displacement of (100), followed by shear deformation of (100). Next was lateral separation of molecules in (100), and finally a deformation of these planes together with extension of the molecules in the orientation direction. Even though a distribution of molecular orientation may have given rise to some complications these measurements indicate that planar orientation is an important factor in determining mechanical properties.

9.9 NYLONS 6·6 AND 6

The values of elastic compliances[38] and the form of the dynamic loss peaks are each affected by absorbed water. Even when humidity control has given a self-consistent set of data there may be difficulty in linking these results to other experiments performed under different conditions.

Apart from the early measurements of Poisson's ratio of drawn filaments already mentioned,[20,21] which were not used to throw any light on mechanical processes or structure, the only static test data on nylon 6·6 are those of Hadley et al.,[38] which suggest that the drawing process is essentially a simple phenomenon of orientating pre-existing structural elements (Fig. 14).

Dynamic measurements by Thomas[79] showed that the α peak, in the region of 100°C, moved to higher temperatures with increasing draw

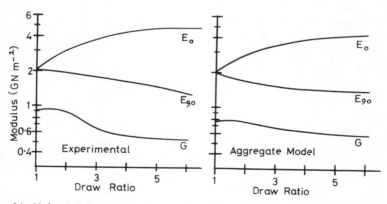

Fig. 14. Nylon 6·6. Room temperature measurements of the variation with draw ratio of extensional, transverse and torsion moduli of monofilaments; and comparison with Ward aggregate model assuming continuity of strain. (Adapted from Hadley et al.[38])

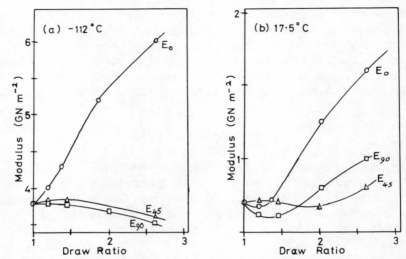

Fig. 15. *Nylon 6. Moduli of strips from cold drawn sheet, measured at* $-112°C$ *and* $17·5°C$. *(Adapted from Owen and Ward.*[82]*)*

ratio, suggesting that at these temperatures orientation restricted molecular mobility. This relaxation was affected by the overall sample crystallinity,[80] and was thought to represent a non-crystalline process. Murayama *et al.*[81] examined the dynamic elastic behaviour of an annealed oriented fibre, as part of a programme involving creep and stress relaxation studies, and concluded that the α relaxation can be caused by the response of only a few chain elements at any one time. They discussed the similarity with the mechanisms responsible for molecular motion in glassy polymers.

Cold drawn specimens of nylon 6 have recently been investigated by Owen and Ward[82] who measured static moduli between $-110°C$ and $+20°C$, and dynamic tan δ between $-70°C$ and $+110°C$ in a vibrating reed instrument. As can be seen from Fig. 15 the pattern of anisotropy changes with temperature, there being a minimum in E_0 and E_{90}, near room temperature, around the draw ratios at which X-ray diffraction indicates a transformation from an α form to a more stable γ form.[83] These structural changes, and the changes responsible for the rise in the temperature of the α relaxation with orientation, prevented a detailed understanding of the mechanical deformation processes.

9.10 NON-CRYSTALLINE POLYMERS

Although many polymers are non-crystalline only a relatively small amount of data are available on such materials, primarily because they lack any readily observable structure which might be affected by orientation.

Hennig[84] quotes room temperature values of extensional, transverse and torsional moduli for polyvinyl chloride, polymethylmethacrylate and polystyrene. Extensional data were obtained from dynamic testing at 320 Hz, while torsional measurements were made at 1 Hz. These values, together with those of Robertson and Buenker[85] on bisphenol A polycarbonate are summarised in Table 3.

As discussed in Chapter 8 Hennig attempted to fit these data to a theory of orienting rods, and Kausch-Blecken von Schmeling[86] used them to test a deforming network model.

A few data on slightly crystalline polyvinyl chloride are given by

TABLE 3

MODULI OF ORIENTED AMORPHOUS POLYMERS, IN UNITS OF GN m^{-2}.
(Adapted from Hennig.[84])

Material	Draw ratio	E_0	E_{90}	$1/S_{44}$
Polyvinyl chloride	1	3·20	3·20	1·22
	1·5	3·62	3·13	1·26
	2·0	3·92	3·05	1·28
	2·5	4·11	2·97	1·30
	2·8	4·21	2·93	1·31
	∞	4·90	2·64	1·37
Polymethyl-methacrylate	1	4·68	4·68	1·88
	1·5	4·80	4·66	1·91
	2·0	4·90	4·65	1·93
	2·5	5·00	4·62	1·96
	3·0	5·10	4·61	1·98
Polystyrene	1	3·30	3·30	1·30
	2·0	3·38	3·29	1·30
	3·0	3·46	3·28	1·30
Polycarbonate	1	2·66	2·66	0·95
	1·3	3·18	2·45	1·02
	1·6	3·73	2·32	1·08

Müller[87] to illustrate the large effect on mechanical properties of a small amount of crystalline material.

Recently Wright et al.[88] have obtained room temperature values of all five elastic constants for uniaxially drawn polymethylmethacrylate (Perspex) and polystyrene (Carinex) by measuring the critical angle for total reflection of an ultrasonic beam incident on immersed samples. The specimens were stretched by similar extents at different temperatures, with orientation assessed by measurements of optical birefringence.

From Fig. 16 it can be seen that stiffnesses were only slightly dependent on orientation, the largest change being an increase of 30% in Young's modulus in the axial direction for polymethylmethacrylate. Static loading experiments on strips cut at various angles to the stretch direction enabled some check values to be obtained. The trend was similar to that in the ultrasonic measurements, but absolute values of stiffness were lower, presumably due to a frequency dependent effect. The authors comment that despite the small effect of orientation on low strain

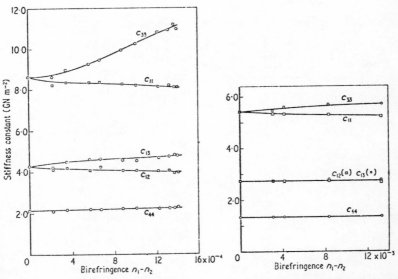

Fig. 16.(a) Polymethylmethacrylate (Perspex) and (b) polystyrene. Room temperature measurements of stiffness constants for uniaxially drawn specimens. (From Wright et al.[88])

properties a slight degree of orientation produces very pronounced anisotropy in fracture properties.[89]

Kashiwagi et al.[90] have shown that Wright's data on Perspex are consistent with the predictions of a Ward aggregate model, using orientation functions obtained by broad line nuclear magnetic resonance.

9.11 CELLULOSIC MATERIALS

Shinohara and Tanzawa[33] examined the effect of orientation on the dynamic properties of viscose rayon films and fibres. (Only an abstract of their studies appears to be available in English.) Making the assumption that the structure was polycrystalline they deduced a relation between extensional and shear moduli in terms of an orientation factor; tan δ for each modulus was found to be independent of orientation.

The extensional moduli parallel and perpendicular to the stretch direction of drawn films of cellulose acetate were measured by Wilson.[34] Experiments were performed at 20°C for relative humidities of 65% and 100%. In each case the parallel modulus (E_0) increased markedly with draw ratio while E_{90} showed a slight decrease.

9.12 CONCLUSIONS

The studies discussed in this chapter emphasise the necessity of performing mechanical measurements, using well characterised specimens, in parallel with the fullest possible examination of molecular and supermolecular structure and its variation with orientation. Only then can plausible suggestions be made concerning the processes governing elastic properties, and even so it is necessary to distinguish between mechanical and molecular processes and mechanisms. The relative significance of particular stiffness components can alter dramatically over a small range of temperature, and experiments at the most accessible temperature may not be revealing.

At the present time it seems possible that for non-crystalline polymers the observed mechanical anisotropy is consistent with a network deformation process, but detailed understanding awaits further evidence on possible supermolecular structure occurring in such materials.

Cold drawn semi-crystalline polymers are characterised generally by oriented unordered fractions, through which pass strained intercrystallite molecular ties. In these circumstances it appears that frequently an orientation mechanism is a major factor in determining low strain properties, although other plastic deformations processes may intervene.

Annealing and hot drawing have the effect of relaxing the material between crystallites, and frequently a clear lamellar structure is produced. When this occurs mechanical properties are sensitive to the lamellar geometry, and a full understanding is dependent on more detailed knowledge of lamellar structure.

APPENDIX. A BRIEF SURVEY OF THE MOLECULAR STRUCTURE OF LOW DENSITY POLYETHYLENE

Typical material with 2-3% branching has a density of 0·92 Mg m^{-3} and an X-ray crystallinity of 60-50%, compared with a density of 0·95 Mg m^{-3} and crystallinity > 80% for high density polyethylene with only one side branch per 1000 main chain carbon atoms.

Cold drawn specimens usually show an indefinite image from small angle X-ray scattering suggesting a single phase material with many defects. Indications of a lamellar structure have, however, been recorded,[91] with lamellae in material quenched from the bulk having the molecules almost perpendicular to the basal planes, but after drawing lamellae were inclined at 45° to the stretch direction, with molecular chains then at 45° to the basal planes. Drawing appeared to cause shear of the lamellae, reducing the lamellar periodicity but not the molecular length within lamellae.[49]

Unordered regions in cold drawn polyethylene are partially oriented, and are probably traversed by strained tie molecules linking crystallites together.

Annealed samples show clear lamellar periodicity, with lamellae approximately at 45° to the stretch direction. After high enough annealing temperatures all orientation in the inter-lamellar regions is lost.

Cold drawing followed by rolling produces a twinned double texture, which after progressively higher annealing temperatures gives macroscopic analogues of single crystal and solution grown platelet textures.[3] In the first case the c-axis remains parallel with the draw direction, with the b-axis in the plane of the sheet and the a-axis perpendicular to it (b–c sheet). A structure of crystalline lamellae, with interfaces at about 45° to the draw direction, is obtained. Molecular chains are thus at about 45° to the basal planes of the lamellae.

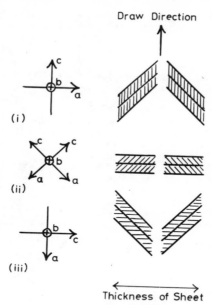

Fig. 17. Low density polyethylene. Schematic morphology of drawn, rolled and annealed sheets; (i) b–c sheet; (ii) parallel lamellae sheet; (iii) a–b sheet.

An increase in annealing temperature causes rotation of lamellae about the b-axis until the basal planes are perpendicular to the original draw direction (parallel lamellae sheet). Further annealing causes further rotation so that the c-axis becomes perpendicular to the plane of the sheet (a–b sheet).

Idealised schematic representations of lamellar and molecular orientation in the Hay and Keller sheets are shown in Fig. 17.

REFERENCES

1. McCrum, N. G., Read, B. E. and Williams, G. (1967). *Anelastic and Dielectric Effects in Polymeric Solids,* Wiley, London.
2. Point, J. J. (1953). *J. Chim. Phys.,* **50,** 76.
3. Hay, I. L. and Keller, A. (1966). *J. Mater. Sci.,* **1,** 41; (1967) **2,** 538.
4. Buckley, C. P., Gray, R. W. and McCrum, N. G. (1970). *J. Polymer Sci., B,* **8,** 341.
5. Ladizesky, N. H. and Ward, I. M. (1973). *J. Mater. Sci.,* **8,** 980.

6. Biot, M. A. (1955). Int. U. Theor. Appl. Mechs. Colloq., Springer, Berlin, p. 251.
7. Ladizesky, N. H. and Ward, I. M. (1971). *J. Macromol. Sci., Phys.*, **B5**, 661.
8. Ladizesky, N. H. and Ward, I. M. (1971). *J. Macromol. Sci., Phys.*, **B5**, 745.
9. Ladizesky, N. H. and Ward, I. M. (1974). *J. Macromol. Sci., Phys.*, **B9**, 565.
10. Lockett, F. J. and Turner, S. (1971). *J. Mech. Phys. Solids*, **19**, 201.
11. Hadley, D. W., Ward, I. M. and Ward, J. (1965). *Proc. Roy. Soc.*, **A285**, 275.
12. Lekhnitskii, S. G. (1963). *Theory of Elasticity of an Anisotropic Elastic Body*, Holden-Day, San Francisco, p. 197.
13. Biot, M. A. (1939). *J. Appl. Phys.*, **10**, 860.
14. Ward, I. M. (1971). *Mechanical Properties of Solid Polymers*, Wiley, London, p. 232.
15. Nielsen, L. E. (1962). *Mechanical Properties of Polymers*, Reinhold, New York.
16. Ferry, J. D. (1970). *Viscoelastic Properties of Polymers*, Wiley, New York.
17. Hillier, K. W. (1961). *Prog. Solid Mechs.*, **2**, 201.
18. Gupta, V. B. and Ward, I. M. (1967). *J. Macromol. Sci., Phys.*, **B1**, 373.
19. Pinnock, P. R., Ward, I. M. and Wolfe, J. M. (1966). *Proc. Roy. Soc.*, **A291**, 267.
20. Davis, V. (1959). *J. Text. Inst.*, **50**, T688.
21. Frank, R. I. and Ruoff, A. L. (1958). *Text. Res. J.*, **28**, 213.
22. Raumann, G. (1962). *Proc. Phys. Soc.*, **79**, 1221.
23. Takayanagi, M. (1965). *Proc. 4th Int. Cong. Rheol.*, Interscience, New York, Part I, p. 161.
24. Pinnock, P. R. and Ward, I. M. (1966). *Polymer*, **7**, 255.
25. Lord, P. and Wetton, R. E. (1961). *J. Sci. Instrum.*, **38**, 385.
26. Meredith, R. (1954). *J. Text. Inst.*, **45**, T489.
27. Heijboer, J., Dekking, P. and Staverman, A. J. (1954). *Proc. 2nd Int. Cong. Rheol.*, Academic Press, New York, p. 123.
28. Inoue, Y. and Kobatake, Y. (1958). *Kolloid-Z.*, **159**, 18.
29. Nolle, A. W. (1948). *J. Appl. Phys.*, **19**, 753.
30. Robinson, D. W. (1955). *J. Sci. Instrum.*, **32**, 2.
31. Wakelin, J. H., Voong, E. T. L., Montgomery, D. J. and Dusenburg, J. H. (1955). *J. Appl. Phys.*, **26**, 786.
32. Kawaguchi, T. (1957). *Chem. High Polym., Japan*, **14**, 176.
33. Shinohara, Y. and Tanzawa, H. (1957). *Chem. High, Polym., Japan*, **14**, 488.
34. Wilson, N. (1960). *J. Polymer Sci.*, **43**, 257.
35. Raumann, G. and Saunders, D. W. (1961). *Proc. Phys. Soc.*, **77**, 1028.
36. Ward, I. M. (1962). *Proc. Phys. Soc.*, **80**, 1176.
37. Raumann, G. (1963). *Brit. J. Appl. Phys.*, **14**, 795.
38. Hadley, D. W., Pinnock, P. R. and Ward, I. M. (1969). *J. Mater. Sci.*, **4**, 152.
39. Gupta, V. B. and Ward, I. M. (1968). *J. Macromol. Sci., Phys.*, **B2**, 89.
40. Hillier, K. W. and Kolsky, H. (1949). *Proc. Phys. Soc., B*, **62**, 111.
41. Gupta, V. B., Keller, A. and Ward, I. M. (1968). *J. Macromol. Sci., Phys.*, **B2**, 139.
42. Frank, F. C., Gupta, V. B. and Ward, I. M. (1970). *Phil. Mag.*, **21**, 1127.
43. Gupta, V. B. and Ward, I. M. (1970). *J. Macromol. Sci. Phys.*, **B4**, 453.
44. Gupta, V. B. and Ward, I. M. (1971). *J. Macromol. Sci., Phys.*, **B5**, 629.
45. Hellwege, K. H., Kaiser, R. and Kuphal, K. (1956). *Kolloid-Z.*, **147**, 155.
46. Hellwege, K. H., Kaiser, R. and Kuphal, K. (1957). *Kolloid-Z.*, **157**, 27.
47. Stachurski, Z. H. and Ward, I. M. (1968). *J. Polymer Sci., A-2*, **6**, 1083.
48. Stachurski, Z. H. and Ward, I. M. (1968). *J. Polymer Sci., A-2*, **6**, 1817.
49. Stachurski, Z. H. and Ward, I. M. (1969). *J. Macromol. Sci., Phys.*, **B3**, 427.
50. Stachurski, Z. H. and Ward, I. M. (1969). *J. Macromol. Sci., Phys.*, **B3**, 445.
51. Davies, G. R. and Ward, I. M. (1969). *J. Polymer Sci. B*, **7**, 353.
52. Davies, G. R., Owen, A. J., Ward, I. M. and Gupta, V. B. (1972). *J. Macromol. Sci., Phys.*, **B6**, 215.
53. Darlington, M. W. and Saunders, D. W. (1970). *J. Phys. E.*, **3**, 511.

54. Saunders, D. W. and Darlington, M. W. (1968). *Nature*, **218**, 561.
55. Darlington, M. W. and Saunders, D. W. (1970). *J. Phys. D*, **3**, 535.
56. Owen, A. J. and Ward, I. M. (1971). *J. Mater. Sci.*, **6**, 485.
57. Ladizesky, N. H. and Ward, I. M. (1974). *J. Macromol. Sci., Phys.*, **B7**, 565.
58. Sinnott, K. M. (1966). *J. Appl. Phys.*, **37**, 3385.
59. Takayanagi, M. and Matsuo, T. (1966). *J. Macromol. Sci., Phys.*, **B1**, 407.
60. Crissman, J. M. and Passaglia, E. (1966). *J. Res. Nat. Bur. Std.*, **70A**, 225.
61. Eby, R. K. and Colson, J. P. (1966). *J. Acoust. Soc. Amer.*, **39**, 506.
62. McCrum, N. G. and Morris, E. L. (1966). *Proc. Roy. Soc.*, **A292**, 506.
63. Iwayanagi, S. (1962). *Rep. Prog. Polym. Phys. Japan*, **5**, 135.
64. Takayanagi, M., Imada, K. and Kajiyama, T. (1966). *J. Polymer Sci. C*, **15**, 263.
65. Hosemann, R. (1963). *J. Appl. Phys.*, **34**, 25.
66. Buckley, C. P., Gray, R. W. and McCrum, N. G. (1969). *J. Polymer Sci. B*, **7**, 835.
67. Buckley, C. P., Gray, R. W. and McCrum, N. G. (1970). *J. Polymer Sci. B*, **8**, 341.
68. Carder, D. (1972). *Ph.D Thesis*, University of Reading.
69. Bassett, D. C. and Carder, D. (1973). *Phil. Mag.*, **28**, 535.
70. Pinnock, P. R. and Ward, I. M. (1966). *Brit. J. Appl. Phys.*, **17**, 575.
71. Ward, I. M. (1960). *Trans. Faraday Soc.*, **56**, 648.
72. Owen, A. J. and Ward, I. M. (1973). *J. Macromol. Sci., Phys.*, **B7**, 417.
73. Thompson, A. B. and Woods, D. W. (1956). *Trans. Faraday Soc.*, **52**, 1383.
74. Pinnock, P. R. and Ward, I. M. (1963). *Proc. Phys. Soc.*, **81**, 260.
75. Ladizesky, N. H. and Ward, I. M. (1973). *J. Mater. Sci.*, **8**, 980.
76. Pinnock, P. R. and Ward, I. M. (1964). *Brit. J. Appl. Phys.*, **15**, 1559.
77. Allison, S. W. and Ward, I. M. (1967). *Brit. J. Appl. Phys.*, **18**, 1151.
78. Farrow, G. and Ward, I. M. (1960). *Polymer*, **1**, 330.
79. Thomas, A. M. (1957). *Nature*, **179**, 862.
80. Takayanagi, M. (1963). *Mem. Fac. Eng. Kyushu Univ.*, **23**, 1.
81. Murayama, T., Dumbleton, J. H. and Williams, M. L. (1967). *J. Macromol. Sci., Phys.*, **B1**, 1.
82. Owen, A. J. and Ward, I. M. (1973). *J. Macromol. Sci., Phys.*, **B7**, 279.
83. Miyasaki, K. and Ishikawa, K. (1968). *J. Polymer Sci. A-2*, **6**, 1317.
84. Hennig, J. (1967). *Kunststoffe*, **57**, 385.
85. Robertson, R. E. and Buenker, R. J. (1964). *J. Polymer Sci.*, *A-2*, **2**, 4889.
86. Kausch-Blecken von Schmeling, H. H. (1970). *Kolloid-Z.*, **237**, 251.
87. Müller, F. H. (1967). *J. Polymer Sci. C*, **20**, 61.
88. Wright, H., Faraday, C. S. N., White, E. F. T. and Treloar, L. R. G. (1971). *J. Phys. D*, **4**, 2002.
89. Curtis, J. W. (1970). *J. Phys. D*, **3**, 1413.
90. Kashiwagi, M., Folkes, M. J. and Ward, I. M. (1971). *Polymer*, **12**, 697.
91. Keller, A. and Sawada, S. (1964). *J. Polymer Sci.*, **74**, 190.
92. Young, R. J., Bowden, R. B., Ritchie, J. M. and Rider, J. G. (1973). *J. Mater. Sci.*, **8**, 23.
93. Keller, A. and Pope, D. P. (1971). *J. Mater. Sci.*, **6**, 453.

CHAPTER 10

ANISOTROPIC CREEP BEHAVIOUR

M. W. DARLINGTON and D. W. SAUNDERS

10.1 INTRODUCTION

This chapter will be concerned with developing a full description of the anisotropic character of the creep behaviour of polymeric materials which have some degree of molecular orientation. Many of the ideas and methods employed are extensions of those used in the very highly developed studies of creep in isotropic materials; here attention will be given to those aspects of the subject which are special to materials showing anisotropic behaviour. Further, no attempt will be made to discuss the large amount of work concerned with how the creep behaviour in a single particular direction depends upon the molecular orientation, as for example in drawn synthetic fibres. Discussion will be concentrated on those studies which are concerned with the *anisotropy* of creep behaviour and are therefore not limited to studies along one direction only.

In many articles fabricated from polymeric materials molecular orientation arises either accidentally, as in say injection moulding, or intentionally, as in say film blowing, as a result of the flow processes involved in the fabrication procedure. Such orientation affects properties but the patterns of orientation obtained are often highly complex and interpretation of properties in terms of structure is a very difficult task which has not yet been attempted in any detail. It was observed by Nielsen[1] that 'little work had been carried out on the effect of orientation on creep and relaxation behaviour of rigid polymers'. This is still largely the case but there is currently an expanding interest and a growing recognition that the effects of orientation can be of real significance in determining engineering behaviour. This work is currently mainly aimed at a full description and understanding of samples specially prepared in the laboratory to have a well controlled and homogeneous pattern of orientation. Further, *full* characterisation of the creep properties of such oriented materials, whilst essentially more complicated than for isotropic materials, promises to lead more readily to a critical and rigorous interpretation of creep and deformation mechanisms in terms of structure. This is apparent already from the work discussed in Chapters 8 and 9 on dynamic properties of well characterised oriented structures and is equally true for creep. Accordingly the study of anisotropy of creep behaviour in oriented materials is of major importance to both the physicist and the engineer and represents also a significant challenge both experimentally and theoretically.

10.2 THEORETICAL BACKGROUND

The classical elastic theory of anisotropic materials at infinitesimal strains has been discussed in an earlier chapter. The starting point is the generalised Hooke's law which can be written as

$$e_{ij} = S_{ijkl}\,\sigma_{kl} \tag{1}$$

in which e_{ij} and σ_{kl} are the tensor components of strain and stress respectively and S_{ijkl} are the tensor components of elastic compliance.

For convenience we shall base our discussion on the more usual matrix formulation of this equation which can be written down in the (mixed) notations

$$e_{xx} = S_{11}\,\sigma_{xx} + S_{12}\,\sigma_{yy} + S_{13}\,\sigma_{zz} + S_{14}\,\sigma_{yz} + S_{15}\,\sigma_{zx} + S_{16}\,\sigma_{xy}$$

$$e_{yy} = S_{21}\,\sigma_{xx} + S_{22}\,\sigma_{yy} + S_{23}\,\sigma_{zz} + S_{24}\,\sigma_{yz} + S_{25}\,\sigma_{zx} + S_{26}\,\sigma_{xy}$$

$$e_{zz} = S_{31}\,\sigma_{xx} + S_{32}\,\sigma_{yy} + S_{33}\,\sigma_{zz} + S_{34}\,\sigma_{yz} + S_{35}\,\sigma_{zx} + S_{36}\,\sigma_{xy}$$

$$e_{yz} = S_{41}\,\sigma_{xx} + S_{42}\,\sigma_{yy} + S_{43}\,\sigma_{zz} + S_{44}\,\sigma_{yz} + S_{45}\,\sigma_{zx} + S_{46}\,\sigma_{xy} \tag{2}$$

$$e_{zx} = S_{51}\,\sigma_{xx} + S_{52}\,\sigma_{yy} + S_{53}\,\sigma_{zz} + S_{54}\,\sigma_{yz} + S_{55}\,\sigma_{zx} + S_{56}\,\sigma_{xy}$$

$$e_{xy} = S_{61}\,\sigma_{xx} + S_{62}\,\sigma_{yy} + S_{63}\,\sigma_{zz} + S_{64}\,\sigma_{yz} + S_{65}\,\sigma_{zx} + S_{66}\,\sigma_{xy}$$

in which $e_{xx}, e_{yy}, \ldots, e_{xy}, \sigma_{xx}, \sigma_{yy}, \ldots, \sigma_{xy}$ are respectively the conventional components of strain and stress used in engineering. $S_{11}, S_{12}, S_{13}, \ldots, S_{66}$ are the corresponding elastic compliances for the material such that $S_{ij} = S_{ji}$. Full discussions of this elastic theory are set out in Nye[2] and Hearmon.[3]

It follows that in the most general case of anisotropy there are 21 independent elastic compliance constants necessary for the complete characterisation of simple elastic behaviour. This number is however progressively reduced, as the number of symmetry elements of the system increases, to a minimum of 2 for the fully isotropic case.

The two particular symmetries of concern here are:

(a) *Fibre symmetry* involving a plane of isotropy perpendicular to a single axis, the *fibre axis*. This axis will be taken to be parallel to the Oz direction and the compliance matrix then takes the form

$$S_{ij} = \begin{matrix} S_{11} & S_{12} & S_{13} & 0 & 0 & 0 \\ S_{21} & S_{11} & S_{13} & 0 & 0 & 0 \\ S_{31} & S_{31} & S_{33} & 0 & 0 & 0 \\ 0 & 0 & 0 & S_{44} & 0 & 0 \\ 0 & 0 & 0 & 0 & S_{44} & 0 \\ 0 & 0 & 0 & 0 & 0 & 2(S_{11}-S_{12}) \end{matrix} \quad (3)$$

which involves only 5 independent compliance constants (since $S_{ij} = S_{ji}$).

In many cases this symmetry will result from drawing the isotropic material under a single uniaxial tension allowing free contraction in perpendicular directions. It will then be convenient to take the draw ratio, defined as the ratio of lengths parallel to Oz before and after drawing, as one useful index of orientation.

(b) *Orthorhombic symmetry* in which case the compliance matrix takes the form

$$S_{ij} = \begin{matrix} S_{11} & S_{12} & S_{13} & 0 & 0 & 0 \\ S_{21} & S_{22} & S_{23} & 0 & 0 & 0 \\ S_{31} & S_{32} & S_{33} & 0 & 0 & 0 \\ 0 & 0 & 0 & S_{44} & 0 & 0 \\ 0 & 0 & 0 & 0 & S_{55} & 0 \\ 0 & 0 & 0 & 0 & 0 & S_{66} \end{matrix} \quad (4)$$

with 9 independent compliance constants.

Polymeric materials with such symmetry will result from deformation of isotropic material in which no two of the principal strains are equal. Procedures which have been used to produce orthorhombic symmetry include pure shear, simple shear and combinations of uniaxial drawing and rolling.

In the above it is implied that the symmetry of the oriented material is simply related to the geometry of the drawing deformation imposed upon the isotropic material. If this implication is accepted then it follows that the most general anisotropy to be expected in polymeric materials in which the orientation results only from a simple deformation process, which is a usual case, would have orthorhombic symmetry requiring only 9 independent compliance constants and not the more general anisotropy typical of a triclinic crystal which requires the full 21 independent constants.

It is important to note that in polymeric materials one frequently encounters marked inhomogeneity of anisotropy. This occurs especially in articles in which the orientation results from solidification of a melt in which flow orientation has not had time to relax out. In such cases the variation of anisotropy within the body is related to the inhomogeneity of the deformation during processing which produces orientation and/or to the variation of relaxation of orientation from point to point within the body combined with the variations in thermal history. Such inhomogeneity represents the most difficult aspect of characterisation of mechanical behaviour in so-called 'simple' mouldings. Discussion will be primarily related to idealised systems in which great care has been taken to ensure homogeneity of drawing deformation.

The theory outlined above is rigorous only for infinitesimal elastic deformation. Creep of polymeric materials is explicitly concerned with time dependence and implicitly with finite strains and therefore non-linear behaviour. The nature of the non-linear behaviour is complex and varies not only from material to material but also with direction within a given sample of material, *i.e.* the non-linearity of behaviour is anisotropic. It is found, therefore, that on a particular definition of strain the behaviour of a sample may appear to be linear in one direction and significantly non-linear in another. Such a phenomenon is demonstrated in results presented below.

There appears to be no useful formalism capable of giving a rigorous and succinct description of the complicated behaviour observed in oriented polymers and therefore the formalism of classical elastic theory is retained for describing time dependent finite strain deformation; accepting the lack of rigour and investigating its utility. The individual compliance constants S_{ij} are allowed to become functions of time and of stress and/or strain in any one experiment.

Concentrating initially on time dependence for infinitesimal strains, the correspondence principle relating viscoelastic and elastic behaviour, well established for isotropic systems,[4] may be simply extended to apply to the anisotropic case.[5] There is, however, a difficulty in showing that the compliance matrix S_{ij} will necessarily have the same symmetry properties in the viscoelastic case as in the classically elastic case.[6] This difficulty arises from the thermodynamic nature of part of the argument used in proving symmetry. In the viscoelastic case the proof would depend upon the less well established principles of irreversible thermodynamics. No discussion on this point will be attempted; the symmetry properties of S_{ij} as determined in elastic theory will be accepted and its validity examined in the light of the experimental data available. This data shows that there may be systematic deviations from the assumptions in work at finite strains and further work is needed in this area. However, the manner in which these deviations occur does not detract significantly from the utility of the simple formalism in many cases.

The linear formalism implies,[7] that for many classes of problems in deformation, solutions in the viscoelastic case can be simply obtained from elastic solutions by Laplace transform methods as for isotropic materials. Further in particularly simple cases this reduces to simple replacement of constants in elastic solutions by their time dependent analogues in creep. Most of the work of determining significant creep parameters for oriented materials falls into this latter class.

The experimental work discussed in this chapter consists, in the main, of simple tensile creep measurements on specimens cut in various directions with respect to the symmetry axes of the oriented material. Let us consider a sheet of material in which the principal directions of the deformation used to produce the orientation were parallel to Ox, Oy and Oz with Ox parallel to the sheet thickness direction.

Inspection of eqn. (2) shows that for application of a simple stress σ_{zz} in the Oz direction we obtain a deformation given by

$$e_{xx} = S_{13}\,\sigma_{zz}$$

$$e_{yy} = S_{23}\,\sigma_{zz}$$

$$e_{zz} = S_{33}\,\sigma_{zz}$$

$$e_{yz} = S_{43}\,\sigma_{zz}$$

$$e_{zx} = S_{53}\,\sigma_{zz}$$

$$e_{xy} = S_{63}\,\sigma_{zz}$$

$$(5)$$

since all other stress components are zero. S_{33} is therefore the reciprocal of the Young's modulus, E_{33} in the Oz direction. For the creep case we write

$$e_{zz}(t) = S_{33}(t)\sigma_{zz}$$

where $S_{33}(t)$ is the creep compliance and

$$\frac{1}{S_{33}(t)} = E_{33}(t)$$

the creep modulus for the Oz direction.

Similarly

$$S_{13}(t) = \frac{e_{xx}(t)}{\sigma_{zz}}$$

is termed the *lateral creep compliance* relating contraction in the Ox direction to the applied tensile stress along Oz and $S_{23}(t)$ is the other lateral creep compliance for this loading. We note further that for the materials symmetries considered, eqns. (3) and (4), the other materials parameters in eqn. (5) are all zero.

Analogous procedures for tensile stresses parallel to Ox and Oy separately define the nature of $S_{11}(t)$, $S_{22}(t)$ and $S_{12}(t)$.

Application of shear stresses σ_{yz}, σ_{zx} and σ_{xy} separately lead to the relations such as

$$e_{xx} = S_{14}\,\sigma_{yz}$$
$$e_{yy} = S_{24}\,\sigma_{yz}$$
$$e_{zz} = S_{34}\,\sigma_{yz}$$
$$e_{yz} = S_{44}\,\sigma_{yz} \qquad (6)$$
$$e_{zx} = S_{54}\,\sigma_{yz}$$
$$e_{xy} = S_{64}\,\sigma_{yz}$$

in which, for the symmetries of interest, only S_{44} is non-zero and it follows that $S_{44}(t)$ is therefore a shear creep compliance, the reciprocal of a shear creep modulus $G_{44}(t)$ with analogous significance for $S_{55}(t)$ and $S_{66}(t)$.

It is apparent that in the case of orthorhombic symmetries 9 independent creep functions are necessary to characterise fully the creep behaviour of the material. For materials with fibre symmetry only 5 such functions are needed.

Creep contraction ratios, which are time dependent functions analogous to the Poisson's ratios of elastic theory can be defined.

Thus

$$v_{13}(t) = -\frac{e_{xx}(t)}{e_{zz}(t)} \text{ for loading in the } Oz \text{ direction}$$

$$= -\frac{S_{13}(t)}{S_{33}(t)}$$

whereas

$$v_{31}(t) = -\frac{e_{zz}(t)}{e_{xx}(t)} \text{ for loading in the } Ox \text{ direction}$$

$$= -\frac{S_{31}(t)}{S_{11}(t)} = -\frac{S_{13}(t)}{S_{11}(t)}$$

and it follows that $v_{13}(t) \neq v_{31}(t)$. Other contraction ratios $v_{23}(t)$, $v_{32}(t)$, $v_{12}(t)$ and $v_{21}(t)$ are similarly defined.

For both fibre and orthorhombic symmetries it can be shown from eqn. (2) (see Hearmon,[3] or Nye,[2]) that for tensile loading in the plane of the sheet, yOz, at an angle θ to Oz the Young's modulus $E(\theta)$ is given by

$$\frac{1}{E(\theta)} = S_{22} \sin^4 \theta + (2S_{23} + S_{44}) \sin^2 \theta \cos^2 \theta + S_{33} \cos^4 \theta \qquad (7)$$

In the creep case the $E(\theta)$, S_{22}, S_{23}, S_{44} and S_{33} are replaced by their time dependent analogues and thus for loading at $\theta = 45°$

$$\frac{4}{E(45°, t)} = S_{22}(t) + 2S_{23}(t) + S_{44}(t) + S_{33}(t) \qquad (8)$$

in which $E(\theta, t)$ is the creep modulus at angle θ and time t.

It should be noted that in the case of fibre symmetry (in which there are only 5 independent creep functions) tensile creep measurements on specimens cut at only three angles to the fibre axis (say 0°, 45° and 90°) will give three independent combinations of the five, say $S_{22}(t)$, $[2S_{23}(t) + S_{44}(t)]$ and $S_{33}(t)$. Whilst if lateral contraction measurements are also carried out during creep then all 5 functions, including the shear compliance $S_{44}(t)$, can be obtained without recourse to measurements in shear.

For orthorhombic symmetry on the other hand, tensile creep and lateral compliance measurements on specimens cut from oriented sheet will yield only 6 of the 9 required creep functions; those not accessible by this method being $S_{11}(t)$, $S_{55}(t)$ and $S_{66}(t)$. The two shear compliances $S_{55}(t)$ and $S_{66}(t)$ can be obtained by torsional creep experiments, but these need to be carefully designed and involve complex experimental procedures.[8,9] The only possibility for measurement of $S_{11}(t)$ on sheet appears to be by compressive creep techniques, however, one would expect substantial experimental difficulties largely associated with strain measurement and specimen geometry. There appears to be no reported evaluation of the full characterisation of creep for the case of orthorhombic symmetry.

So far the parameters defined, compliances, moduli and contraction ratios, are in forms which can be rigorously interpreted for infinitesimal time dependent deformation under constant stress, thus creep moduli $E(\theta, t)$ are functions of angle and time. Extension to accommodate non-linear behaviour at finite strain is obtained by allowing the quantities to become also functions of stress or strain. Thus modulus has the form $E(\theta, t, \overset{\sigma}{e})$ and compliance functions $S_{ij}(t, \overset{\sigma}{e})$. Such an extension is not rigorous but is useful.

There is a need therefore to obtain and present results over significant ranges of two or three variables. The scope of the problem and the presentation adopted is discussed below.

10.3 EXPERIMENTAL METHODS

Fundamental studies of creep behaviour in oriented polymers have been largely confined to sheet material homogeneously oriented by carefully controlled deformation. Such restriction is essential if meaningful *materials parameters* are to be evaluated. Other work which has been reported, for example that on injection-moulded samples, suffers from a basic difficulty in interpretation since orientation is non-uniform throughout the sample and analysis of the non-uniformity is prohibitive. Clearly the behaviour of non-uniform systems can be highly complex and may include features which could be falsely attributed to more general symmetries than can occur as a materials property. Reported results of measurements on such samples whilst possibly of use in some applications have no basic significance since at best they represent an unspecified average of behaviour over the non-uniformity present.

The most common technique employed to date has been that of creep in uniaxial tension. It was shown above that with the inclusion of lateral strain measurements this is a powerful technique giving access to up to 6 independent creep compliance functions. This is more than for any other known method. It further has the overwhelming advantage over many methods, such as say torsional or flexural creep, that the stress is sensibly uniform over the working volume of the specimen. This advantage is paramount in studies of materials displaying non-linear behaviour in creep since analysis of the non-uniform stress situation in non-linear systems is not well developed. Attempts to overcome the non-uniform stress situation in torsion, by recourse to, say, torsion of thin walled tubes, lead to severe difficulties in specimen preparation in oriented materials, when anisotropy of behaviour is to be studied.

Tensile creep, with or without lateral strain measurements, has been augmented by torsional creep, but usually creep is then limited to very small strains hopefully to avoid the non-linear behaviour. The results are therefore limited. Torsional creep measurements do seem, however, to constitute the only known method at this time for evaluating $S_{55}(t)$ and $S_{66}(t)$ for oriented sheet materials with orthorhombic symmetry.

Tensile and torsional creep methods will be discussed below. For discussion of the very specialised techniques which have been used for studies of anisotropy of compliance in fibres and monofilaments, such as the Hertzian contact technique by Hadley et al.[10] and Pinnock et al.,[11] readers are referred to the original papers and to Ward.[12] It should be noted that such methods are not well adapted to creep measurement and are mainly used for determination of isochronous parameters. They suffer from all the limitations, referred to above, associated with non-uniform stress situations.

10.3.1 Creep in uniaxial tension

Experimental techniques for creep measurements on isotropic materials are highly developed and have been discussed in detail by Turner and coworkers.[13,14] Discussion here will be limited to the special requirements associated with studies of anisotropy.

The usual method of preparing homogeneously oriented specimens from thermoplastics involves subjecting sensibly isotropic sheet to large permanent deformations by hot or cold drawing and/or rolling techniques. These deformations may involve strains of several hundred per cent which, since they occur essentially at constant volume, involve large reductions in some dimensions. Many investigations have involved tensile strains in excess of 400% and consequent reductions in width and/or thickness by factors greater than 2. In some cases even larger figures are involved.

Since it is necessary to cut samples in a variety of directions in the plane of the sheet the overall length of such samples can be severely limited by the dimensions of the drawn sheet. If drawing procedures are to be kept within the scope of normal laboratory facilities the thickness of the drawn sheet may be less than 1 mm and the width or length less than 5 cm. It follows that creep specimens cut from drawn sheet may be both short and lacking in rigidity when compared with conventional isotropic specimens. The lack of rigidity is aggravated for relatively soft materials such as low density polyethylene (LDPE).

The following constraints are therefore relevant to the design of tensile creep apparatus if it is to be versatile in application to a wide range of materials and orientations:

(a) In all creep measurements it is of prime importance that strains due to extraneous loads, such as creep preloads, extensometers, etc., shall be small compared with the creep strains under study. In the case of non-rigid specimens referred to above this requirement becomes a severe constraint, requiring great care in ensuring friction-free movements in the loading system and extensometers, and effectively precludes the use of extensometers whose mass would have to be supported by the specimen.

(b) The requirement for axial loading is also made more severe by the inability of the specimen to contribute to the maintenance of this condition. A completely friction-free guidance system for the moving grip becomes highly desirable.

(c) The limited specimen length necessitates a highly sensitive displacement measurement system if there is the requirement for measurement of small strains. In detailed studies of polymer systems, capability for accurate measurements of strains of 0·1% is desirable.

(d) The limited specimen length also militates against grip separation measurements of strain even though such measurements may be made with high sensitivity. Suitable extensometry working on a gauge length clear of grips and with carefully designed attachment techniques becomes desirable.

The nature of end effects in anisotropic materials falls into two categories. (i) The problem of slippage, pull out and stress distribution at the grips. All these depend upon the stiffness of the specimen and therefore have different significance for specimens of the same material cut at different directions, making comparisons difficult.

(ii) In specimens cut at finite angles to the symmetry axes $Oxyz$ it can be shown from eqn. (2) that simple tensile loading leads to a shear strain in the yOz plane superposed on the expected tensile deformations (shear-coupling). The resulting sideways displacement of the 'free' end of the specimen is in practice largely inhibited by the constraints imposed by the loading system and the grips, or completely restrained by a good grip-guidance system. Accordingly there is an area of inhomogeneous strain extending into the specimen from the grips which would not be present in isotropic specimens or in specimens loaded parallel to Ox, Oy or Oz. It is clearly desirable to make strain measurements remote from those regions which in short specimens can occupy a significant proportion of the overall length. The problem is minimised by the use of extensometers. Discussion of this behaviour (in highly anisotropic composites) has been given by Halpin and Pagano.[15]

An apparatus specially developed to meet these conditions which has proved to work reliably and with high sensitivity for a wide range of polymers, from LDPE to highly rigid short fibre reinforced thermoplastics, over a wide temperature range is that described by Darlington and Saunders[16] and Darlington (1971, unpublished).

A general view of the apparatus is shown in Fig. 1. The loading assembly consists of a 5:1 ratio lever-loading arm (1) supported on knife edge bearings, with facility for balance adjustment and application of accurately determined preloads by moveable counterweights. Load is transferred to the specimen via small clamps fixed to the specimen which engage with hooks (2) on the loading assembly. The lower hook is fixed rigidly to the main frame whilst the upper hook is constrained to move axially on the specimen axis by a linear spring guide (3). This movement is entirely free of friction and since there are no sliding surfaces is unaffected by temperature and corrosion.

Tensile and lateral strains are monitored by an extensometer, the weight of which is supported by the main frame of the machine, and the design is such that the maximum loads applied to the specimen by the extensometer are negligibly small compared with the creep load. In Fig. 1 the tensile extensometer (4) and the extensometer measuring thickness strain (5) are shown. A third extensometer measuring width strain can be accommodated simultaneously from a mounting attached to the front face of the frame.

The basic mode of action of the extensometers can be seen from Fig. 2.

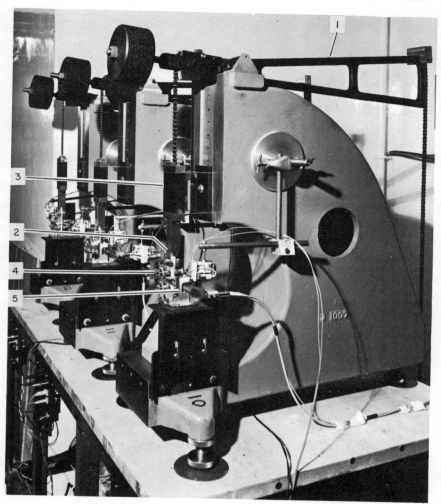

Fig. 1. General view of uniaxial tensile creep machine (Darlington, 1971, unpublished). 1, Lever arm; 2, upper hooks; 3 linear spring guide; 4, tensile extensometer; 5, lateral extensometer (for thickness strain).

Fig. 2. Extensometer details. 1, Arms of tensile extensometer; 2, screw pins of tensile extensometer; 3, transducer; 4, arms of lateral extensometer; 5, brass contact pieces of lateral extensometer; 6, glass plates.

The tensile extensometer consists of two arms (1) supported on fine bearings at their mid-points to rotate freely in a vertical plane. The arms are attached at one end to the specimen by light contact between the specimen surface and screw pins (2) attached to the arms, so that the gauge length for the measurement of tensile strain is the distance between the screw pins. As the specimen elongates in creep this distance increases, the arms rotate through a small angle about their pivots and the separation between their far ends is monitored by the sensitive and stable linear displacement capacitance transducer (3). The output from the transducer is processed to give digital display and print-out of the required displacement (the transducer and associated electronics are manufactured by Rank Taylor Hobson). The lateral extensometer shown in Fig. 2 is positioned for thickness strain measurement. Its principle of operation is identical to that of the tensile extensometer. The two arms (4) rotate in a horizontal plane and form effectively a caliper with rounded brass domes (5) forming the contact between the arms and the specimen. Contact is maintained by light spring pressure. Thin glass microscope cover slides (6) are inserted between the specimen faces and the brass domes to obviate indentation by the brass domes.

Full details of the apparatus are given in the above references.

Specimens with gauge lengths as small as 12 mm can be accommodated with simultaneous measurement of length and thickness strains. All three principal strains may be measured within a gauge length of 26 mm. Specimens are carefully machined from oriented sheet by high speed routing to produce the necessary surface finish.

Sensitivity is such that movements of <0.25 μm can be readily detected and accurate tensile strain measurements made over the range 0.1 to $>5\%$ if required. The system has been proved over a number of years by inter-laboratory comparisons, especially with the highly sophisticated apparatus for creep in isotropic materials developed by Turner.[13] Stability for long term creep measurements has been proved for periods up to 6 months. Full details of the apparatus proving trials are given by Clayton et al.[17]

An alternative method of measuring tensile and lateral strains simultaneously, avoiding end effects and the use of extensometers, is the photographic method developed by Ladizesky and Ward.[18] A fine pattern of grid lines is deposited on the polymer surface by vacuum deposition of aluminium through an electron microscope specimen grid. The grid is carefully oriented with respect to specimen axes and loading direction and photographed repeatedly during creep deformation. Special methods were used for the photographic processing followed by travelling microscope measurements to high accuracy of the dimension of the grid images on the plates. The authors claim an accuracy of better than $\pm 3\%$ in the measurement of strains of magnitude 1% and corresponding accuracy of $\pm 5\%$ in the measure of creep contraction ratios. The design of the loading and guidance system was relatively crude involving a simple frame with sliding cross members guided by four linear bearings.

The authors appear to present no measurements of the time dependence of either lateral or axial strains generated on this apparatus and it may be that the performance of the loading frame was not suitable for accurate creep work. They do give 10 second isochronous compliance data, both lateral and axial, obtained on this apparatus and there is in principle no reason why the optical technique should not be used on a more refined loading frame to generate time dependent data.

Ladizesky and Ward did present a few examples of time dependence of tensile compliances $S_{33}(t)$ and $S_{22}(t)$ and S(isotropic) but it seems likely that these were obtained on the simple creep apparatus described by Gupta and Ward.[19] This apparatus employed simple direct dead weight loading and took the observed movement of the free grip relative to the base frame of the apparatus as the basis of strain measurement. An inductance bridge method in conjunction with a micrometer screw was used to monitor the movement. The creep measurements reported are limited to c. 1000 s and a sensitivity of only 2.5×10^{-4} cm is claimed. It is apparent that such a system involves no features of particular relevance to the measurement of anisotropy of creep behaviour and suffers all the disadvantages listed above.

Buckley and McCrum[20] have recently published work on the anisotropy of creep obtained using a tensile creep apparatus based on precision measurement of clamp displacement. Creep strains up to a maximum of 0.1% were used. The creep compliance was subject to error limits of $\pm 5\%$ but the scatter of points on a given creep curve was always less than 0.5% provided creep was terminated after 60 s below room temperature or after 180 s above room temperature. Details of the apparatus have not yet been published.

Reference has been made earlier to the well known creep apparatus developed by Turner[13,14] and used in many laboratories for accurate creep measurements on a wide range of isotropic polymers over a wide temperature range. It is apparent that although not suitable for measurements on the small samples usually encountered in work on homogeneously oriented sheets this apparatus, with its guidance system and optical extensometry, is admirably suited for axial strain measurements on anisotropic materials when large rigid samples are readily available, and examples of its use for inhomogeneously anisotropic injection mouldings will be discussed below. No measurements of lateral strain with this apparatus have yet been reported. It is apparent that there are other creep apparatuses in regular use which could be employed in this way.

10.3.2 Creep in torsion
The requirements that the apparatus can handle small, weak specimens with negligible friction in the moving parts (as discussed in the previous

section) are more easily satisfied in torsion than in tension, and apparatus designs already in existence for use with isotropic materials are easily adapted.

A versatile and accurate apparatus has been described by Morrison *et al.*[21] in which the torque is applied to the specimen by passing a constant current through a coil suspended in a radial magnetic field. Modifications of this system have been used successfully to study the creep of oriented polymers by McCrum and Morris[22] and by Clayton *et al.*[17]

Simpler torsional creep apparatus used in studies on oriented polymers has been described by Raumann[23] and by Ladizesky and Ward.[8]

All the above methods suffer the disadvantage that strain measurement is accomplished by monitoring the relative grip rotation. If the specimens are both small and cut from highly anisotropic material, end effects may be significant and difficult to eliminate. The extraction of meaningful materials parameters from torsion measurements is difficult, even for the case of linear elastic anisotropic materials, except in special cases. The techniques used depend upon the theory of St. Venant and demand careful selection of specimen geometry and orientation. Even then they do not appear to be completely rigorous. The extension of such methods to the time-dependent case, and further to the non-linear viscoelastic case, increases the difficulties which have been discussed by Ladizesky and Ward.[8] Probably because of these difficulties, little use has yet been made of torsional creep techniques in studies on oriented polymers.

10.4 PROBLEMS OF DATA PRESENTATION

A major problem in data presentation for anisotropy of creep behaviour arises from the dependence of behaviour on many parameters. In Section 10.2 the modulus for a given degree of molecular orientation under given environmental conditions for a single material was given as a function of angle, time and either stress or strain whilst the material compliance functions for these conditions were functions of time and stress or strain. Changes of temperature, composition, orientation, structure, etc., will of course affect the whole pattern of behaviour.

It is necessary therefore to establish systematic methods for obtaining and presenting the data so as to give a clear overall description of behaviour involving the minimum of experimental work. The importance of such methods for isotropic material has been reviewed by Turner.[24] In the anisotropic case the problems are more extensive.

The value of the simplified technique for producing isochronous stress–strain curves for non-linear isotropic materials by successive loading and unloading of a single sample have been amply demonstrated over many years and fully described elsewhere.[14,25] These techniques become even more valuable in studies of anisotropy, where samples may be difficult to obtain in large numbers and where the scope of the problem is much larger. A considerable proportion of work on oriented materials reported in the literature is essentially confined to this measurement and does not include studies of time dependence of behaviour. Detailed work has been carried out validating this procedure for oriented materials by comparison of the isochronous stress–strain data with isochronous sections from families of creep curves.[26,27]

In studying time dependence, *i.e.* creep behaviour, it is necessary to carry out tests at several stress levels in each direction of interest in the oriented material. This can involve a prohibitive amount of experimental work and, in practice, little is generally lost by reducing the tests to, say, one creep curve and one isochronous stress–strain curve in each direction. The problem then becomes one of selection of the absolute value of stress for each of the creep curves and is most severe when non-linearity and its anisotropy are well developed. The choice of stress levels is arbitrary but interesting special cases are (a) equal stress levels at all angles and (b) equal strain ranges at all angles.

In the case of (a), since there can be substantial variations in both compliance and strength (creep rupture) with angle, this may result in creep in some directions involving extremely low strains, and therefore presenting severe measurement problems, whilst in other directions very rapid large creep or rupture may occur thus limiting the information available. It has therefore been found preferable to employ (b) and to choose stress levels in different directions so as to produce equal strains after a specified creep time in all directions. Furthermore if correlation of creep behaviour with deformation mechanisms is sought it may well be desirable to compare the polymer response when the different mechanisms produce similar strains. Selection of appropriate stress-levels is achieved by use of the isochronous stress–strain curves.

10.5 UTILITY OF ELASTICITY THEORY

A brief outline of the extension of the formalism of the classical theory of elasticity for the description of non-linear viscoelastic behaviour was presented in Section 10.2. In this section the utility of certain aspects of this approach will be examined in the light of experimental evidence.

10.5.1 Angular variation of behaviour

A high degree of anisotropy of both isochronous compliance and non-linearity of behaviour is found in highly drawn LDPE having fibre symmetry, as shown by Raumann and Saunders[28] and Darlington and Saunders[26,29] and thus provides a suitable system in which to examine the utility of the theory.

The variation of the isochronous modulus at 100 s with the magnitude of the creep strain at 100 s for strains in the region 0·1–10% in samples cut at various angles to the fibre axis is shown in Fig. 3. The data were obtained using the isochronous stress–strain procedure, previously referred to, on LDPE drawn at 20°C so as to produce fibre symmetry with a draw ratio of 4·2. In this figure horizontal straight lines would indicate linear viscoelastic behaviour. The strain at which significant deviation from

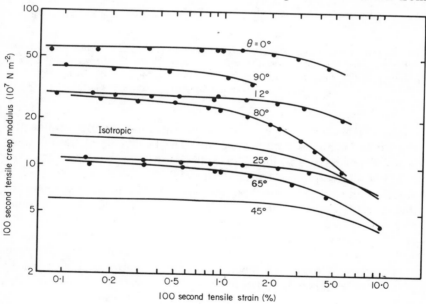

Fig. 3. Isochronous modulus-strain data for specimens cut at various angles, θ, to the fibre axis of cold-drawn LDPE, draw ratio 4·2. In all diagrams, except where explicitly stated otherwise, test temperature was 20°C. (After Darlington and Saunders, 1970).[26]

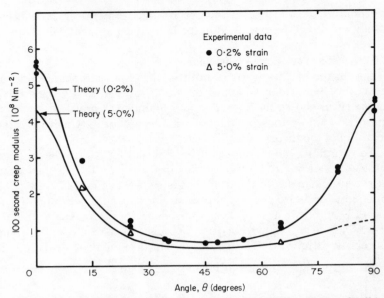

Fig. 4. *Isochronous modulus-angle data, at 100 second tensile strains of 0·2% and 5·0%, for specimens cut at various angles, θ, to the fibre axis of cold-drawn LDPE, draw ratio 4·2. (Reproduced from* SPE Technical Papers, **19** *(May 1973) by permission of the Society of Plastics Engineers.)*

linearity first occurs changes radically with direction and the non-linearity is greatest for angles approaching 90°. The behaviour of the undrawn (isotropic) material is included for comparison.

In Fig. 4 cross-plots at tensile strains of 0·2% and 5% are presented as points (with some additional points). The full lines in this figure are the theoretical lines derived from eqn. (7) fitted, in the case of 0·2% strains, at $\theta = 0$, 45 and 90° and in the case of 5% strain at $\theta = 0, 45$ and 80°. No experimental value of $E(90)$ was available above a tensile strain of 1·7% owing to early fracture being a feature of behaviour in this direction.

It is apparent from the agreement between the theoretical lines and the points for the remaining values of θ that the angular variation of modulus suggested by eqn. (7) applies over a wide range of tensile strains.

During the above experiments measurements of the lateral strain in the sheet thickness direction were also made. The ratio of this strain to the tensile strain, designated $v_{th}(t, \theta)$, has been derived for $t = 100$ s and various values of θ. This data for tensile strains in the range 1–6% is presented, as points, in Fig. 5. The theoretical relation for this angular variation is given by

$$v_{th}(t, \theta) = \frac{\dfrac{v_{th}(0)}{E(0)}\cos^2\theta + \dfrac{v_{th}(90)}{E(90)}\sin^2\theta}{\dfrac{\sin^4\theta}{E(90)} + \sin^2\theta\cos^2\theta\left(\dfrac{4}{E(45)} - \dfrac{1}{E(90)} - \dfrac{1}{E(0)}\right) + \dfrac{\cos^4\theta}{E(0)}} \quad (9)$$

and is shown as the full line in the figure. This line is completely determined by the 100 second values of $E(0)$, $E(45)$, $E(90)$, $v_{th}(0)$ and $v_{th}(90)$. It is apparent from the agreement between the theoretical line and the points at all angles other than 0° and 90° that the theoretical relation applies over a wide range of tensile strain. In considering the scatter of the points in this figure it should be remembered that the general level of lateral displacement associated with lateral strain in specimens of this thickness (c. 1 mm) is very small (c. 10^{-5} cm) and demands extreme sensitivity in measurement.

It should be noted that the basis for these predicted variations with angle is founded in infinitesimal strain theory, but it may be concluded that the formalism derived from classical theory is useful for finite strain non-linear behaviour.

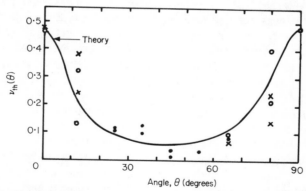

Fig. 5. *Variation of the 100 second thickness contraction ratio $v_{th}(\theta)$ with angle θ.* ○, *measured using two, single transducer extensometers (1% tensile strain)*; ×, *as above (5% tensile strain)*; ●, *measured using travelling microscope plus one, double-transducer extensometer (tensile strains in range 3% to 6%) (from Darlington and Saunders[26]).*

10.5.2 Symmetry of the compliance matrix

In Section 10.2 the effect of materials symmetry on the number of independent compliance constants S_{ij} for linear elastic behaviour was presented. For the case of fibre symmetry, eqn. (3), we have in particular, $S_{13} = S_{31} = S_{23} = S_{32}$. For the linear viscoelastic case Rogers and Pipkin[6] were able to show theoretically that without recourse to the arguments of irreversible thermodynamics it was not possible to show that $S_{13} = S_{31}$ and $S_{23} = S_{32}$. Further the validity of all these equalities must be in doubt in non-linear behaviour at finite strains.

The techniques of creep measurement described above allow of some experimental examination of these equalities.[27]

It should be noted that S_{23} is the compliance relating a strain along the Oy direction to a stress along the Oz direction whilst S_{32} relates a strain along Oz to a stress along Oy. These two quantities may therefore be obtained in separate experiments on the same material by lateral contraction measurements (width) on samples cut at 0° and 90° to the fibre axis. Similarly S_{13} and S_{12} may be obtained separately from lateral contraction measurements (thickness) on those two specimens.

The results of such measurements are shown in Fig. 6. The measurements were carried out at 20°C on samples cut from a sheet of polymethylmethacrylate (PMMA) drawn with fibre symmetry at 124°C to a draw ratio of 3·3 and birefringence $\Delta n = 0\cdot001$. This represents nearly the highest orientation obtainable in this material by the methods available. The figure shows that within experimental error $S_{13} = S_{23}$ over the full timescale of the measurements (c. 30 to 5×10^5 s) whereas S_{32}, although equal to S_{23} at short times, increases more rapidly than S_{23} at longer times. The maximum tensile strains in these experiments were 1·14% and 1·27% for the 0° and 90° specimens respectively The stress levels were chosen to give identical 100 second strains of 0·75% and were 25·4 and 20·8 MN/m^2 for 0° and 90° specimens respectively. It is apparent that if the comparisons had been made at equal stresses in this non-linear material the divergence of S_{32} and S_{23} would have been even greater.

It therefore appears that at these long times and/or high creep stresses or strains the symmetry of the matrix can not be assumed. It should be noted that the anisotropy of the tensile compliances was not very high, covering only a 20% spread at short times, but that the anisotropy was becoming more marked as time progressed (see Section 10.7). This preliminary work shows clearly that the matrix symmetry cannot be assumed, but more detailed investigations are required to establish whether the divergence results from finite strain effects at high

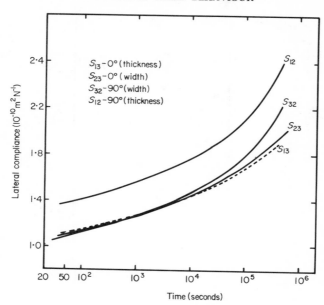

Fig. 6. Variation of lateral compliance with time during tensile creep of specimens cut at 0° and 90° to the fibre axis of a sheet of oriented polymethylmethacrylate (drawn at 124°C, birefringence 0·001).[27] Note that the values of the lateral compliances are all negative.

stresses or from time effects. Nor is the generality of the effect established. Halpin and Pagano[15] have examined this problem in a composite nylon fibre reinforced rubber and claim to have established $S_{ji} = S_{ij}$ at finite strains but in a linear system. However, they do not give sufficient experimental detail to allow a critical assessment to be made.

10.6 ANISOTROPIC CREEP IN SEMI-CRYSTALLINE POLYMERS

The first detailed study on anisotropy of stiffness behaviour in LDPE was reported by Raumann and Saunders[28] for material drawn at room temperature to a range of draw ratios from 1·0 (undrawn, isotropic) to 4·6 (very highly drawn). These initial measurements, reproduced in Fig. 7, were carried out in cyclic repeated loading and highlighted at least two

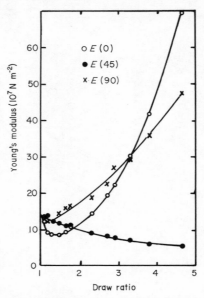

Fig. 7. Variation of E*(0),* E*(45) and* E*(90) with draw ratio for cold-drawn LDPE (from Raumann and Saunders*[28]*).*

unexpected features in the form and development of anisotropy with molecular orientation: (a) a decrease at low draw ratios in the tensile moduli $E(0)$ and $E(90)$ such that $E(0) < E(90) \sim E(45)$ and (b) anisotropy at high draw ratios such that $E(0) > E(90) \gg E(45)$. This work was not concerned with time or stress effects and clearly indicated the need for further work in creep and dynamic behaviour. These authors recognised that a dominating feature at high draw ratios, viz. the low value of $E(45)$, was simply related to a large shear compliance, S_{44}, corresponding to easy-shear parallel or perpendicular to the fibre axis and that further studies of such a feature could lead to basic understanding of fundamental deformation mechanisms. Such creep studies have been reported by Gupta and Ward,[19] Gupta et al.[30] and Darlington and Saunders.[26,29]

The creep behaviour of LDPE drawn, at room temperature, to high draw ratio was described by Darlington and Saunders.[26,29] The 100 s isochronous data has already been given in Figs. 3, 4 and 5, and discussed in Section 10.5. The creep response at stresses chosen to give 100 second tensile strains of c. 1% and 5% is shown in Figs. 8 and 9

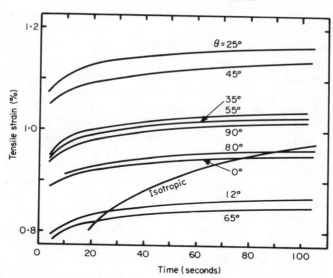

Fig. 8. Creep curves for specimens cut at varous angles, θ, to the fibre axis of cold-drawn LDPE, draw ratio 4·2 (100 second tensile strains in the range 0·85–1·15%) (Darlington and Saunders[26]).

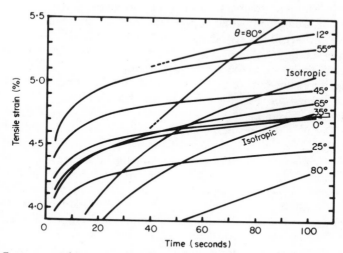

Fig. 9. Creep curves for specimens cut at various angles, θ, to the fibre axis of cold-drawn LDPE, draw ratio 4·2 (100 second tensile strains in the range 4·5–5·3%) (Darlington and Saunders[26]).

respectively. The creep curves for the isotropic material are included for comparison and show a higher creep rate than those for any direction in the drawn material, with the exception of the $\theta = 80°$ specimen at large stresses. It is of interest to note that examination of recovery behaviour following removal of the creep loads showed that for creep strains up to at least 2%, the oriented specimens recovered 98% of their final creep strain within a recovery time of ten times the creep time. Furthermore, the indications were that at even higher creep strains (i.e. in the region of 5%) the specimens showed virtually complete recovery given sufficient recovery time.

Mention has been made above of the possible importance of an 'easy shear' mechanism, associated with a high value of S_{44}, and corresponding to laminar shear parallel or perpendicular to the fibre axis. The relevance of this to the data presented merits discussion.

Wide angle X-ray diffraction studies on the highly drawn sheets show that the crystallographic c-axes are highly aligned parallel to the fibre axis and the easy shear therefore occurs parallel or perpendicular to the c-axes, although its exact nature is not known from these studies.

In tensile loading of a specimen, the maximum shear stresses occur on planes with their normals at 45° to the tensile stress direction. In tensile loading of a 45° specimen we are interested therefore in laminar shear on planes with their normals parallel to the sheet surface which either contain or are perpendicular to the fibre axis (i.e. are at 45° to the tensile stress direction). Since easy shear on either set of planes is consistent with the observed high value of S_{44}, additional measurements (such as X-ray diffraction) must be made during deformation in order to determine the relative importance of the possible molecular deformation modes. Such measurements were not attempted in the above study.

If the easy shear mechanism is the dominant feature of deformation, since it occurs on planes perpendicular to the sheet surface it follows that the deformation will involve no change in sheet thickness. In Fig. 5, the lateral (thickness) strain measurements during creep are shown as lateral contraction ratio vs. θ. It is clear that at 45°, where the easy shear mechanism is dominant, the thickness changes are indeed very small but increase gradually as the angle departs from 45°; the increase becoming rapid near to 0° and 90°, thus confirming the proposed mechanism. Measurements over a 100 second tensile strain range from 0·1% to 10% suggest that this one simple mechanism operates over the wide strain range.

The quantitative significance of the easy shear mechanism in the highly drawn material was discussed by Darlington and Saunders.[26] The differences in creep rate at angles of 45°, 55° and 65°, apparent in Fig. 9, were shown to be simply related to the increase in resolved shear stress, on the easy shear planes, due to rotation of the shear planes with increasing tensile strain during creep at finite strains (see Section 10.3.1).

The low creep rates for the oriented material in Fig. 8 indicated that the easy shear process which produced the high anisotropy of modulus occurred before the start of the creep measurements. This is consistent with the dynamic mechanical data of Stachurski and Ward[31] on similar cold drawn LDPE sheets, which showed a highly anisotropic relaxation, attributed to shear parallel to the oriented crystalline chains, at a frequency of 50 Hz and temperature of 0°C. From the above it would be expected that high creep rates would be obtained on specimens cut at 45° to the fibre axis, when the temperature is lowered into the relaxation region. Baker and Darlington (1973, to be published) have confirmed that such high creep rates are observed at times between 5 and 2×10^3 s, when the temperature is lowered to $-10°C$.

The increasing non-linearity of behaviour at angles approaching 90°, apparent in Fig. 3, was tentatively associated by Darlington and Saunders[26] with a progressive change in deformation mechanism with increasing tensile strain. It was suggested that the fall in stiffness with increasing 100 s strain resulted from easy shear parallel to chain axes deviating slightly from the fibre axis. During deformation such c-axes appear to rotate towards the tensile stress axis, thus progressively increasing the resolved shear stress on the easy shear planes. An attempt to quantify such a concept was made by Ladizesky and Ward.[18]

It should be noted that, when discussing trends in degree of linearity as a function of angle in highly anisotropic materials, the observed patterns of behaviour may be significantly different for comparisons made at equal stress or at equal strain.[32]

The time dependence of S_{33}, S_{22}, S_{13}, S_{12} and $1/E(45°)$ in LDPE drawn at room temperature to a range of draw ratios has been obtained by Clayton et al.[32,33]

At low draw ratios, the tensile compliances (as distinct from the lateral compliances) exhibited appreciable time-dependence, with a marked anisotropy in their rate of change with time. As the draw ratio increased, the time dependence, and its anisotropy, decreased until, at very high draw ratio, the low creep rates presented earlier were obtained with no significant anisotropy of time-dependence. Data at draw ratios of 1·4 and 2·0 are presented in Figs. 10 and 11 (from Clayton[34]).

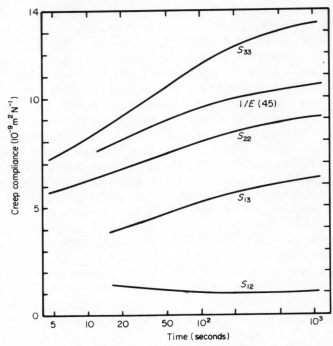

*Fig. 10. Creep data for cold-drawn LDPE with fibre symmetry. Draw ratio 1·4.
(From Clayton.[34]) Note that the values of S_{13} and S_{12} are negative.*

The behaviour of the lateral compliance, S_{13}, is similar to that of S_{33} but the behaviour of S_{12} at low draw ratio is worthy of note. Thus, on applying the creep load, an initial decrease in specimen thickness is followed first by a slight time dependent thickening and then the more usual time dependent decrease in thickness. No explanation has yet been offered for this unexpected effect. It has also been observed in LDPE drawn at 90°C and is completely reproducible and experimentally significant.[35]

From a combination of X-ray diffraction, isochronous modulus and dynamic mechanical studies Ward and coworkers explained the anomalous behaviour of $E(0)$ at low draw ratio (see Fig. 7) in terms of shear parallel to oriented crystalline chains.[19,30,31,36] Thus, at low draw ratio, the crystalline units become preferentially oriented with their c-axes conically disposed about the fibre axis, the half-angle of the cone being approximately 35°. Tensile loads on specimens cut parallel to the fibre axis then result in high resolved shear stresses in the direction of the aligned chains. If the easy shear mechanism is activated this will give a low value for $E(0)$. Tensile loads on specimens cut at other angles to the fibre axis activate this mechanism to a lesser extent, depending on angle, and hence are associated with higher moduli. As the draw ratio increases above 1·4 so the cone-angle progressively decreases. Thus the resolved shear stress on the easy shear planes decreases and $E(0)$ rises steadily.

From the dynamic mechanical measurements of Stachurski and Ward[36] this process is seen to occur at 50°C in the region of 150 Hz. Clayton et al. concluded therefore that the tensile creep behaviour observed at low draw ratio could be the result of the cone easy shear process occurring in their time scale at 20°C. The observed anisotropy of time dependence of S_{33} and S_{22} (i.e. $dS_{33}/dt > dS_{22}/dt$) leads to a low value of $E(0)$ compared with $E(90)$, which is consistent with the above conclusion. Furthermore, calculation of the variation of volume strain with time during tensile creep at 0° and 90° (made possible by the measurement of both lateral and axial strains) showed that at 0° the

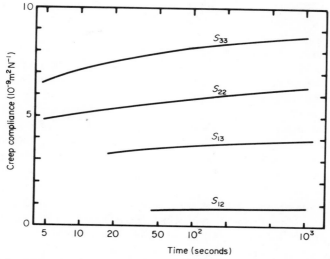

Fig. 11. Creep data for cold-drawn LDPE with fibre symmetry. Draw ratio 2·0. (From Clayton.[34]) Note that the values of S_{13} and S_{12} are negative.

creep deformation occurred at constant volume, so supporting the idea of a simple shear process. However, at 90°, the volume strain increased with time, indicating that the deformation at 90° included a significant contribution from a different mechanism to that at 0°. The anomalous behaviour of S_{12} is not inconsistent with this view.

The time dependence of the shear compliance $S_{44}(t)$ was studied over a range of draw ratios in cold-drawn LDPE by Clayton[34] using torsional creep apparatus. In general, good agreement was obtained between this directly measured shear compliance and that calculated by inserting the tensile creep data above into eqn. (8) (see Clayton et al.[17]). It should be mentioned that the measurement of S_{44} for a material with fibre symmetry is one of the cases where the difficulties mentioned in Section 10.3.2 are least severe, since torsion occurs about an axis which is perpendicular to a plane of isotropy in the specimen.

A detailed examination of the problems of direct measurement of $S_{44}(t)$ and $S_{66}(t)$ in highly cold-drawn LDPE with fibre symmetry was carried out by Ladizesky and Ward[8] using torsional creep apparatus. They found only a small dependence of $S_{44}(t)$ upon time at high draw ratio, in agreement with the above mentioned studies on tensile and torsional creep. The direct determination of $S_{66}(t)$, by experiments in torsion was, however, shown to require a complicated double extrapolation procedure; reasonable agreement then being obtained between the measured values and those derived from the measurements of S_{11} and S_{12} during tensile creep.

The double extrapolation procedure invoked involves firstly extrapolation of measurements of torque on specimens with various known axial tensions to zero axial tension. These extrapolated values were then used in the St. Venant expression for torsion of non-circular cylinders to obtain apparent values of S_{66} for rectangular prisms of various aspect ratios (width : thickness). If the theory was rigorous these values of S_{66} would be independent of aspect ratio. However, this was not the case and a second extrapolation was made to obtain a value for S_{66} at the limiting aspect ratio at which the theory could be regarded as most rigorous.

This complicated procedure, together with the inhomogeneity of stress and strain inherent to torsion of solid bodies, which precludes quantitative interpretation and observation of non-linear effects, must be regarded as a severe restriction on the utility of torsional testing in creep studies such as those discussed here. However, as already indicated, torsional measurements appear currently to offer the only route to the determination of all three shear compliances in materials with orthorhombic symmetry.

Creep studies have also been carried out on LDPE drawn at 55°C, drawn at 95°C and drawn at room temperature followed in this case by various thermal treatments at 55°C; all the sheets having fibre symmetry (see Darlington et al.[35]). The sheets drawn at 55°C and room temperature had high values of draw ratio whilst those drawn at 95°C were produced with a full range of draw ratios.

The 45° specimens cut from the sheet drawn to a ratio of 3·8 at 95°C (hot-drawn) exhibited two significant differences in creep behaviour when compared with the previous 45° specimens from a cold drawn sheet with similar draw ratio: (1) significantly higher creep rate in the hot drawn material when compared at stresses producing identical tensile strains at 100 s, and (2) significantly higher degree of non-linearity (i.e. sensitivity of isochronous modulus to stress level) in the hot drawn material. In contrast 0° and 90° specimens in both hot and cold drawn materials had markedly similar behaviour in creep rate and non-linearity. This behaviour is shown in Figs. 12 and 13.

Despite these different behaviours for 45° specimens from hot drawn and cold drawn sheets, lateral strain measurements in sheet thickness directions showed that in the hot drawn material, as in the cold drawn material, the tensile deformation occurred without thickness change. This

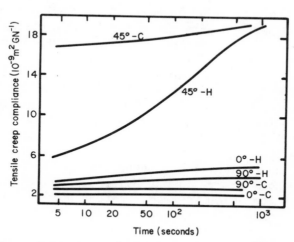

Fig. 12. Variation of tensile creep compliance with time for specimens cut at angles of 0°, 45° and 90° to the fibre axis of hot-drawn (H) and cold-drawn (C) LDPE (from Darlington et al.[35]).

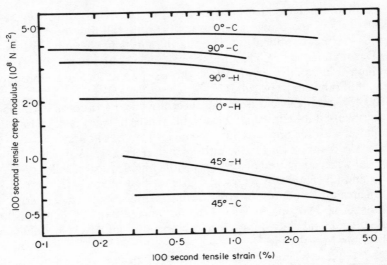

Fig. 13. Isochronous modulus-strain data for specimens cut at angles of 0°, 45° and 90° to the fibre axis of hot-drawn (H) and cold-drawn (C) LDPE. (From Darlington et al.[35])

observation, together with the form of the developing anisotropy, suggests that the easy shear mechanism parallel to the oriented crystalline chains is, in the hot-drawn specimens, occurring *during* the time scale of the creep measurements. Such a shift of time scale for the easy shear mechanism, from that of cold-drawn LDPE, is confirmed by the dynamic mechanical studies of Stachurski and Ward.[37] Thus, in LDPE drawn at room temperature and annealed at 105°C, they observed a highly anisotropic relaxation at a temperature of 80°C and a frequency in the region of 100 Hz which they associated with the easy shear mechanism. The shift in time scale between cold-drawn and cold-drawn and annealed specimens was attributed to the structural changes during annealing.

The significant difference in the stress sensitivity of isochronous modulus between the two 45° specimens would appear at first sight to be difficult to explain if it is accepted that the same basic deformation mechanism occurs in both specimens (although at different times). However, an explanation may be found in the recent work of Baker and Darlington (1973, to be published). They carried out creep tests over a range of stress levels on 45° specimens from the hot-drawn LDPE. The tests were carried out at three temperatures, chosen such that the experimental time scale coincided with the start, middle and end of the relaxation process. Typical results are presented in Fig. 14 as the variation of log creep compliance with log time. When cross-plots of the data are taken at a selected creep time, variations of isochronous compliance with stress as shown in Fig. 15 are obtained. It is apparent that the stress sensitivity of the compliance (and hence also of modulus) is greatest in the central region, and least at *either* end, of the relaxation process.

The isochronous behaviour of the 45° specimen from the hot drawn LDPE is now seen to be highly non-linear at 20°C because the relaxation process is occurring during the time scale of the creep measurements. However, as noted earlier, in the case of the cold-drawn material at 20°C, the relaxation process has occurred well before the time scale of the creep measurements and therefore the isochronous modulus shows little depend-

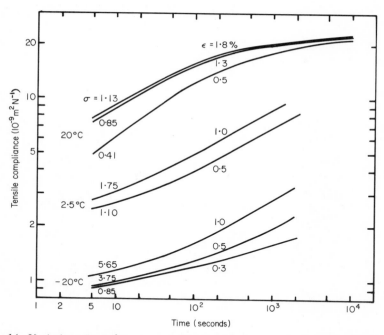

Fig. 14. *Variation of tensile creep compliance with time, at three temperatures, for specimens cut at 45° to the fibre axis of hot-drawn LDPE. Draw ratio = 3·9. σ is the applied stress in MN/m². ε is the 100 second tensile strain. (Baker and Darlington, 1973, to be published.)*

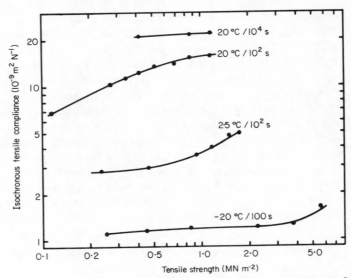

Fig. 15. Isochronous tensile compliance-stress data at three temperatures, for specimens cut at 45° to the fibre axis of hot-drawn LDPE. Draw ratio = 3·9. (Baker and Darlington, 1973, to be published).

ence on stress (at least up to 100 second tensile strains of 1%). Creep studies on the 45° specimens from cold-drawn LDPE at a temperature of −10°C confirm that the isochronous modulus also shows high stress sensitivity near the centre of the relaxation process.

It is readily apparent that great care must be taken when comparing the degree of non-linearity in the viscoelastic response of different materials, particularly when isochronous data is viewed in isolation.

The tensile creep behaviour of specially oriented sheets of LDPE was studied by Gupta and Ward.[38] However, their data took the form of a dependence of isochronous modulus on temperature; time dependence and non-linearity being ignored. The interpretation of such data is reviewed elsewhere in this book.

The tensile creep behaviour of oriented high-density polyethylene has been studied by McCrum and coworkers[20,39,40] and by Ward and co-workers.[38,41] No creep curves as such are given, but the variation of isochronous modulus with temperature for specimens cut at various angles to the symmetry axes is presented. McCrum and coworkers carried out all their measurements at very low tensile strains with the intension of keeping within the linear viscoelastic region. Although not containing creep curves, these papers do serve to demonstrate the power of stiffness measurements on oriented polymers when attempting to elucidate deformation mechanisms.

Similar isochronous studies have been carried out on polyethylene terephthalate sheets with orthorhombic symmetry by Ladizesky and Ward,[9] on cold-drawn Nylon 6 and 66 by Owen and Ward[42] and on oriented polypropylene by Owen and Ward.[43]

The anisotropy of stiffness in a polypropylene injection moulding was studied by Ogorkiewicz and Weidmann.[44] The moulding took the form of a centre-gated tray and, as expected, the radial direction in the tray bottom was a principal axis. The variation of tensile modulus with angle to the radial direction was studied in the plane of the tray, using 100 second isochronous creep procedures at low strains (to avoid non-linear visco-elastic effects). It should be noted that, because of the geometry of flow in the tray bottom, there would be through-thickness variations in molecular orientation and hence in the material anisotropy. The authors neglected such effects and the parameters quoted by them must therefore be recognised as average values rather than absolute materials parameters.

Two sets of moulding conditions were used in the above study. For the set producing the greater anisotropy, the 'modulus' varied from a maximum of 1.13 GN/m^2 in the radial direction to a minimum of 0.825 GN/m^2 at $50°$ to the radial direction. At $90°$ the 'modulus' was 0.92 GN/m^2. The full variation of 'modulus' in the plane of the tray was calculated using eqn. (7) and the 'modulus' values at $0°$, $45°$ and $90°$ to the radial direction. Modulus data at intermediate angles were shown to lie close to the curve so obtained. A similar exercise in torsion of these specimens showed greater scatter which was attributed by the authors to the variations in stiffness through the thickness of the moulding.

The creep anisotropy of polypropylene when formed into wheels by the solid-phase forming, or forging, process has been studied by Abrahams et al.[45] A displacement factor was defined as $(t_0 - t_1)/t_0 \times 100\%$, where t_0 was the original billet thickness and t_1 the final thickness of the wheel web. Specimens were cut parallel $(\theta = 0°)$ and perpendicular $(\theta = 90°)$ to the radial direction. Isochronous creep tests showed little change in stiffness at $90°$ to the radial direction for a range of displacement factors from 0 to 58%. However, significant increases in stiffness were found at $0°$, above a displacement factor of 50%. Creep tests on other specimens appeared to indicate that, at both 0° and 90°, the creep rate was higher than for the unforged material, but interpretation of the results is complicated by the choice of specimen sizes and positions and by the complex nature of the forming process.

10.7 ANISOTROPIC CREEP IN AMORPHOUS THERMOPLASTICS

The anisotropy of creep behaviour of oriented amorphous thermoplastics appears to have received even less attention than that of oriented semi-crystalline thermoplastics. This may well be associated with the fact that early measurements based on standard tests demonstrated a marked

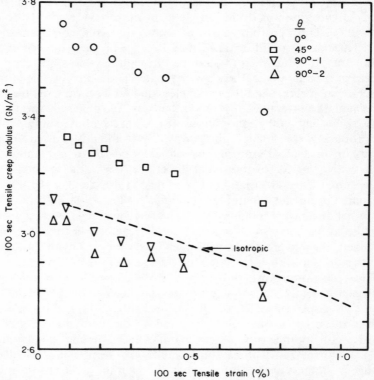

Fig. 16. Isochronous modulus-strain data for specimens cut at various angles, θ, to the fibre axis of an oriented PMMA sheet (drawn at 124°C, birefringence 0·001).[27]

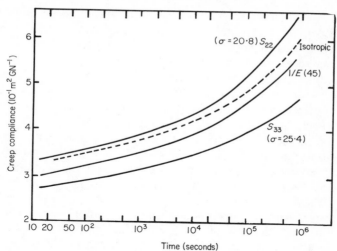

Fig. 17. Variation of tensile compliance with time for isotropic and oriented PMMA (drawn at 124°C, birefringence 0·001)[27] (σ is the applied stress in MN/m²).

dependence of strength properties, but small dependence of stiffness properties, on the level of molecular orientation in amorphous thermoplastics (see Nielsen[1]).

The anisotropic creep behaviour of highly drawn polymethylmethacrylate (PMMA) possessing fibre symmetry was studied by Darlington.[27] The oriented sheet was produced by drawing an initially isotropic sheet in uniaxial tension at a temperature of 124°C; the drawn sheet being cooled at constant length. The final draw ratio was 3·36 and the birefringence was 0·001. All three strains were measured during creep in uniaxial tension of specimens cut at 0°, 45° and 90° to the fibre axis.

The variation of the 100 second tensile creep modulus with 100 second tensile strain is presented in Fig. 16. The behaviour of a specimen cut from an isotropic sheet (which had been subjected to the same thermal cycle as the drawn sheet) is included for comparison. It is apparent that all specimens exhibited non-linear viscoelastic behaviour, but there is little anisotropy of non-linearity. Furthermore the degree of non-linearity exhibited by the specimens from the drawn sheet is similar to that of a specimen from the isotropic sheet. At any chosen creep strain the anisotropy of modulus for the drawn sheet is relatively low.

The variations of the tensile compliances S_{33}, S_{22} and $1/E(45)$ with time are presented in Fig. 17. The behaviour of a specimen from the heat-treated isotropic sheet is again included for comparison. For each specimen the stress level was chosen to give a 100 second tensile strain of 0·75%. The actual stress levels are indicated in Fig. 17. The variations of the lateral compliances with time during creep of the above specimens has already been presented in Fig. 6 and discussed in Section 10.5.2.

It is apparent from Fig. 17 that there is anisotropy of creep rate for the drawn sheet. However, the degree of anisotropy is relatively low; the difference between S_{22} and S_{33} rising from some 20% at a creep time of 10 s to some 30% at a creep time of 4×10^5 s. It should be noted that, as the material exhibits non-linear viscoelastic behaviour, comparisons of S_{22} and S_{33} from tests carried out at equal stress would give slightly higher values. These preliminary results suggest that, although the level of anisotropy is relatively small at short times at room temperature, it may develop to a significant level at long times and/or higher temperatures.

The results of creep rupture tests on a drawn PMMA sheet with a lower level of molecular orientation (birefringence of only 0·0006) are presented in Fig. 18.[46] This shows a high level of anisotropy of creep rupture which contrasts with the relatively low level of anisotropy of stiffness found in the more highly oriented sample. The level of molecular orientation, even in the sample with lower birefringence, must therefore be regarded as significant, and the low anisotropy of stiffness must be

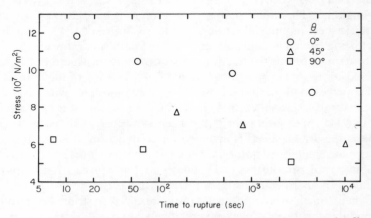

Fig. 18. Creep rupture data for specimens cut at various angles, θ, to the fibre axis of an oriented PMMA sheet (birefringence 0·0006).[46]

regarded as due to a low sensitivity of this property (in this time/temperature range) to orientation rather than to a low value of orientation.

Wright et al.[47] have also examined the stiffness behaviour of drawn PMMA using a uniaxial tensile test in which the load was alternately applied and removed at 30 second intervals. The 'modulus' data from these tests showed reasonable agreement with the 100 second isochronous data presented above, when compared at equal birefringence.

10.8 GENERAL CONCLUSIONS

Creep testing of oriented polymers, intended to fully characterise the anisotropy of stiffness behaviour, presents formidable difficulties. For materials with fibre symmetry, techniques are now available which allow complete characterisation. These techniques are considerably more sophisticated than simple creep testing in isotropic materials. For lower symmetries it is still not possible to achieve full characterisation.

There appears to be no rigorous theoretical scheme for describing anisotropy of creep behaviour in these materials. However, simple extensions of linear viscoelastic theory are presented and shown to be useful though not completely rigorous. Further development is clearly desirable.

Oriented thermoplastics can show large anisotropy in creep behaviour, expecially in partially crystalline polymers. Significantly different patterns of behaviour occur in different materials. Not only is there anisotropy of isochronous stiffness, but also of creep rate and non-linearity. If stiffness is regarded as a function of time, direction and stress or strain, the behaviour is such that the variables are not normally separable.

In partially crystalline polymers the creep behaviour of the oriented material varies systematically with structure. Anisotropic creep studies on oriented materials, whilst considerably more complicated, can more readily lead to understanding of deformation mechanisms than do creep studies in isotropic materials.

Anisotropy of creep behaviour in oriented glasses seems to be less well developed than in partially crystalline materials, but the anisotropy increases with time and presumably temperature. The low level of anisotropy in the time/temperature region investigated may well be a consequence of the polymer being well below its glass transition temperature. Clearly there is scope for systematic investigation of the contributions of the various relaxation phenomena to anisotropy in oriented glassy polymers.

REFERENCES

1. Nielsen, L. E. (1962). *Mechanical Properties of Polymers*, Reinhold, New York.
2. Nye, J. F. (1957). *Physical Properties of Crystals*, Clarendon Press, Oxford.
3. Hearmon, R. F. S. (1961). *An Introduction to Applied Anisotropic Elasticity*, Clarendon Press, Oxford.
4. Lee, E. H. (1955). *Quarterly Applied Maths.*, **13**, 183.
5. Biot, M. A. (1965). *Mechanics of Incremental Deformations*, Wiley, New York.
6. Rogers, T. G. and Pipkin, A. C. (1963). *J. Appl. Maths. Phys.*, **14**, 334.
7. Schapery, R. A. (1967). *J. Comp. Mater.*, **1**, 228.
8. Ladizesky, N. H. and Ward, I. M. (1971). *J. Macromol. Sci., Phys.*, **B5**, 745.
9. Ladizesky, N. H. and Ward, I. M. (1973). *J. Mater. Sci.*, **8**, 980.
10. Hadley, D. W., Ward, I. M. and Ward, J. (1965). *Proc. Roy. Soc.*, **A285**, 275.
11. Pinnock, P. R., Ward, I. M. and Wolfe, J. M. (1966). *Proc. Roy. Soc.*, **A291**, 267.
12. Ward, I. M. (1971). *Mechanical Properties of Solid Polymers*, Wiley—Interscience, London.
13. Dunn, C. M. R., Mills, W. H. and Turner, S. (1964). *Brit. Plastics*, **37**, 386.
14. Thomas, D. A. and Turner, S. (1969). Chapter 2 in *Testing of Polymers*, Vol. 4 (Ed. W. E. Brown), Interscience, New York.
15. Halpin, J. C. and Pagano, N. J. (1968). *J. Comp. Mater.*, **2**, 18 and 68.
16. Darlington, M. W. and Saunders, D. W. (1970). *J. Phys. E:Sci. Instrum.*, **3**, 511.
17. Clayton, D., Darlington, M. W. and Hall, M. M. (1973). *J. Phys. E: Sci. Instrum.*, **6**, 218.
18. Ladizesky, N. H. and Ward, I. M. (1971). *J. Macromol. Sci.—Phys.*, **B5**, 661.
19. Gupta, V. B. and Ward, I. M. (1967). *J. Macromol. Sci., Phys.*, **B1**, 373.
20. Buckley, C. P. and McCrum, N. G. (1973). *J. Mater. Sci.*, **8**, 928.
21. Morrison, T. A., Zapas, L. J. and DeWitt, T. W. (1955). *Rev. Sci. Instrum.*, **26**, 357.
22. McCrum, N. G. and Morris, E. L. (1966). *Proc. Roy. Soc.*, **A292**, 506.
23. Raumann, G. (1962). *Proc. Phys. Soc.*, **79**, 1221.
24. Turner, S. (1969). Chapter 1 in *Testing of Polymers*, Vol. 4 (Ed. W. E. Brown), Interscience, New York.
25. British Standard 4618, Parts 1.1 and 1.1.1 (1970).
26. Darlington, M. W. and Saunders, D. W. (1970). *J. Phys. D: Appl. Phys.*, **3**, 535.
27. Darlington, M. W. (1969). Cranfield Institute of Technology, Memo. No. 67.
28. Raumann, G. and Saunders, D. W. (1961). *Proc. Phys. Soc.*, **77**, 1028.
29. Saunders, D. W. and Darlington, M. W. (1968). *Nature*, **218**, 561.
30. Gupta, V. B., Keller, A. and Ward, I. M. (1968). *J. Macromol. Sci., Phys.*, **B2**, 139.
31. Stachurski, Z. H. and Ward, I. M. (1969). *J. Macromol. Sci., Phys.*, **B3**, 445.
32. Darlington, M. W. and Saunders, D. W. (1971). *J. Macromol. Sci., Phys.*, **B5**, 207.
33. Clayton, D., Darlington, M. W. and Hall, M. M. (1970). *International Conference on Yield, Deformation and Fracture of Polymers*, Inst. of Physics. Cambridge, U.K.
34. Clayton, D. (1971). *Ph.D. Thesis*, Cranfield Institute of Technology.
35. Darlington, M. W., McConkey, B. H. and Saunders, D. W. (1971). *J. Mater. Sci.*, **6**, 1447.
36. Stachurski, Z. H. and Ward, I. M. (1969). *J. Macromol. Sci., Phys.*, **B3**, 427.
37. Stachurski, Z. H. and Ward, I. M. (1968). *J. Polymer Sci.*, A-2, **6**, 1817.
38. Gupta, V. B. and Ward, I. M. (1968). *J. Macromol. Sci., Phys.*, **B2**, 89.
39. Buckley, C. P., Gray, R. W. and McCrum, N. G. (1969). *J. Polymer Sci.*, A-2, **6**, 1817.
40. Buckley, C. P., Gray, R. W. and McCrum, N. G. (1970). *J. Polymer Sci.*, **B8**, 341.
41. Davies, G. R., Owen, A. J., Ward, I. M. and Gupta, V. B. (1972). *J. Macromol. Sci., Phys.*, **B6**, 215.
42. Owen, A. J. and Ward, I. M. (1973). *J. Macromol. Sci., Phys.*, **B7**, 279.
43. Owen, A. J. and Ward, I. M. (1973). *J. Macromol. Sci., Phys.*, **B7**, 417.
44. Ogorkiewicz, R. M. and Weidmann, G. W. (1972). *Plast. Polymers*, **40**, 337.

45. Abrahams, M., Spedding, C. E. and Marsh, B. J. (1970). *Plast. Polymers,* **38,** 124.
46. Darlington, M. W. (1969). College of Aeronautics Memo No. 184.
47. Wright, H., Faraday, C. S. N., White, E. F. T. and Treloar, L. R. G. (1971). *J. Phys. D: Appl. Phys.,* **4,** 2002.

CHAPTER 11

ANISOTROPIC YIELD BEHAVIOUR

R. A. DUCKETT

11.1 PRELIMINARY DEFINITIONS

We will discuss here the anisotropic yield behaviour of oriented polymers but there is a need for a few preliminary remarks regarding the topic of yield in general. In describing the deformation of many crystalline materials, especially metals and ceramics, it is often convenient to introduce the idealisation of an 'elastic–plastic transition'. The term 'elastic' is used to describe the components of the strain which are proportional to the applied stresses, and which are completely recovered on removal of the stresses. Plastic strains are observed only for stresses greater than or equal to the 'yield stress' and are not recovered on removal of the stress. The yield stress defines the elastic–plastic transition.

It has been possible, for metals and ceramic materials, to demonstrate by direct observation the existence of lattice defects called dislocations, using the techniques of transmission electron microscopy. These studies have shown that it is often adequate to assume that dislocation motion is responsible for the observed plastic, or permanent, deformation, and that this motion is negligible at stresses below the yield stress. Although very refined microstrain measurements and internal friction experiments have failed to define a stress range in which dislocation motion is completely absent, there is still a clear distinction for these materials between elastic and plastic strain, both on a macroscopic level, in terms of permanency of deformation, and on a microscopic level in terms of large scale dislocation motion.[1,2]

Yield stresses are often recognised as features of the stress–strain curves at constant strain-rate. For example in Fig. 1(a) three common working

366

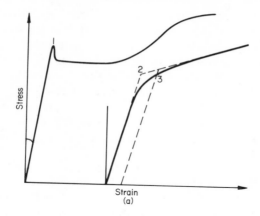

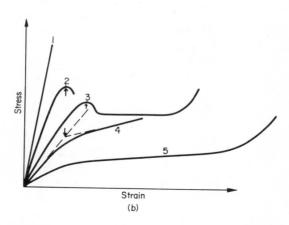

Fig. 1.(a) Three common working definitions of the yield point for metals. (1) Load maximum, (2) tangent method, (3) 'proof-stress' or 'strain-offset' method. (The proof-strain is commonly taken to be 0·1%, but is quite arbitrary.) (b) Load elongation curves for polymers. (1) Brittle, (2) strain softening, (3) cold-drawing, (4) strain-hardening, (5) rubbery. Typical definitions of the yield point are marked by arrows on curves (2), (3) and (4). Any one polymer can show behaviour ranging from (1) to (5) depending on test conditions, e.g. temperature, strain-rate, tension or compression.

definitions for the yield stress of a metal are illustrated. The first definition is based on the stress at the first load-maximum and is reasonably unambiguous. The second and third definitions are often used where such a clear load-maximum is not observed. Although they are completely arbitrary, they are often used in situations not requiring high numerical accuracy and, as such, are adequate measures of the onset of permanent deformation.

If a polymer is deformed at constant strain-rate it is often possible to draw analogies between the observed response and the yield behaviour of metals, see Fig. 1(b).

Load–elongation curves covering the complete range of behaviour from 'brittle' (1), to rubbers (5), can be obtained for most polymers by varying the temperature and/or strain-rate of the test in the manner indicated. The curves (2)–(4) all indicate a critical stress (corresponding to the points marked on Fig. 1), similar to that observed in metal, which can be taken as a measure of the 'yield stress'. The analogy with the yield behaviour of metals is strengthened by the observation of 'necking phenomena' and particularly deformation bands at stresses around those marked, which are generally associated with 'permanent' deformation, although the time-dependent nature of polymer deformation renders any definition of yield in terms of permanency of deformation somewhat impracticable.

For the purposes of the discussion to follow we will wherever possible define the yield stress as the first maximum in stress on a curve such as (2) or (3), although tests under different conditions such as temperature, strain-rate or orientation may perhaps produce curves of type (4) requiring a different definition of 'yield' in the same material. At this stage it is appropriate to recognise that tests at large strains in tension, compression or torsion are often accompanied by the development of mechanical instabilities such as deformation or slip bands, buckling, etc., which obscure the shape of the true stress–strain curve and which can only be inferred from indirect observation.[3,4] In general the numerical differences resulting from the use of these various definitions of the yield stress are small compared with the effects to be described here, and further discussion of the significance of yield or plastic flow of polymers will be reserved until Section 11.4 below.

11.2 THE PHENOMENOLOGY OF YIELD

There have been two main contrasting approaches to the establishment of a yield criterion for oriented polymers. On the one hand we have those who seek a yield criterion which is an extension of those applicable to isotropic materials, treating the oriented polymer as a continuum. On the other hand we find many who draw analogy with the plasticity of metal single crystals. We will present arguments here in support of each of these approaches recognising that at this stage the links between them have yet to be formed.

11.2.1 The continuum approach for isotropic materials

In order to appreciate the development of the continuum approach for oriented polymers it is helpful to discuss briefly the progress made in dealing with plasticity of isotropic polymers.

Most attempts at dealing with the yield criteria of isotropic polymers derive from the criteria due to Hencky–von Mises or Tresca (see Ref. 5, Chapter 2 for an excellent introduction to this topic). The Hencky–von Mises criterion can be written in the alternative forms

$$(\sigma_y - \sigma_z)^2 + (\sigma_z - \sigma_x)^2 + (\sigma_x - \sigma_y)^2 + 6(\sigma_{yz}^2 + \sigma_{zx}^2 + \sigma_{xy}^2) = 6k^2 \tag{1}$$

or
$$(\sigma_1 - \sigma_2)^2 + (\sigma_2 - \sigma_3)^2 + (\sigma_3 - \sigma_1)^2 = 6k^2 \tag{2}$$

where σ_1, σ_2 and σ_3 are the principal components of the general stress tensor

$$\begin{bmatrix} \sigma_x & \sigma_{xy} & \sigma_{xz} \\ \sigma_{xy} & \sigma_y & \sigma_{yz} \\ \sigma_{xz} & \sigma_{yz} & \sigma_z \end{bmatrix}$$

These equations state that yield will occur when the function of the stress components represented by the left-hand side of these equations reach a critical value, $6k^2$. If we consider an isotropic material with $\sigma_1 = -\sigma_2 = k$ and $\sigma_3 = 0$, it is clear that this stress configuration satisfies (2), and so we can identify k with the yield stress in pure shear.

Note that both (1) and (2) are symmetrical with respect to the subscripts x, y, z and 1, 2, 3 respectively, reflecting the isotropy of the material. Note further, that both of these equations are unchanged by increasing each normal stress by a constant amount, p, i.e. if we write $\sigma_x \rightarrow \sigma_x + p, \sigma_y \rightarrow \sigma_y + p, \sigma_z \rightarrow \sigma_z + p$, then eqn. (1) is unchanged. This feature, implying that hydrostatic pressure does not affect yield, is a necessary ingredient for a yield criterion for metals which deform mainly by slip processes at constant volume. It however serves only as a first approximation to the yield behaviour of isotropic polymers (see Ref. 6, for which a considerable body of experimental data has now been collected to show the significant effect of hydrostatic pressure.[7-12] In fact it must also be recognised that the yield stress of polymers is strongly strain-rate, temperature and pressure dependent (see for example Ref. 4), *i.e.*

$$(\sigma_1 - \sigma_2)^2 + (\sigma_2 - \sigma_3)^2 + (\sigma_3 - \sigma_1)^2 = f(\dot\varepsilon, T, p) \qquad (3)$$

Such a dependence on hydrostatic pressure naturally implies a difference between tensile and compressive yield stresses as is observed for most polymers.

Two other yield criteria which have been applied with more or less success to the yield behaviour of isotropic polymers are due to Tresca and Coulomb. According to Tresca,[5] yield occurs when

$$|\sigma_1 - \sigma_3| = 2k, \quad \text{for} \quad \sigma_1 > \sigma_2 > \sigma_3 \qquad (4)$$

where k can again be identified with the maximum shear stress which can be supported by the material. Naturally this criterion is also deficient in dealing with the pressure and strain-rate dependence of polymer yield behaviour and could be improved by writing as

$$|\sigma_1 - \sigma_3| = f(\dot\varepsilon, T, p) \qquad (5)$$

A criterion which is somewhat similar to (5) was proposed by Coulomb to describe the failure of soils and has subsequently been applied to the yield behaviour of polymers:[4]

$$\tau = \tau_0 - \mu\sigma_N \qquad (6)$$

Here τ_0 and μ are material parameters which define the shear stress τ required to produce yield on a plane on which the normal stress is σ_N. (Note that this criterion is written in terms of σ_N and not p.) It should, in principle, be possible to distinguish between eqns. (3), (5) and (6), by experiments of high accuracy, covering a wide range of stress-states, but this has yet to be achieved.

Several ingenious explicit versions of, and alternatives to eqns. (3), (5) and (6) have been proposed (Refs. 4, 13–16) which need not be discussed further here. It is sufficient to remark that most of them are purely phenomenological, and apply equally well to amorphous polymers below their glass transition, T_g, *e.g.* polycarbonate, polymethylmethacrylate, polystyrene, and semi-crystalline polymers both below and above T_g (*e.g.* polypropylene, polyethylene, polyethylene terephthalate).

11.2.2 The continuum approach for oriented polymers

When an initially isotropic polymer is drawn or extruded to a high deformation ratio under suitable conditions, it develops appreciable anisotropy which is apparent in mechanical tests at all stresses up to and beyond the yield stress. A typical example of the anisotropy of yield observed is shown in Fig. 2 in which the tensile yield stress of oriented polyethylene terephthalate (PET) sheets is shown as a function of the angle θ between the tensile axis (TA) and the initial draw direction (IDD). This large anisotropy is somewhat similar to that observed in cold-rolled metal sheets, for which theories of anisotropic plasticity were suggested by Hill,[17] Yoshimura[18] and others. A modification of the theory

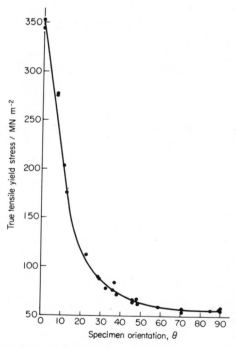

Fig. 2. The variation of tensile yield stress with θ, the angle between the initial draw direction and the tensile axis. The material is oriented polyethylene terephthalate sheet, draw ratio 5:1. Equation (8) was used to construct the full-line and is a good fit to the available data (after Brown et al.[19]).

due to Hill was found to deal satisfactorily with many features of the yield data from PET,[19,20] and has also been applied to polyvinyl chloride (PVC),[21] and polypropylene (PP).[22]

Hill's yield criterion for anisotropic materials[17] (see also Ref. 5, Chapter 10) is based on the following equation

$$f(\sigma_{ij}) = F(\sigma_y - \sigma_z)^2 + G(\sigma_z - \sigma_x)^2 + H(\sigma_x - \sigma_y)^2$$
$$+ 2(L\sigma_{yz}^2 + M\sigma_{zx}^2 + N\sigma_{xy}^2) = 1 \qquad (7)$$

It is assumed that the material possesses three mutually orthogonal planes of symmetry, whose lines of intersection are the axes $0x, 0y, 0z$, used to define the components of stress. The parameters, F, G, H, L, M, N characterise the anisotropy, and in the limit of vanishing anisotropy $F \rightarrow G \rightarrow H \rightarrow 1/6k^2$, $L \rightarrow M \rightarrow N \rightarrow 1/2k^2$ and eqn. (7) reverts to the Hencky–von Mises eqn. (1). The anisotropy is therefore treated as a perturbation on the normal isotropic behaviour. Equation (7) follows (1) in stating that yield is independent of the hydrostatic component of stress, $p = -\frac{1}{3}(\sigma_x + \sigma_y + \sigma_z)$ and also that the tensile and compressive yield stresses are equal. These two points will be examined further below.

In order to check the validity of eqn. (7) in describing the tensile yield data in Fig. 2 it is necessary to label the symmetry axes of the oriented material in terms of the cartesian axes, $0x, 0y, 0z$. We take $0z$ parallel to the initial drawing direction (IDD) and $0x$ as the sheet normal. Hence for a tensile test at an angle θ to the IDD resulting in an axial yield stress σ, we have $\sigma_z = \sigma \cos^2 \theta, \sigma_y = \sigma \sin^2 \theta, \sigma_{zy} = \sigma \sin \theta \cos \theta$ and all other stress components are zero. Equation (7) then reduces to

$$\sigma^2 \{ (F + G)\cos^4 \theta + 2(L - F)\sin^2 \theta \cos^2 \theta + (H + F)\sin^4 \theta \} = 1 \qquad (8)$$

(Note: This convention does not follow previous publications,[19,20] but is in common with other contributions to this volume.) If equation (8) is fitted to the data of Fig. 2 at any three points, say $\theta = 0°$, $45°$ and $90°$, the coefficients $(F + G)$, $(H + F)$ and $2(L - F)$ can be evaluated, enabling the full line in the figure to be computed. It is clear that an excellent fit to the PET tensile yield data is afforded by this representation.

In order for a theory of this type to be useful it is necessary that it is not only descriptive but also predictive. The Hill criterion offers such a possibility in that it allows a prediction of the plastic strain increments at yield, provided sufficient yield stresses have been determined to evaluate F, G, H, L, M and N explicitly. By making the conventional assumption of normality (see Ref. 5, Chapters 2 and 10), it follows that the components of plastic strain increments at yield are deduced from the yield criterion as follows

$$de_x = \frac{\partial f}{\partial \sigma_x} d\lambda \qquad de_y = \frac{\partial f}{\partial \sigma_y} d\lambda \qquad de_z = \frac{\partial f}{\partial \sigma_z} d\lambda$$

$$de_{yz} = \frac{\partial f}{\partial \sigma_{yz}} d\lambda \qquad de_{zx} = \frac{\partial f}{\partial \sigma_{zx}} d\lambda \qquad de_{xy} = \frac{\partial f}{\partial \sigma_{xy}} d\lambda \qquad (9)$$

where $d\lambda$ is a scalar factor of proportionality with the dimensions of stress. Thus knowing F, G, H, and L, M, N and the stress components corresponding to a particular test, it is possible to predict the plastic strain increments directly. A suitable experimental measure of the plastic strain increments is afforded by measurement of the directions of the deformation bands which are associated with the yield event in many oriented polymers. These bands must form parallel to directions of zero elongation, which can be calculated from the plastic strain increment

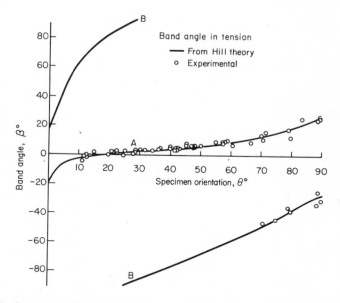

Fig. 3. *Deformation band directions in tensile specimens of oriented PET sheet. The full curves are predicted from the Hill theory using values of the constants obtained from Fig. 2. Most experimental data lie on branch A (slippy bands), but some kinky bands are observed at high θ (branch B). (After R. A. Duckett, Ph.D. thesis, Bristol University, 1968.)*

components. Such a test of the representation has been performed for data from PET, PVC and PP with varying degrees of success, see Fig. 3.

Note that, in general, the theory predicts two possible band directions whereas only one type of band is observed for most specimen orientations, corresponding to branch A in Fig. 3. The strains associated with these bands, which we call slippy bands, cause a line in the material initially parallel to the IDD to rotate away from the tensile axis. By contrast in the case of kinky bands (branch B in Fig. 3) a line initially parallel to the IDD, is rotated towards the tensile direction. These points are illustrated further in Figs. 4, 15 and 16 and will be amplified in Section 11.3.1 below.

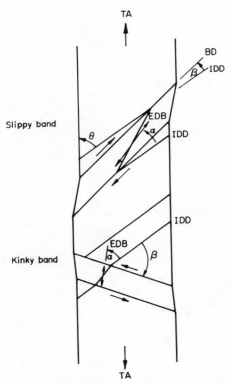

Fig. 4. Two types of deformation band are possible in tensile specimens of oriented polymer, and are illustrated in this figure. The slippy band is the most common for specimens oriented at angles less than 70° to the tensile axis.

It was immediately recognised from somewhat indirect evidence that two of the assumptions made in deriving the Hill criterion were invalid for polymers; namely the assumptions of invariance under stress reversal and superposed hydrostatic pressure. Tensile tests in different directions on oriented polymers produce widely varying yield stresses, and therefore differing hydrostatic components. Examination of Fig. 2 reveals that over the complete range of orientations the hydrostatic component changes from 120 MN m^{-2} at $\theta = 0°$ to about 20 MN m^{-2} at $\theta = 90°$. A change in hydrostatic pressure of this magnitude changes the shear yield behaviour of isotropic PET by only 10%[11] and so we may assume, until direct measurements become available, that the effects of pressure on the tensile yield behaviour of oriented PET arising from the anisotropy are negligible compared with the direct effects of orientation. It is also significant that if these indirect pressure effects could be eliminated the effects of anisotropy would be even greater.

The second assumption regarding explicitly the equality of tensile and compressive yield stresses cannot be so easily dismissed. Robertson and Joynson[23,24] and subsequently Brown et al.,[19] Bridle et al.[20] made measurements in simple shear on several oriented polymers which indicated that the tensile yield stress parallel to the IDD was considerably greater than the compressive yield stress in the same direction. An attempt was made to modify the Hill yield criterion to take this behaviour into account.[19-21] This was achieved formally by assuming the existence of an internal compressive stress in the direction of initial drawing. This internal stress, σ_i, referred to as the Bauschinger term, has to be overcome by an applied tensile stress parallel to the IDD, and aids an external compressive stress in the same direction. Thus we write $\sigma_z = \sigma \cos^2 \theta - \sigma_i$, in (7), with all the other stress components unchanged. Note that the yield criterion is still insensitive to a change in the hydrostatic component of the applied stress. With the 'internal' stress included the variation of yield stress with θ is now given by

$$\sigma^2\{(F+G)\cos^4 \theta + 2(L-F)\sin^2 \theta \cos^2 \theta + (H+F)\sin^4 \theta\}$$

$$+ 2\sigma\sigma_i\{F \sin^2 \theta - (F+G)\cos^2 \theta\} + \sigma_i^2(F+G) = 1$$

Although this changes the form of eqn. (8) somewhat the overall fit to yield stress and band angles in tension is not much affected, whereas the fit to the simple shear data can be significantly improved.[19-21]

In the work so far described no direct measure of the yield stress in compression along the IDD was available and so σ_i was necessarily treated as a disposable parameter. It has been possible subsequently to investigate directly the yield behaviour in compression of several oriented polymers. Rabinowitz et al.[25] measured tensile and compressive yield stresses of oriented PET sheet as a function of draw ratio using an admittedly unfavourable thin-walled tubular geometry (Fig. 5(a)). Their results suggested that the compressive yield stress parallel to the IDD is relatively independent of draw ratio despite the large increase in tensile yield stress at $\theta = 0$ (implying much larger Bauschinger terms in highly oriented PET than hitherto suspected). They have since been supported by the measurements of Duckett et al.[26,22] and Shinozaki and Groves[27] on specimens of oriented polypropylene for which there is no possibility of buckling occurring below the true compressive yield stress (see Figs. 5(a), 5(b) and 6). It is clear that a large Bauschinger effect is an intrinsic feature of the yield anisotropy for oriented polypropylene, polyethylene terephthalate and also polyethylene.[22] This suggests that an *ad hoc* modification of a theory of anisotropic plasticity to include the Bauschinger effect is not likely to lead to a physical understanding of the anisotropy, and this point must be taken up later. In particular, it must be noted that it is not possible for any material to support an homogeneous internal stress and all attempts to interpret the Bauschinger effect in terms of an internal stress have been unsuccessful. It is probably advisable to consider σ_i only as a convenient curve fitting parameter, enabling the yield surface to be modelled closely, which is the main prerequisite for the prediction of band angles.

11.2.3 Single crystal plasticity

In this section we will deal with attempts to describe the yield criterion for anisotropic polymers in terms of a critical resolved shear stress by analogy with the deformation of metal single crystals. Further discussion of the 'single crystal' approach follows in section 11.3.2 where we discuss the structural reorganisation occurring within deformation bands.

We have seen that the continuum theories of plasticity go a long way towards dealing with the relationship between yield stress and plastic strain increments for anisotropic polymers. In particular the directions in which deformation bands form are the directions of zero elongation, and are predicted using the formalism developed to describe necking and 'earing' of cold-rolled polycrystalline metal sheets. We now turn briefly to theories which draw analogy between the deformation bands in polymers and similar phenomena observed in plastically deformed metal *single* crystals. The discussion will refer many times to 'slip' and 'kink'

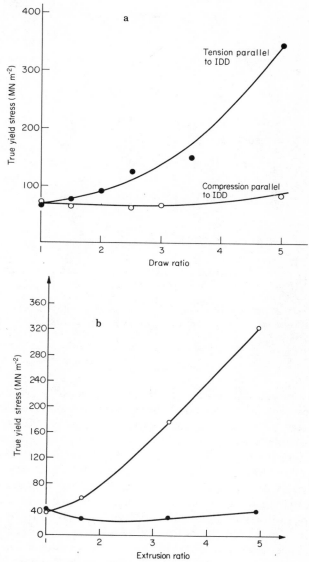

Fig. 5. The tensile and compressive yield stresses of (a) oriented polyethylene terephthalate,[25] (b) oriented polypropylene plotted as a function of extension ratio.[26] The stress axis is parallel to the orientation direction. It is clear that for both materials, only the tensile yield stress is appreciably dependent on degree of orientation.

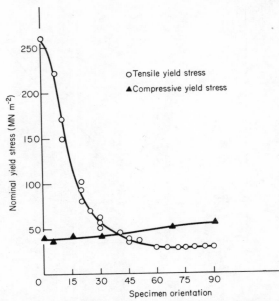

Fig. 6. Tensile and compressive yield stresses of oriented polypropylene plotted versus the angle θ between the stress axis and the orientation direction (after Shinozaki and Groves[27]).

bands. These are regions of localised shear strain and correspond to the two types of deformation band discussed in the previous section. We will see later that the two extremes, with $\beta \sim 0°$ and $\beta \sim 90°$ are physically distinguishable. There is in between the full spectrum of values of band angle and a smooth transition from one type to the other. It is, however, not necessary to distinguish between them at this stage of the discussion.

Slip bands and kink bands were first studied in compression of oriented nylon 6,6 and 6,10 by Zaukelies,[28] and subsequently in tensile specimens of oriented high density polyethylene (HDPE) by Kurakawa and Ban[29] and Keller and Rider[30] Zaukelies interpreted the angle between kink bands in oriented nylon and the compression axis in terms of Orowan's theory of crystal kinking,[31] postulating dislocation mechanisms for the process. Keller and Rider[30] and Kurakawa and Ban[29] were impressed by the appearance of deformation bands in high density polyethylene in directions close to the IDD. In this polymer system the

IDD is synonymous with the preferred orientation of the crystal c-axes suggesting an interpretation of the deformation in terms of the extended molecular chains sliding past each other parallel to their main axis which is called the 'c-slip process'. Keller and Rider supported this proposal by observing that the yield stress varied with specimen orientation in a way which could, for an appreciable range of orientations, be interpreted by a modified critical resolved shear stress criterion. Their proposal was that yield occurs when the applied shear stress parallel to the preferred c-axis direction reaches a critical value, which depends on the normal stress on that slip plane. For a specimen cut with the tensile axis at an angle θ to the IDD this criterion can be written

$$\sigma \sin \theta \cos \theta = \tau_c - k\sigma \sin^2 \theta \qquad (10)$$

where τ_c and k are treated as disposable parameters in order to fit the data. If we rewrite this equation as

$$\sigma(\sin \theta \cos \theta + k \sin^2 \theta) = \tau_c$$

then the connection with the more widely applicable Hill criterion becomes apparent

$$\sigma^2 \{(F + G)\cos^4 \theta + 2(L - F)\sin^2 \theta \cos^2 \theta + (H + F)\sin^4 \theta\} = 1 \qquad (8)$$

There is a formal similarity between (10) and the Coulomb theory of yield mentioned above for isotropic materials. On the latter theory yield occurs on the plane which makes an angle α with the tensile axis, for which the shear component of the applied stress reaches a critical value. This shear component at yield decreases linearly with the normal component of stress on that plane. For an axial stress σ

$$\sigma \sin \alpha \cos \alpha = \tau_c - k\sigma \sin^2 \alpha \quad \text{or} \quad \sigma(\sin \alpha \cos \alpha + k \sin^2 \alpha) = \tau_c$$

The shear plane is the one which maximises the value of $\sin \alpha \cos \alpha + k \sin^2 \alpha$, and it is easily shown that this plane makes an angle $\phi = \frac{1}{2}\tan^{-1} k$ with the plane of maximum shear stress. On the Coulomb theory α is uniquely defined by a material constant, whereas in (10) θ is an independent variable defining the specimen orientation.

Hinton and Rider[32] extended the notion of c-slip to a study of homogeneous plastic deformation in cold-drawn high density polyethylene. They noted that the direction of maximum refractive index (which they labelled as the current c-axis) swung round towards the tensile axis during the tensile deformation of specimens cut at a range of angles θ to the IDD.

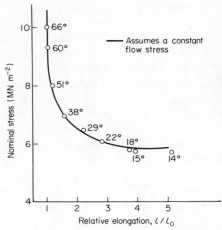

Fig. 7. The drop in nominal stress with increasing elongation can be attributed to the reduction in cross-sectional area as the slip plane (parallel to the c-axis) rotates towards the tensile axis. The shear stress on the slip plane changes by only 8% during 500% extension. The points are labelled in terms of the current angle between the c-axis and the tensile axis. (After Hinton and Rider.[32])

The magnitude of the rotation could be related approximately to the overall strain according to the ideas of Schmidt-Boas[94] by assuming an incremental slip, always parallel to the current c-axis. Furthermore they were able to explain the observed load–elongation curve by assuming the existence of a constant flow stress on the c-slip system. This requires decreasing axial loads as the slip plane rotates towards the tensile axis, because of the rapidly reducing cross-sectional area (see Fig. 7).

Several recent papers have maintained interest in crystalline deformation processes. Shinozaki and Groves[27] suggest that a theory proposed for fibre reinforced composites by Kelly and Davies[33] is helpful in describing the tensile yield behaviour of oriented polypropylene. Three distinct failure mechanisms are proposed, as depicted in Fig. 8.

Tensile specimens in the range $30° < \theta < 45°$ are considered to deform essentially by c-slip, from the observation that the yield stress in that range is governed approximately by the equation $\sigma \sin \theta \cos \theta = $ const. Their proposal is supported by wide and low angle X-ray data taken from specimens under load.[39] The X-ray measurements are of considerable interest but take us into the ill-defined territory between small strain viscoelasticity and 'yield'—a topic to be considered further in a later section. It is perhaps worthwhile to note that Owen and Ward[35] conclude from a study of the 'elastic' behaviour of a similar hot drawn polypropylene that the c-shear process is not operative at room temperature, an observation also relevant to high density polyethylene.[101] We will be careful therefore to restrict the term 'c-shear' to elastic deformation, and use the more general terms 'c-slip' and 'intermolecular slip' at higher strains.

It is also interesting that in compression tests on oriented polypropylene at low temperatures it is not in general appropriate to describe the deformation in terms of c-slip.[36] Deformation bands in these specimens form in general at approximately 45° to the compression axis, *i.e.* near to planes of maximum shear stress, independent of specimen orientation. Bands in specimens for $\theta < 45°$ and $\theta > 60°$ are highly localised, indicating considerable 'strain-softening' after yield. By contrast at $\theta = 45°$, which

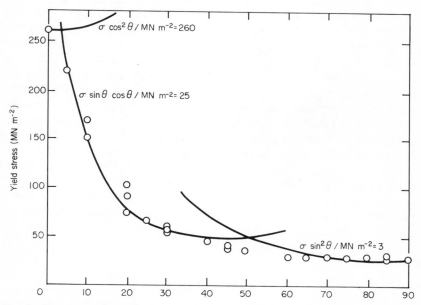

Fig. 8. Illustrating the Kelly–Davies criterion proposed by Shinozaki and Groves[27] to explain the anisotropic tensile yield data of oriented polypropylene.

would provide optimum conditions for c-slip, no well defined slip bands occur, which is perhaps indicative of some slight strain-hardening. Both Shinozaki and Groves[27] and Zihlif[22] observe that the compressive yield stress of oriented polypropylene is approximately independent of specimen orientation at room temperature. This fact is difficult to rationalise with the c-slip process suggested for the tensile deformation.

A similar thesis to Ref. 34 has also been proposed by Young *et al.*[37] to describe the deformation of high density polyethylene in compression. On the basis of *in situ* wide and low angle X-ray scattering data from oriented HDPE in compression at angles of approximately 20°, 40° and 60° to the IDD, Young *et al.* suggest an interesting distinction between 'elastic' and 'plastic' strain, at least for highly annealed material. Their measurements indicate that 'elastic' deformation occurs primarily by inter-lamellar slip and that 'plastic' deformation arises from 001 chain slip (equivalent to the c-slip or intermolecular slip discussed above). Young *et al.*[37] also propose a semi-quantitative description of the subsequent strain hardening observed in the compression tests. They propose that the c-slip process has a well defined flow stress, independent of plastic strain. The axial stress has to increase suitably to maintain the shear stress parallel to the crystalline c-axes constant as they rotate away from the compression axis (Refs. 32 and 34). It appears that such a clear distinction between elastic and plastic strains cannot be made for materials which have not been annealed and do not have such a well-defined morphology. Both Young *et al.*[37] and Shinozaki and Groves[27,34] were careful to point out that their proposed deformation mechanisms were restricted to specimens oriented so that the applied stress had a significant shear component on their slip systems. Their ideas must be considered as developments of earlier work due to Hay and Keller,[38] Seto and Tajima,[39] Keller and Pope[40] and Bowden and Young.[41]

Finally we must note that yield surfaces dominated by the c-slip shear stress term ($\sigma \sin \theta \cos \theta$) are not restricted to highly oriented highly crystalline polymers such as HDPE, but are also observed for low draw-ratio, low crystallinity PET, for which materials a crystalline slip process cannot be relevant.[96]

11.3 STRUCTURAL CHANGES RESULTING FROM PLASTIC STRAIN

11.3.1 Continuum approach

We now discuss in more detail structural changes arising during plastic deformation of oriented polymers. Ward and co-workers have investigated the optical properties of deformation bands in oriented PET sheet.[43−46] This polymer possesses a high degree of transparency in contrast to oriented HDPE and thus allows a study of the deformation processes by following the optical birefringence.

When a thin sheet of an anisotropic material is observed normally between crossed polarisers and rotated about the observation direction, the intensity of the transmitted light is found to pass through four equispaced minima (ideally zeros) separated by four equal maxima. The minima of intensity arise when the principal directions of refractive index in the plane of the sheet are parallel to the polariser or the analyser directions. It is convenient to label the direction of maximum refractive index the 'extinction' direction and note that a minimum of transmitted intensity arises whenever the extinction direction is parallel to either the polariser or the analyser directions.

A thin specimen of an oriented polymer containing a deformation band observed between crossed polarisers is impossible to orient so that the material in the band and in the undeformed material outside the band (the 'matrix') are at extinction simultaneously. The parameter of immediate interest is the angle α needed to rotate from extinction in the matrix to extinction in the band. (Note that the direction of maximum refractive index in many polymers coincides with the average direction of chain orientation.)

In Fig. 4 we have seen typical deformation bands in PET. The observed extinction directions in the matrix (IDD) and in the bands (EDB) are marked showing the angle α between them. From Fig. 9 it can be seen that a homogeneous shear could not explain the observed reorientation if all the chains were aligned parallel to the IDD, as the direction of maximum refractive index would then rotate in the opposite sense to that observed.

An explanation was proposed[44] for the PET results which has since been shown to apply equally to oriented polypropylene,[47] high density polyethylene[48,22] and nylon.[49] The polymer sheet is considered as an oriented continuum characterised by three principal extension ratios $(\lambda_1, \lambda_2, \lambda_3)$. If the isotropic sheet is considered as the state of zero strain, then the oriented polymer has extension ratios λ_3 in the draw or orientation direction (IDD) and it is convenient to take λ_1 in the direction of the sheet normal. Thus λ_3 and λ_2 define the projection of the strain ellipsoid in the plane of the sheet. When a deformation band forms in the oriented polymer the deformation can be described in terms of two shear

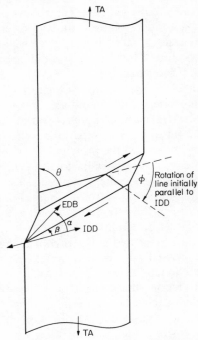

Fig. 9. This illustrates that in a slippy band a line initially parallel to the initial drawing direction, IDD, and therefore parallel to the average chain direction in the matrix, rotates in the opposite sense to the observed rotation of extinction direction.

strains: (a) a simple shear γ on a plane whose normal is N with displacements parallel to the band direction, making an angle $(\theta - \beta)$ with the tensile axis; (b) a pure shear with principal extension ratios e, 1 and $1/e$ in the directions N, BD and the thickness direction respectively, see Fig. 10. (These strains can be measured from the deformation of grids either shadowed or lightly scratched on the specimen surfaces.)

In terms of these parameters it is straightforward to calculate the angle α_t between the directions of maximum elongation outside the band (the IDD) and inside the band, from the equation

$$\tan 2(\alpha_t - \beta) = \frac{2\gamma(\cos^2 \beta + (\lambda_3^2/\lambda_2^2)\sin^2 \beta) - e^{-1}(\lambda_3^2/\lambda_2^2 - 1)\sin 2\beta}{\{(\gamma^2 - 1)(\cos^2 \beta + (\lambda_3^2/\lambda_2^2)\sin^2 \beta) - \gamma e^{-1}(\lambda_3^2/\lambda_2^2 - 1)\sin 2\beta + e^{-2}(\sin^2 \beta + (\lambda_3^2/\lambda_2^2)\cos^2 \beta)\}} \quad (11)$$

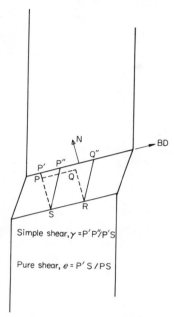

Fig. 10. The simple and pure shears γ and e respectively are defined in terms of the deformation of the element PQRS into $P''Q''RS$. $\gamma = P'P''/P'S$, $e = P'S/PS$.

These geometric aspects are shown in Fig. 11.

Brown *et al.*[44] made the intuitive step of relating the direction of maximum elongation to the direction of maximum refractive index, regardless of strain-history. Thus the angle α_t calculated above should relate exactly to the angle α between the extinction directions for material inside and outside the deformation bands.

This prediction was found to be satisfied for tensile specimens of all orientations (implying a range of values of β, γ and e) for PET sheet with initial draw ratios from 1·5 up to 5.[44,46] Table 1 includes representative data from PET, and also from polypropylene[47] and high density polyethylene.[48]

Attempts have been made to apply this theory to homogeneous

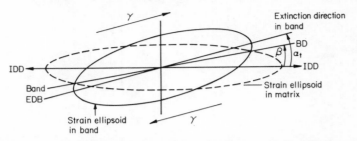

Fig. 11. The strain ellipsoids for material in the matrix and in a deformation band (zero strain referring to the isotropic state).

deformation in oriented PVC[50] and high density PE[32] with much less success. However, in each case features of the experimental arrangements were not ideal for this work. Rider and Hargreaves[21,50] performed the initial drawing process for their PVC sheet at 80°C, followed by a second deformation at 50°C. They found that the initial draw process had to be described by an effective draw ratio which was less than that actually applied, and that this could be attributed to the differing strain-optical characteristics of PVC at the temperatures 50°C and 80°C. Hinton and Rider[32] modified the theory in an attempt to describe their proposal of incremental slip parallel to the IDD (or current c-axis, *i.e.* $\beta = 0$). They calculated that the observed rotation of the c-axis ($0.02°$ per 1% shear) would arise from a sheet initially drawn to a value of λ_3/λ_2 of 5:1, whereas their material corresponded to $\lambda_3/\lambda_2 = 20$. It is worth noting that if one assumes a value for β of only $1°$ their results can be explained by this model. Direct measurements of β from deformation bands would support a value of $\beta \sim 1$–$5°$ for their material, although they were working in a temperature range in which homogeneous deformation is observed.

Thus this simple model accurately described optical aspects of the molecular reorientation observed in deformation bands, in PET polypropylene (PP) and high density polyethylene (PE), in terms which are completely independent of the chemical structure of the chains, and independent of the fact that one of the materials, PET, is of low crystallinity tested below its glass-transition T_g, whereas the other two are appreciably crystalline, tested above their T_g. The success of this simple phenomenological description of the reorientation should not be considered to imply that the molecular deformation processes are themselves simple. Taken at face value, the model suggests that the microscopic structural units

TABLE 1

| Material | | Mode of Testing | Geometry | | Deformation | | Optical Reorientation | |
Polymer	Draw ratio		Orientation $\theta°$	Band angle $\beta°$	Simple shear γ	Pure shear e	Predicted from eqn. (11) $\alpha_t°$	Measured $\alpha°$
PET	2·5	Tension	8	−33·1	0·42	1·15	−1·1	−1·8
PET	2·5	Tension	30	−6·5	1·29	1·13	8·8	8·4
	3·5	Tension	36	0·3	1·03	1·10	7·3	6·3
	3·5	Tension	72	12·9	1·33	1·50	25·1	25·8
	5	Tension	33	3	3·43	1·00	7·9	8·5
	5	Tension	80	53·0	0·18	1·10	79·7	79·5
PP	5·36	Tension	28	1·4	2·77	1·14	6·3	7·0
	5·36	Tension	70	15·7	1·94	1·89	53·7	54·2
HDPE	9·4	Tension	61·1	4·5	5·59	1·95	13·75	16·1
	11·1	Tension	34·5	4·5	8·60	1·15	9·45	8·8
	11·7	Tension	71·7	69·5	0·57	1·01	32·15	31·7

reorient to be, on average, parallel to the direction of maximum extension. Such behaviour might be expected from a structure containing an effective three dimensional network. The fact that the deformation of oriented PET obeys this scheme is consistent with the success of the aggregate models (pseudo-affine and 'rubber-like') in describing the change of structure and properties of this material on cold drawing (see Chapter 3).

It is not necessary to make any assumptions regarding specific deformation mechanisms in order to describe the optical reorientation. Note that it is implicit even in a superficially simple deformation of a network that elements in different situations (including differences in orientation) will experience completely different deformations.

When we move on therefore, to a partially oriented semi-crystalline polymer such as HDPE and PP for which there are as yet many unsolved problems regarding the initial morphology, it is somewhat surprising that the simple scheme for molecular reorientation is applicable. It is not clear how a simple morphology of alternating crystalline and amorphous regions could always reform so that the direction of maximum refractive index was parallel to the direction of maximum elongation, for all specimen orientations and initial draw ratios, as is required by this scheme (Fig. 12).

Such suspicions of extreme complexity at the molecular level have been confirmed by recent work[48] in which the optical reorientation measure-

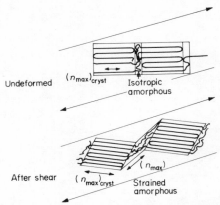

Fig. 12. A simple alternating crystalline/amorphous block model could not in general predict that the extinction direction coincides with the direction of maximum elongation.

ments have been complemented by measurements of fibrillar orientation and wide-angle (WAXS) and small-angle (SAXS) X-ray diffraction in deformation bands in HDPE, PP and PET. Although ideally these measurements should all be made in the same region of the deformation band, this is not always possible due to severe practical difficulties. Very thin specimens are required in order to define clear extinction directions, and the bands themselves are in general very narrow (approximately 50 μm in width). It is necessary therefore to piece together an overall picture from somewhat fragmentary data.

Optical microscopy of deformation bands in HDPE and PP reveals immediately evidence of 'fibrillar-like' structures similar to those which have been used[55-57] to monitor the deformation within the bands. Although the scale of the markings is very much larger, the optical micrograph in Fig. 13 bears a striking resemblance to the electron micrograph of a deformation band shown by Robertson[55] (his Fig. 2). Duckett et al.[48] were able to measure the strain in the band independently by means of a grid printed on the specimen surface and this has revealed a surprising feature. From a knowledge of the strain components it is possible to calculate the new direction of a line initially parallel to the fibrils (i.e. initially parallel to the IDD). This new direction is not in general parallel to the 'fibrillar' markings in the band as shown in Fig. 13. The real significance of these markings within the band has not yet been established but it is clear, first, that they are not related directly to those outside the band, and secondly, that the deformation cannot be attributed solely to 'inter-fibrillar shear' (see Sections 11.2.3 and 11.3.2). It is not clear whether the markings observed within the deformation bands by optical microscopy differ only in scale or in kind from those observed by electron microscopy, and so for clarity we shall henceforth refer to them as 'pseudo-fibrils'.

Optical microscopy has therefore shown (a) that the direction of maximum refractive index in the band (EDB) is parallel to the direction of maximum extension. (b) A pseudo-fibrillar texture is observed within the band which is not apparently related to the EDB nor to the direction a fibril in the drawn material would have if it followed an 'affine' deformation in the band. WAXS measurements reveal, further, that the preferred crystallographic c-axis direction is neither parallel to the 'pseudo-fibrils' nor to the extinction direction in the band (EDB). The WAXS photographs from material within the bands provide a clue to the explanation why the crystalline c-axes are not on average parallel to the maximum refractive index direction (EDB). For all materials examined (PET, PP

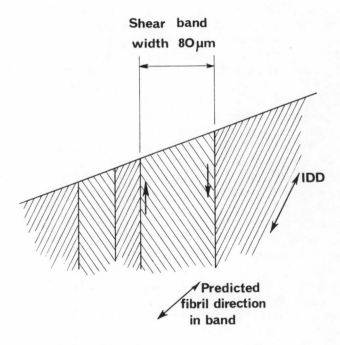

Fig. 13. *An optical micrograph showing the pseudo-fibrillar texture in a deformation band in a compression specimen of HDPE.*[48]

and HDPE) a considerable loss in crystallinity is observed within the deformation bands. This may be due to the formation of kinks and jogs within the crystallites by mechanisms similar to those suggested by Robertson[51] (see Fig. 14). These molecular kinks would contribute to the lattice distortion and reduce the crystallinity (see Refs. 52–54). Alternatively the effect of a compressive component of stress parallel to the chain axes may be to pull away fragments of the crystallites in the manner suggested in Fig. 15. This results in a considerable decrease in crystal size and loss in orientation which could together be interpreted as a loss in crystallinity.

Both of these suggested mechanisms would be selective in reducing crystallinity mainly in those crystals oriented such as to experience large compressive strains during the deformation, whilst the majority of the structure must deform in a near 'affine' manner to reproduce the predicted extinction directions. The WAXS is dominated by the regions of higher

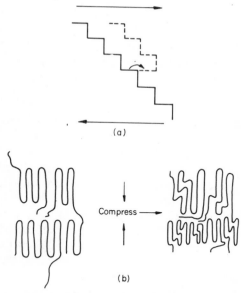

Fig. 14. (a) A possible mechanism of jog formation due to the compressive stress along the molecular axis (after Robertson[51]). (b) The manner in which compressive strains may reduce crystallinity by increasing the jog content and therefore increasing the lattice dislocation.

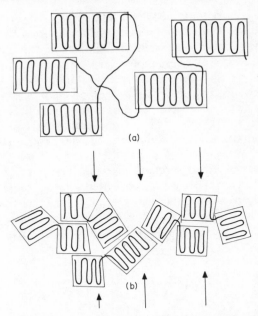

Fig. 15. *Compressive stresses along the draw direction may reduce the crystal size by pulling away fragments of relatively perfect crystal.*

crystallinity which may not be parallel to the overall orientation direction.

Similar observations of the WAXS from material within kink bands in oriented HDPE were made by Robertson.[55] He noticed not only that the diffraction spots from the sheared material within the band became more diffuse, but also the existence of what he called 'advance' reflections from material which rotated by up to 60° more than the fibrils. Robertson's model of the morphological rearrangement within a kink band (Fig. 16) is not completely consistent with the more extensive, recent data of Zihlif *et al.*[22,48] However, he does show an electron micrograph (his Fig. 3) which lends support to ideas of loss of organisation of crystal blocks within the kink band, and there is considerable general agreement between his data and that in Refs. 22 and 48. Both sets of data emphasise the dangers inherent in basing an interpretation of the complex deformation processes on measurements say of only one of optical reorientation, X-ray diffraction or surface topography.

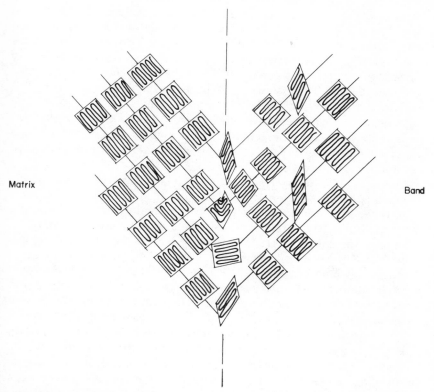

Matrix

Band

Fig. 16. A model to explain the existence of 'advance reflections' in the WAXS photographs of slippy bands in HDPE (after Robertson[55]). Although this explains qualitatively some features of the X-ray diffraction it does not predict the correct extinction direction.

11.3.2 Single crystal plasticity approach

It is not yet possible to establish a clear link between the approach of Ward and co-workers to the problem of molecular reorientation, which is discussed above at length, with many theories in the literature which take as their starting point the crystalline nature of the material. Some of these ideas have already been discussed in relation to the yield criterion for anisotropic polymers, but it is worth looking at them again to assess how they relate to the structure of the material which has been deformed in the process. We have seen in Section 11.2.3 that one of the earliest papers on this topic by Zaukelies[28] described kink bands in compressed nylon bristles by the theory of crystalline kinking due to Orowan.[31] This theory makes the quantitative prediction that there should be reflection symmetry between the crystal lattices on either side of the kink plane. Robertson[56,57] interprets the fibrillar texture of kink bands in HDPE as supporting this explanation. Robertson[55] does not comment that the X-ray diffraction data break this apparent symmetry. We have seen evidence that the pseudo-fibrillar texture observed by optical microscopy does not relate directly to the overall molecular orientation direction, nor to the preferred crystalline c-axis direction.

A considerable body of evidence is available now that with highly oriented polymers which have not been annealed the deformation for a wide range of specimen orientations can be mainly ascribed to a slip process parallel to the IDD or c-axis. This is apparently true only for situations where the material deforms homogeneously on a macroscopic scale.[27,32,34,37] Such homogeneity of deformation implies an absence of the strain softening, in contrast to the formation of a localised deformation band. It is hopeful that such features will eventually be incorporated into a molecular theory of yield. Note that even in situations where deformation bands form it has been shown that the band angle β (see Fig. 11) decreases with increasing degrees of orientation (data on PET sheet[58]). Examination of Fig. 11 shows that the deviations between the extinction direction, the band direction, and the initial draw direction all reduce significantly as the strain ellipsoid becomes more elongated. Thus the disagreement between the crystallographic theories and the continuum approach becomes more academic at high degrees of orientation.

Specific crystallographic deformation mechanisms have been proposed by many other workers, e.g. Refs. 29, 38, 42, 43, 55, but the distinction between elastic and permanent deformation has not always been made. Indeed the question of whether this is in general possible, and whether the yield behaviour is related in any way to the elastic behaviour is still to be settled.

Another possible link between the two schools may be established by observations of the X-ray photographs published by several workers (Refs. 22, 34, 37, 48, 59–62) which show indications, at least, of a loss of crystallinity with plastic strain. Tajima and Seto[59] measured the change in lamellar orientation and c-axis orientation on redrawing oriented HDPE. They commented that the WAXS scattering from specimens experiencing a further tensile strain in the draw direction became too diffuse to measure

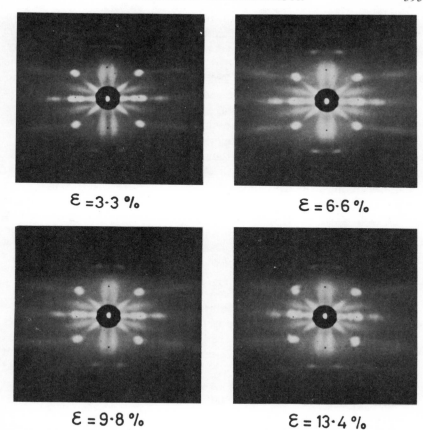

$\mathcal{E} = 3.3\%$ $\mathcal{E} = 6.6\%$

$\mathcal{E} = 9.8\%$ $\mathcal{E} = 13.4\%$

Fig. 17. The result of increasing axial compressive strain on the WAXS photographs from oriented polypropylene. Increasing disorientation and distortion of the Bragg reflection is apparent at strains greater than 10% (after Zihlif[22]).

at large (unspecified) elongations. SAXS photographs on strained oriented HDPE and polypropylene[34] show a similar tendency for the low angle scattering to become more diffuse. These observations are in fact from tensile specimens at $\theta = 30°$, the optimum orientation for intermolecular slip. Ishikawa *et al.*[60] took WAXS and SAXS photographs from fibres of HDPE under both compressive and tensile axial stresses. The WAXS photographs both during and after compression to 10–20% strain show an appreciable disorientation and an increase in the amorphous halo. These observations are also applicable to the WAXS photographs from oriented HDPE in compression at $\theta = 17\cdot5°$ (Ref. 37). Figure 17 is a set of WAXS photographs of an extruded polypropylene specimen under increasing axial compression, after Zihlif.[22] (The nominal strain values are shown for each case.) It can be seen that the overall crystallinity is appreciably reduced at compressive strains of 10% and above. These findings are in general agreement with those of Kaufman and Schultz[62] from their studies of the tensile deformation of isotropic polyethylene.

The break-up of crystallites and the reformation of the lamellar fragments into microfibrils is the basis of a theory for the cold-drawing of isotropic semi-crystalline polymers due to Peterlin.[63-66] (See also Hosemann et al.[67] and Robertson[68].) Both Peterlin and Hosemann assert that the main mechanism is the break-up of each crystallite into approximately twenty smaller units which lie like pearls on a string with their chain axes parallel to the IDD. Many aspects of these theories would seem to be relevant to the deformation of oriented polymers of modest draw ratios.

11.3.3 Kinetic fracture

One further topic merits discussion in this section in view of its success in dealing with the mechanical properties of oriented fibres, which are after all anisotropic polymers. That is the theory of kinetic fracture, developed mainly by Zhurkov and co-workers. Evidence has been presented from electron spin resonance (e.s.r.),[69-73] mass spectrometry,[74] and infra-red spectroscopy[75] that when highly oriented fibres or heavily cross-linked rubbers experience a tensile stress (along the axis for fibres) an appreciable fraction of main-chain bonds are broken by the applied stress. These scission events are observed to occur more or less homogeneously throughout the fibre and are not localised in the fracture plane. Many sets of data show that the lifetime t_b of a fibre under stress is described approximately by the following equation

$$t_b = t_0 \exp \frac{(\Delta U - \sigma v)}{kT}$$

Where t_0, ΔU, v are constants for a given fibre, σ is the applied stress and T is the absolute temperature. Zhurkov[69] achieved an excellent correlation between ΔU calculated from such stress-rupture data and ΔU_n, the activation energy for thermal degradation for a wide range of fibres. It was also possible to identify the free radicals and degradation products formed under stress with those produced by thermal degradation,[69,74] lending weight to the argument that under these conditions of tension parallel to the fibre axis main chain bonds are broken.

Campbell and Peterlin[70] and Peterlin[76] concluded from e.s.r. measurements on isotropic and highly drawn nylon 6 and 6.6 fibres that no detectable free radicals were formed in the isotropic state, whereas approximately 1 chain in 250 was fractured in a fibre under high axial tension at failure. These fractured chains were later identified with the 'tie molecules' linking adjacent crystallites together in the fibre direction. Quantitative theories have since been developed by Kausch et al.[77,78] and more recently by DeVries et al.[79] which attempt to correlate creep, creep-rupture, and stress-relaxation in fibres in terms of the measured main chain scission.

DeVries,[79] Peterlin[76] and Zhurkov[75] all specifically identify the dominating mechanism in the tensile deformation of fibres as the scission of 'tie molecules' which are assumed to have a distribution of lengths.

It should be pointed out that no strong evidence exists for chain scission in unoriented polymers or in oriented polymers in, say, compression or transverse tension or compression. Ito[80] proposed an explanation of the yield drop observed in many polymers as due to the stress required to break secondary bonds superposed on a rubber-like stress–strain curve. However, Steg and Ishai[81] in a study of a plasticised epoxy system concluded that there was no significant change in cross-link density on passing through yield.

It is clear that there are still many important experiments to be done in order to rationalise, select from and to reconcile aspects of these many and varied ideas regarding the molecular aspects of yield in polymers.

11.4 THE NATURE OF THE YIELD POINT

We have discussed at some length theories which attempt to describe the stresses needed to cause yield and the molecular reorganisation resulting from yield. The yield criteria have related to measurements of somewhat arbitrary features of the stress–strain curves, with the implicit assumption that these features coincide with the onset of permanent deformation. Various models of 'elastic–plastic' deformation have been used in describing the polymer behaviour.

Other treatments of yield, not discussed in detail here, assume that at yield the deformation is 'pseudo-viscous'. The polymer flows momentarily at a constant stress, the yield stress, which is a function of the applied strain-rate. On this basis it is not necessary to associate the yield stress with permanent deformation but to treat it as a transient effect which perhaps is related to the general nonlinear viscoelastic behaviour of the polymer. We now describe briefly a new approach to non-linear viscoelasticity which appears to offer a reliable and practical link between constant strain-rate tests, including those features associated with yield, with non-linear creep and stress-relaxation tests.

It has long been recognised that the mechanical properties of polymers are time-dependent. The behaviour at very small strains (less than 0·5%) can be described by the theory of linear viscoelasticity. Conventionally the stress σ at time t is related to the strain ε at all previous instants by the equation

$$\sigma(t) = \int_{-\infty}^{t} g(t-\tau)\varepsilon(\tau)\,d\tau \qquad (12)$$

where $G(t) = \int_{-\infty}^{t} g(t-\tau)\,d\tau$ is called the stress-relaxation modulus. The simplest mechanical model of a linear viscoelastic solid that can approximate the behaviour of real polymers in both stress and strain controlled experiments is the standard linear solid[97] for which

$$G(t) = A + Be^{-\lambda t} \qquad (13)$$

It is straightforward to calculate the response of a standard linear solid in a constant strain-rate test, and this is given by

$$\sigma(t) = \dot{\varepsilon}\{At + \frac{B}{\lambda}(1-e^{-\lambda t})\} \qquad (14)$$

where $\dot{\varepsilon}$ is the applied strain-rate.

Inspection of this equation shows that it models reasonably well, on a very superficial level, a stress–strain curve of the type shown in Fig. 1(b), curve (4). In other words it raises the question as to whether the deviations from linear stress–strain relationships observed in constant strain-rate tests might not be merely resulting from the intrinsic time-dependence of the linear viscoelasticity, which can be more clearly studied in creep or stress-relaxation and not due to some new process starting at high stresses. It does not take long to show that at the strain-levels of 3–5% experienced at yield, the response of most polymers is highly non-linear; $\sigma(t)/\dot{\varepsilon}$ is a function of strain-rate $\dot{\varepsilon}$ as well as t, and so eqn. (14) cannot adequately describe the behaviour. However, it is also clear that at least some of the effects ascribed to yield or plasticity might be interpreted in terms of non-linear viscoelasticity, and that such a study might help to pinpoint key features of the stress–strain curves and remove some of the ambiguities discussed in Section 11.1.

Viscoelastic behaviour can be represented in more general terms by either of the two equivalent equations,

$$\sigma(t) = \int_{-\infty}^{t} g_1^*(t-\tau, \varepsilon(\tau))\varepsilon(\tau)\,d\tau \tag{15}$$

or

$$\sigma(t) = \int_{-\infty}^{t} g_2^*(t-\tau, \sigma(\tau))\varepsilon(\tau)\,d\tau \tag{16}$$

where the non-linearity is represented by the stress or strain dependence of the response function. Conventionally eqn. (15) is taken as a starting point, and the response function $g_1^*(t-\tau, \varepsilon(\tau))$ is expanded as a polynomial in $\varepsilon(\tau)$

$$\sigma(t) = \int_{-\infty}^{t} g_1(t-\tau)\varepsilon(\tau)\,d\tau + \int_{-\infty}^{t} \int_{-\infty}^{t} g_2(t-\tau, t-\tau_2)\varepsilon(\tau_1)\varepsilon(\tau_2)\,d\tau_1\,d\tau_2$$
$$+ \int\int\int + \text{etc.} \tag{17}$$

where the infinite set of response functions $g_n(t \ldots . t)$ describe the material response. Analogous arguments lead to a corresponding expression for the strain $\varepsilon(t)$ in terms of the stress-history $\sigma(\tau)$. In Fig. 18 typical isochronal stress–strain curves are shown from creep and constant strain-rate tensile tests on oriented PET (data from Brereton et al.[82]). Several features of these data are important.

(1) The behaviour is time-dependent in both tests, especially at high stresses.

(2) The creep and constant strain-rate data are similar, but not identical.

(3) The behaviour in each test is highly non-linear in that at high stresses a small change in stress is associated with a large change in strain.

This last feature precludes any description of the creep data in terms of a low order polynomial in stress, and led Brereton *et al.*[82] to develop a novel method of representing the non-linear data. The representation is based on an implicit equation relating the stress, strain and time which can be written

$$\int_0^t a(t-\tau)\sigma(\tau)\,d\tau + \int_0^t b(t-\tau)\varepsilon(\tau)\,d\tau$$

$$+ \int_0^t c(t-\tau_1,t-\tau_2)\sigma(\tau_1)\varepsilon(\tau_2)\,d\tau_1\,d\tau_2 = 0 \tag{18}$$

This can be written symbolically in the two alternative forms

$$\sigma = -\left(\frac{b}{a}\varepsilon + \frac{c}{a}\sigma\varepsilon\right) \tag{19}$$

and

$$\varepsilon = -\left(\frac{a}{b}\varepsilon + \frac{c}{b}\sigma\varepsilon\right) \tag{20}$$

By careful analysis of extensive creep and constant strain-rate data from PET and creep data from isotropic PP and PMMA, Brereton showed that eqn. (19) could be usefully written as an integral equation

$$\sigma(t) = \int_0^t g(t-\tau)\varepsilon(\tau)\,d\tau - \int_0^t h(t-\tau)\sigma_L(\tau)\varepsilon(\tau)\,d\tau \tag{21}$$

where

$$\sigma_L = \int_0^t g(t-\tau)\varepsilon(\tau)\,d\tau$$

The response functions $g(t-\tau)$ and $h(t-\tau)$ can be written in terms of a, b and c in (19), and can further be used to describe completely the creep response through the analogue of (20). In other words, the complete non-linear viscoelastic response should be describable in terms of these two functions only. Certain features of this representation can be checked very quickly against the experimental data. For example, it can be shown that for a creep experiment with $\sigma = 0$ for $t < 0$, and $\sigma = \sigma_0$ for $t > 0$, then

$$\sigma_0/\varepsilon(t \to 0) = (j(0))^{-1}(1-\sigma_0 h(0))$$
$$\sigma_0/\varepsilon(t \to \infty) = (j(\infty))^{-1}(1-\sigma_0 h(\infty)) \tag{22}$$

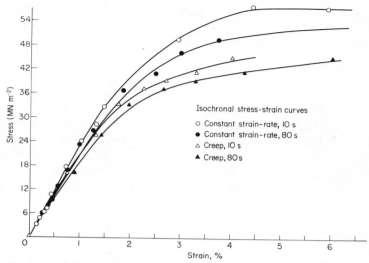

Fig. 18. Isochronal stress–strain curves for oriented PET tested in tension at right angles to the initial draw ($\bigcirc$, $\bullet$) *constant strain-rate data* ($\triangle$, $\blacktriangle$), *creep data (after Croll*[88]*).*

where $j(0)$, $j(\infty)$, $h(0)$ and $h(\infty)$ are limiting values of $j(t)$ and $h(t)$ respectively, and are therefore constants. In other words the creep moduli at these limits, which we call the unrelaxed and relaxed creep moduli respectively, should vary linearly with the applied stress. Figure 19 shows these creep moduli for oriented PET plotted versus the applied stress σ_0. It can be seen from the figure that the functional form of eqn. (22) is obeyed by these data. (Creep data from isotropic PP, PC and PMMA, also fit this scheme.) Similarly it can be shown that for stress-relaxation

$$\varepsilon_0/\sigma(t \rightarrow 0) = j(0) + h(0)\varepsilon_0$$

and
$$\varepsilon_0/\sigma(t \rightarrow \infty) = j(\infty) + h(\infty)\varepsilon_0 \qquad (23)$$

Note that the same parameters $j(0)$, $j(\infty)$, $h(0)$ and $h(\infty)$ appear in (22) as in (23). Figure 20 shows stress-relaxation data for the same oriented PET sheet plotted as unrelaxed and relaxed ($t \rightarrow 0$ and $t \rightarrow \infty$) stress-relaxation compliance v. the applied strain ε_0. The full lines on this figure are a least squares fit to the stress-relaxation data, and the dashed lines are predicted directly from the creep data. This excellent internal consistency is a measure of the adequacy of the implicit eqn. (18) in describing the behaviour.

Inspection of Fig. 19 reveals two critical stresses

$$\sigma_F = \frac{1}{h(\infty)} \quad \text{and} \quad \sigma_B = \frac{1}{h(0)} \qquad (24)$$

at which the relaxed and unrelaxed creep moduli respectively tend to zero. Brereton *et al.* went on to show that by assuming a simple parametric form for $j(t)$ and $h(t)$ eqns. (21) can be solved numerically, indicating the general form of the stress in a constant strain-rate test. This is illustrated in Fig. 21. Note that the stress σ_F has particular significance in each experiment. In creep it can be shown that the polymer deforms to a finite

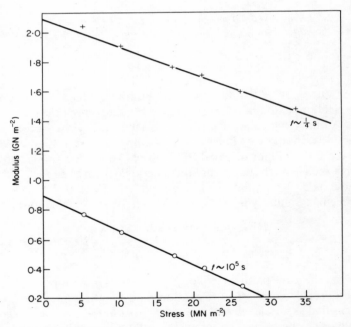

Fig. 19. The stress dependence of the isochronal creep moduli for oriented PET tested in tension perpendicular to the initial draw (+) t ~ 1/4 s, (○) t ~ 7000 s (after Brereton et al.[82]).

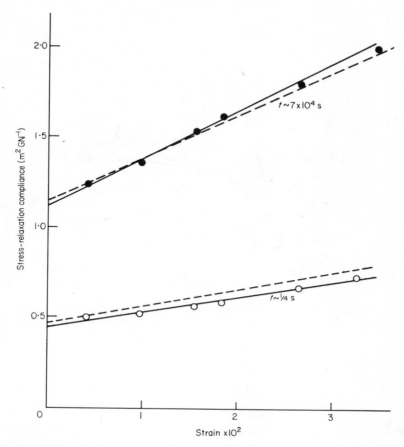

Fig. 20. The strain dependence of the isochronal stress-relaxation compliance for oriented PET tested in tension perpendicular to the initial draw (○) t ~ 1/4 s, (●) t ~ 7000 s. The full lines are a least squares fit to these data, the dashed lines predicted entirely from the creep data of Fig. 19 (after Brereton et al.[82]).

limiting strain for $\sigma_0 < \sigma_F$ but flows with a constant or ever increasing strain-rate for $\sigma_0 > \sigma_F$. Under constant strain-rate conditions the stress always tends to a constant value σ_F at large strains. At strain-rates higher than $\dot{\varepsilon}_F$ (which is the strain-rate observed in creep for $\sigma_0 = \sigma_F$) the stress–strain curves show a maximum σ_y, which increases with increasing strain-

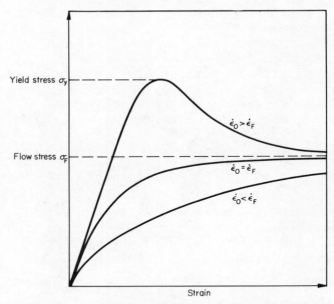

Fig. 21. Schematic stress–strain curves at different strain-rates. The flow stress σ_F at high strains corresponds to the stress at which the long term creep modulus tends to zero (after Brereton et al.[82]).

rate to a limiting value of σ_B at infinite strain-rate. Note then that this representation predicts that true stress maxima can arise in constant strain-rate tests in situations where the strain is completely recoverable. Almost complete recovery is commonly observed in creep even after long duration tests at high stresses,[95] although long recovery times may be required at room temperature. It has also been elegantly demonstrated by Haward[16] that the recovery of 'plastic strain' observed in deformation bands in isotropic PS compression specimens can be so speeded up by annealing just above the glass-transition temperature, that it is effectively complete within a very short period of time.

The question of whether 'plastic' strain is truly permanent, or merely a feature of very slow recovery processes has yet to be settled. The integral equation approach has confirmed that the existence of true stress maxima does not necessarily imply permanent deformation although the latter can readily be included within the representation by the choice of suitable response functions.

Detailed supportive evidence for the implicit equation representation of non-linear viscoelasticity is still very scarce. However, it can be said that for oriented PET tested in tension at $\theta = 90°$ under conditions of constant load (stress) and constant strain-rate, and isotropic PP, PMMA, PC and HDPE in tension at constant load the general features of eqn. (18) are well obeyed at all strains up to the appearance of deformation bands.

The implications and scope of these ideas have yet to be established but the following points are noteworthy.

(a) The form of the non-linear viscoelastic behaviour of oriented PET tested in tension at $\theta = 90°$ is very similar to that of isotropic PP, PMMA, PC and HDPE in tension.

(b) For all of these materials the behaviour is non-linear over the whole range of accessible stresses. The concept of 'yield' can perhaps be related to the creep behaviour as the limiting ($t \to \infty$) creep modulus tends to zero.

(c) The deformation is recoverable, at least for stresses less than the flow stress σ_F. (In tension, necking instabilities prevent the study of both creep and constant strain-rate deformation at higher stresses.)

(d) By describing both creep and constant strain-rate deformation in terms of only two response functions, it should be possible to resolve the frequent conflicts between the measured temperature dependences of these two experiments.

With regard to this latter point Brereton *et al.* point out that the integral equation predicts that the stress-maximum observed at strain-rates greater than $\dot{\varepsilon}_F$, should be approximately linearly dependent on the logarithm of the applied strain-rate. Such a feature is in accord with yield-stress data at different temperatures for a wide range of isotropic polymers, in both tension and compression. Many workers have attempted to describe this linear dependence of the yield stress on log (strain-rate) in terms of various modifications of Eyring's theory of viscosity. For example the equation

$$\dot{\varepsilon} = \dot{\varepsilon}_0 \exp \left\{ \frac{-(\Delta U - v\sigma)}{kT} \right\} \tag{25}$$

where ΔU is an activation energy, v an activation volume, T the absolute temperature, has been shown to describe approximately data from PMMA,[83] PVC,[84] polystyrene,[85] PET[86] and polycarbonate.[87] More accurate fitting of the data has entailed the introduction of increasing complexity into eqn. (25), which still has the striking deficiency of only relating to the yield point. The parameters in this equation obtained from yield data in general bear little or no relationship with those obtained by fitting the same equation to, for example, creep behaviour whereas with the implicit equation one set of parameters can describe both experiments.

Few attempts have been made to apply eqn. (25) to data from oriented polymers. Data over wide temperature and strain-rate ranges from oriented PET[49] and oriented PP[22] can be represented approximately by (25) and show that except for $\theta \sim 0°$ in tension the values of ΔU and v obtained are not very orientation dependent, and are similar to those observed for isotropic polymers.

In a follow-up of the integral equation approach Croll[88] made a preliminary study of the non-linear creep behaviour of the same oriented PET sheet in other orientations, and also of amorphous and crystallised isotropic PET. He discovered that whereas for the oriented material at $\theta = 90°$ and for the amorphous isotropic sheet, the creep modulus versus stress graphs were linear, as suggested by eqn. (22), a more complicated form of non-linearity was evident both for the oriented material at $\theta = 45°$ and $0°$, and for the highly crystalline isotropic material. An example of this is shown schematically in Fig. 22 where the creep modulus/stress graph for oriented PET sheet in tension at $\theta = 0°$, can be seen to have three distinct regimes.

It is interesting to note that in just these situations, *i.e.* tension at $\theta \sim 0°$ for an oriented polymer, and for a highly crystallised material, extended tie molecules may be expected to influence the deformation (cf. Refs. 69–79). No firm interpretation can be offered, however, at this preliminary stage of the study. Croll points out that it is not possible with constant strain-rate data alone to distinguish by visual inspection between polymers which obey eqn. (19) and those which belong to the class typified by Fig. 22 and so for these purposes creep measurements are more definitive.

This short discussion of non-linear viscoelasticity has been included in order to show that most of the features associated with yield in constant strain-rate tests are directly related to aspects of the creep behaviour. It is not yet clear what significance, if any, can be attributed to a measurement of a yield stress from a load-maximum (especially if the latter is complicated by the occurrence of a necking instability), or to an arbitrary proof strain, although the flow stress σ_f, at which the long-term creep

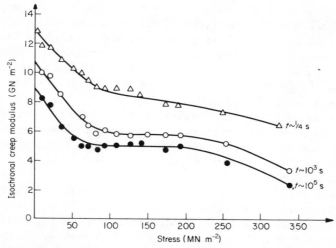

Fig. 22. The stress dependence of the isochronal creep moduli of oriented PET, tested in tension parallel to the initial draw (after Croll[88]).

modulus tends to zero, may have a more ready interpretation. The initiation of 'plastic' or permament deformation has yet to be understood with respect to this scheme.

11.5 GENERAL CONCLUSIONS

We have now discussed in turn, the stresses required to produce yield, the relationship between stress and plastic strain increment, the structural reorientation occurring as a result of yield, and the relationship between 'constant strain-rate yield' and features of non-linear recoverable creep deformation. Theoretical models to describe the behaviour have ranged from single crystal plasticity through to the oriented continuum ideas of plasticity and viscoelasticity. On many points both the experimental data and the interpretations appear almost contradictory and it is therefore helpful to see if any common ground can be established.

   The first point to be established is that the discussion has encompassed data and theories relating to a wide range of polymers and testing conditions. At one end of the spectrum we have PET and PVC drawn approximately 2:1, essentially amorphous materials which have had a rather small amount of order introduced through a low degree of orientation. At the other end of the spectrum we find oriented HDPE drawn in the range 9–12:1. Here the material has appreciable crystallinity, especially after the usual annealing procedures and it is most helpful to visualise the structure in terms of chain folded crystallisation, fringed micelles or near perfect paracrystalline arrays.

It is not unreasonable therefore that the modes of deformation observed with this wide range of polymers should be so diverse. Let us now attempt to understand how the yield anisotropy of a hypothetical polymer might develop as the degree of orientation is increased, starting with the isotropic polymer.

All of the available experimental evidence suggests that the yield surface for both amorphous and crystalline polymers can be approximated by a pressure and rate dependent von Mises criterion (Refs. 4, 6, 11, 13, 15, 16, 22). This implies that the shear components of stress are in the main responsible for the deformation. Many theories have been proposed in the past to explain yield in isotropic polymers ranging from Marshall and Thompson[89] who suggested that the adiabatic heating raised the temperature of the polymer locally to the glass-transition, to the disclination/dislocation theories of Argon[90] and Bowden[91] (see for example Bowden,[92] Ward[4]). These latter theories are based on the assumption that yield in polymers is governed by the motion of defects (disclinations or dislocations) by analogy with the deformation of metals. The motion of such a defect is dominated by the local elastic stress field, much influenced by thermal fluctuations, suggesting a natural link between the shear yield stress and the shear modulus. Allusions to such an association between elastic and plastic deformation have been made by many authors although it is clearly not universal (e.g. Refs. 20, 23, 37, 93, 98–100).

Adiabatic heating has been shown to influence strongly the yielding of polymers at high rates, but Vincent[3] proved that yield drops could also be observed at low rates, at which the heating effect was negligible. Notable amongst theories intermediate between these extremes is that of Robertson[51] in which he assumes that the applied stress induces a conformational state of the chains which is characteristic of a higher temperature and thus a lower viscosity. This theory lays emphasis on the essential flexibility of the polymer chain.

The behaviour of polymers drawn to low and intermediate draw ratios is similar in many ways to that of isotropic polymers, especially under situations which do not involve a substantial tensile stress parallel to the initial draw direction. This similarity is evident not only from the absolute values of yield stress but also from its rate and temperature dependence, for tension at $\theta > 45°$ and compression at all angles.[49,22] The same similarity is observed in the form of the non-linear creep behaviour.[82] Under these circumstances the reorientation observed in deformation bands follows the simple continuum scheme discussed in detail above and is perhaps consistent with the existence of some ill-defined network.

The development of a Bauschinger effect in oriented polymers, which is due to the tensile yield stress at $\theta = 0°$ being much higher than the compressive yield stress at the same angle might have two alternative explanations. First, the compressive yield stress is similar to that of the isotropic material because it involves a similar shear deformation of the network, involving the flexing of many chains. In tension at $\theta = 0°$ few chains are in a suitable orientation to be flexed by the applied stress, which therefore has to reach much higher values to produce yield, i.e. we have one mechanism, differing only qualitatively for different orientations. Alternatively, this same high tensile stress for $\theta = 0°$ may be large enough to introduce a second deformation mechanism—the chain scission detectable by e.s.r., infra-red and mass spectroscopy.[69-79] It appears that chain scission may be of considerable importance in the tensile deformation of fibres and highly cross-linked systems for which the conventional redrawing of a network may be impossible, although it is not likely to affect much the compressive behaviour. Thus the situation of a tensile stress parallel to the draw or orientation direction can be considered to be unique, whereas all other stress configurations produce deformation somewhat similar to that in isotropic materials. This conclusion is supported by the qualitative forms of non-linearity observed in the creep modulus (Figs. 19 and 22).

At very high draw ratios and/or crystallinities the deformation deviates strongly from that of a network and resembles more closely that of a single crystal. In tension, for $20 < \theta < 60°$ where there is a substantial shear stress and a tensile component parallel to the extended molecules, the deformation of HDPE approximates to 'c-slip' or 'intermolecular shear' as described by many authors.[27,29-32] Even in these situations of potentially easy shear, if one lowers the temperature of HDPE, presumably until the high molecular mobility in the less crystalline regions is quenched out, then the network ideas become more relevant, especially in understanding the optical reorientation in deformation bands.[48]

Furthermore if one moves away from the easy slip orientation, other mechanisms become important even in such highly crystalline polymers. For example in tension at $\theta = 0°$ it seems that some scission of extended tie molecules is involved, whereas in compression at the same orientation a low yield stress is observed which is perhaps associated with a high concentration of flexed molecules within both crystals and the amorphous regions and/or crystal break-up (see Refs. 51, 52, 63–66, 71).

Viewed in these simple terms the distinction between isotropic polymers, polymers of low draw ratio, and highly oriented crystalline polymers may be one of degree rather than of kind. For the most part, the oriented polymer is closer to the isotropic material than to the single crystal. It is, however, pertinent to sound a word of warning about too great a reliance on phenomenological similarities between the deformation of these extremes of materials.

Measurements of stress, strain, and to a lesser extent structural studies of birefringence, X-ray diffraction, and most forms of spectroscopy are necessarily averages over the whole structure—different techniques yield different forms of average and provide complementary information. There is a real need for more and varied structural measurements to be undertaken dynamically on stressed polymers in order to elucidate possible deformation schemes.

REFERENCES

1. Cottrell, A. H. (1953). *Dislocations and Plastic Flow in Crystals,* Clarendon Press, Oxford.
2. Cahn, R. (1965). *Physical Metallurgy,* North-Holland, Amsterdam, Chapters 15 and 16.
3. Vincent, P. I. (1960). *Polymer,* **1,** 7.
4. Ward, I. M. (1971). *J. Mater. Sci.,* **6,** 1397.
5. Hill, R. (1950). *The Mathematical Theory of Plasticity,* Clarendon Press, Oxford.
6. Thorkildsen, R. L. (1964). In *Engineering Design for Plastics* (ed. E. Baer), Reinhold, New York, Chapter 5.
7. Holliday, L., Mann, J., Pogany, G. A., Pugh, H. L. D. and Gunn, D. A. (1964). *Nature,* **202,** 381.
8. Ainbinder, G. B., Laka, M. G. and Maiors, I. Y. (1965). *Polymer Mechanics,* **1,** 65.
9. Pae, K. D., Mears, D. R. and Sauer, J. A. (1968). *J. Polymer Sci.,* **B6,** 773.
10. Biglione, G., Baer, E. and Radcliffe, S. V. (1969). In *Fracture '69,* Chapman and Hall, London, p. 503.
11. Rabinowitz, S., Ward, I. M. and Parry, J. S. C. (1970). *J. Mater. Sci.,* **5,** 29.
12. Davis, L. A. and Pampillo, C. A. (1971). *J. Appl. Phys.,* **42,** 4659.
13. Bauwens, J. C. (1970). *J. Polymer Sci., A-2,* **8,** 893.
14. Duckett, R. A., Rabinowitz, S. and Ward, I. M. (1970). *J. Mater. Sci.,* **5,** 909.
15. Rhagava, R., Caddell, R. M. and Yeh, G. S. Y. (1973). *J. Mater. Sci.,* **8,** 225.
16. Haward, R. N. and Thackray, G. (1968). *Proc. Roy. Soc.,* **A302,** 453.

17. Hill, R. (1968). *Proc. Roy. Soc.*, **A193**, 281.
18. Yoshimura, Y. (1959). *Aeronaut. Res. Inst. Tokyo*, No. 349.
19. Brown, N., Duckett, R. A. and Ward, I. M. (1968). *Phil. Mag.*, **18**, 483.
20. Bridle, C., Buckley, A. and Scanlan, J. (1968). *J. Mater. Sci.*, **3**, 622.
21. Rider, J. G. and Hargreaves, E. (1969). *J. Polymer Sci.*, *A-2*, 829.
22. Zihlif, A. (1973). Ph.D. Thesis, University of Leeds.
23. Robertson, R. E. and Joynson, C. W. (1966). *J. Appl. Phys.*, **37**, 3969.
24. Robertson, R. E. (1968). *J. Polymer Sci.*, *A-2*, **6**, 1673.
25. Rabinowitz, S., Duckett, R. A. and Ward, I. M., Unpublished Proceedings of the British Polymer Physics Group Conference on 'Polymer Chain Flexibility', Colchester, (1969).
26. Duckett, R. A., Ward, I. M. and Zihlif, A. (1972). *J. Mater. Sci.*, **7**, 480.
27. Shinozaki, D. M. and Groves, G. W. (1973). *J. Mater. Sci.*, **8**, 71.
28. Zaukelies, D. A. (1961). *J. Appl. Phys.*, **33**, 2797.
29. Kurakawa, M. and Ban, T. (1964). *J. Appl. Polymer Sci.*, **8**, 971.
30. Keller, A. and Rider, J. G. (1966). *J. Mater. Sci.*, **1**, 389.
31. Orowan, E. (1949). *Rep. Prog. Phys.*, **12**, 185.
32. Hinton, T. and Rider, J. G. (1968). *J. Appl. Phys.*, **39**, 4932.
33. Kelly, A and Davies, G. J. (1965). *Met. Rev.*, **10**, 1.
34. Shinozaki, D. M. and Groves, G. W. (1973). *J. Mater. Sci.*, **8**, 1012.
35. Owen, A. J. and Ward, I. M. (1973). *J. Macromol. Sci.-Phys.*, **B7**(3), 417.
36. Duckett, R. A. and Zihlif, A. M. (1974). *J. Mater. Sci.*, **9**, 172.
37. Young, R. J., Bowden, P. B., Ritchie, J. M. and Rider, J. G. (1973). *J. Mater. Sci.*, **8**, 23.
38. Hay, I. L. and Keller, A. (1966). *J. Mater. Sci.*, **1**, 41.
39. Seto, T. and Tajima, Y. (1966). *Jap. J. Appl. Phys.*, **5**, 534.
40. Keller, A. and Pope, D. P. (1971). *J. Mater. Sci.*, **6**, 453.
41. Bowden, P. B. and Young, R. J. (1971). *Nature, Physical Science*, **229**, 23.
42. Simpson, L. A. and Hinton, T. (1971). *J. Mater. Sci.*, **6**, 558.
43. Brown, N. and Ward, I. M. (1968). *Phil. Mag.*, **17**, 961.
44. Brown, N., Duckett, R. A. and Ward, I. M. (1968). *J. Phys. D:Appl. Phys.*, **1**, 1369.
45. Richardson, I. D. and Ward, I. M. (1970). *J. Phys. D: Appl. Phys.*, **3**, 643.
46. Richardson, I. D., Duckett, R. A. and Ward, I. M. (1970). *J. Phys. D: Appl. Phys.*, **3**, 649.
47. Duckett, R. A., Goswami, B. C. and Ward, I. M. (1972). *J. Polymer Sci.*, *A-2, Polymer Physics Edition*, **10**, 2167.
48. Duckett, R. A., Zihlif, A. M., Goswami, B. and Ward, I. M., unpublished.
49. Rabinowitz, S., unpublished.
50. Rider, J. G. and Hargreaves, E. (1970). *J. Phys. D:Appl. Phys.*, **3**, 993.
51. Robertson, R. E. (1960). *J. Chem. Phys.*, **44**, 3950.
52. Pechold, W. (1968). *Kolloid-Z.*, **228**, 1.
53. Hosemann, R. (1963). *J. Appl. Phys.*, **34**, 25.
54. Hosemann, R. (1967). *J. Polymer Sci.*, C, No. 20, 1.
55. Robertson, R. E. (1971). *J. Polymer Sci.*, *A-2*, **9**, 1255.
56. Robertson, R. E. (1969). *J. Polymer Sci.*, *A-2*, **7**, 1315.
57. Robertson, R. E. (1971). *J. Polymer Sci.*, *A-2*, **9**, 453.
58. Rabinowitz, S., Duckett, R. A., unpublished.
59. Tajima, Y. and Seto, T. (1971). *Rep. Prog. Polymer Physics in Japan*, **X**, 205.
60. Ishikawa, K., Miyasaka, K., Maeda, M. and Yamada, M. (1969). *J. Polymer Sci.*, *A-2*, **7**, 1259.
61. Yamada, M., Miyasaka, K. and Ishikawa, K. (1971). *J. Polymer Sci.*, *A-2*, **9**, 1083.
62. Kaufman, W. E. and Schultz, J. M. (1973). *J. Mater. Sci.*, **8**, 41.
63. Peterlin, A. (1966). *J. Polymer Sci.*, C, No. 15, 427.
64. Peterlin, A. and Balta-Calleja, F. J. (1969). *J. Appl. Phys.*, **40**, 4238.
65. Meinel, G., Morosoff, N. and Peterlin, A. (1970). *J. Polymer Sci.*, *A-2*, **8**, 1723.

66. Peterlin, A. (1971). *J. Mater. Sci.,* **16,** 490.
67. Hosemann, R., Loboda–Cackovic, J. and Cackovic, H. (1972). *J. Mater. Sci.,* **7,** 963.
68. Robertson, R. E. (1972). General Electric Report No. 72 CRD028.
69. Zhurkov, S. N., Tomashevsky, E. E. (1966). In *Physical Basis of Yield and Fracture,* Institute of Physics, London, p. 200.
70. Campbell, D. and Peterlin, A. (1968). *J. Polymer Sci., B(Letters),* **6,** 481.
71. Roylance, D. K., DeVries, K. L. and Williams, M. L. (1969). In *Fracture '69,* Chapman and Hall, London, p. 551.
72. Kausch, H. H. (1970). *Reviews in Macromolecular Chemistry,* **5,** Pt. 2, 97.
73. Becht, J. and Fischer, H., (1969). *Kolloid-Z.,* **229,** 167.
74. Regel, V. R., Muinov, T. M. and Pozdnyakov, O. F. (1966). *Physical Basis of Yield and Fracture,* Conference Proceedings, Oxford.
75. Zhurkov, S. N., Vettegren, V. I., Korsukov, V. E. and Novak, I. I. (1969). In *Fracture '69,* Chapman and Hall, London, p. 545.
76. Peterlin, A. (1969). *J. Polymer Sci., A-2,* **7,** 1151.
77. Kausch, H. H., Moghe, S. R. and Hsaio, C. C. (1967). *J. Appl. Phys.,* **38,** 201.
78. Kausch, H. H. and Hsaio, C. C. (1968). *J. Appl. Phys.,* **39,** 4915.
79. DeVries, K. L., Lloyd, B. A. and Williams, M. L. (1972). *J. Appl. Phys.,* **42,** 4644.
80. Ito, K. (1967). *Rept. Prog. Polymer Physics in Japan,* **X,** 361.
81. Steg, I. and Ishai, O. (1967). *J. Appl. Polymer Sci.,* **11,** 2303.
82. Brereton, M. G., Croll, S. G., Duckett, R. A. and Ward, I. M. (1974). *J. Mech. Phys. Solids,* **22,** 97.
        '969). *J. Appl. Polymer Sci.,* **12,** 1653.
84. Bauwens, J. C. (1972). *J. Mater. Sci.,* **7,** 577.
85. Whitney, W. and Andrews, R. D. (1967). *J. Polymer Sci., C,* No. 16, 2981.
86. Rabinowitz, S., Duckett, R. A. and Ward, I. M. (1970). *J. Mater. Sci.,* **5,** 909.
87. Bauwens–Crowet, C. (1973). *J. Mater. Sci.,* **8,** 968.
88. Croll, S. G. (1973). Ph.D. Thesis, University of Leeds.
89. Marshall, I. and Thompson, A. B. (1954). *Proc. Roy. Soc.,* **A221,** 541.
90. Argon, A. S. (1973). *Phil. Mag.,* **28,** 839.
91. Bowden, P. B. (1973). In *The Physics of Glassy Polymers* (ed. R. N. Haward), Applied Science Publishers, London, p. 279.
92. Bowden, P. B., unpublished proceedings of the second conference on 'Yield, Deformation and Fracture of Polymers', Cambridge, (1973).

94. Schmidt, E. and Boas, W. (1935). In *Kristallplastizitat,* Springer, Berlin.
95. Turner, S. (1969). In *Testing of Polymers, Vol.* 4 (ed. W. E. Brown), Interscience, New York.
96. Harris, J. S. and Ward, I. M. (1969). *J. Mater. Sci.,* **5,** 573.
97. Ward, I. M. (1970). *The Mechanical Properties of Solid Polymers,* Wiley, London.
98. Vincent, P. I. (1967). In *Encyclopaedia of Polymer Science and Technology,* Wiley, New York.
99. Robertson, R. E. (1964). General Electric Rept. No. 64-RL-3580C.
100. Allison, S. and Ward, I. M. (1967). *J. Physics D. (Brit. J. Appl. Phys.),* **18,** 1151.
101. Stachurski, Z. H. and Ward, I. M. (1969). *J. Macromol. Sci. Phys.,* **B3,** 445.

CHAPTER 12

ORIENTATION OF FILMS AND FIBRILLATION

G. SCHUUR and A. K. VAN DER VEGT

12.1 INTRODUCTION

In most plastic articles accidental orientation occurs due to the fact that they are usually made from the melt. During the flow and deformation of the melt in extrusion dies and in moulds the material is sheared, with molecular orientation in the shear direction. Part of this orientation relaxes before the material is cooled below the apparent second-order transition point or melting point. Another part, however, remains and is still observable in the final product.

Usually these orientations are undesirable because they result in anisotropic properties. The tensile strength, yield stress, impact strength, etc., are much higher in the direction of orientation than in the perpendicular direction. The tear strength is also anisotropic: a crack easily propagates in the direction of orientation but hardly in the perpendicular direction.

Moreover, molecular orientations often give rise to dimensional instabilities at elevated temperatures. In amorphous polymers, heating to temperatures near the glass-transition point causes a relaxation of the orientation, resulting in a shrinkage in the orientation direction and a swelling in both the other directions. Especially when regions with different orientation directions are present, the article may show a serious distortion. Also internal stresses may develop under these circumstances which weaken the object in certain regions or directions.

In crystalline polymers the phenomena are even more complicated since molecular orientation of the melt increases the rate of nucleation and crystallisation and changes the morphology of the material. Unoriented

413

material shows a spherulitic morphology, but the oriented regions show fewer or no spherulitic structures. These regions are therefore clear or at least less opaque. In microtome sections of the wall of an injection-moulded object three different regions can be observed under the polarising microscope, a highly oriented non-spherulitic skin, a less oriented layer consisting of a row of shear-nucleated spherulites and an unoriented spherulitic core.[1] The thickness of the highly oriented layer varies inversely with the melt—and mould temperature. A lower temperature will result in a higher viscosity and, hence, more shear and orientation and less relaxation.

Also in another respect amorphous and crystalline polymers show a difference in behaviour. An oriented amorphous polymer will tend to shrink in the orientation direction, especially after an increase in temperature. A crystalline polymer may show the same behaviour, but it may also increase in length owing to after-crystallisation in the orientation direction.

The anisotropy of the mechanical properties, caused by unidirectional orientation, is clearly demonstrated by Fig. 1. It shows a microtome section of the wall of an injection-moulded beaker perpendicular to the direction of flow. Bending causes numerous cracks to be formed in the oriented surface layers but they stop abruptly at the unoriented spherulitic core, because this material does not split so easily. Samples taken parallel to the direction of flow do not show these premature cracks on bending, on the contrary, the samples are less brittle than unoriented material.

A second source for accidental orientation is deformation of the material in the solid state, *e.g.* with cold-forming or solid-phase forming, in which a plastic sheet or billet is formed into an object at a temperature below the softening point of the plastic. The relaxation process can hardly take place under these conditions and although the article may be stable at room temperature, the deformation is (almost) completely recoverable at elevated temperatures.[2]

In many instances orientation is produced deliberately. In one-dimensional articles such as fibres and monofils the mechanical properties are, in general, only of importance in one direction and therefore a high anisotropy is desirable. In two-dimensional objects such as sheets and films good properties are required in two or even in only one direction, dependent on the application. Biaxial orientation may be beneficial to improve strength and toughness. A biaxially oriented thick polystyrene sheet, for instance, can be so flexible and tough, that it can be bent through

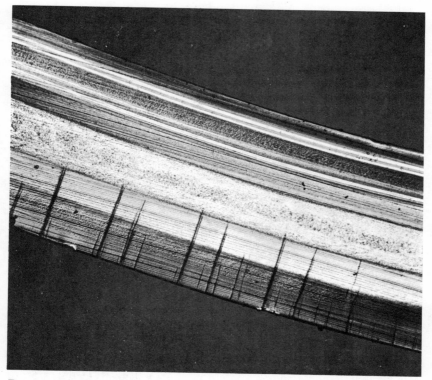

Fig. 1. Photograph of a microtome section of the wall of a polypropylene beaker bent in a direction perpendicular to the direction of flow, between crossed nicols and $\frac{1}{4}\lambda$ glimmer plates. (Magnification ×40.)

180° or punctured by a nail.[3] Unoriented general-purpose polystyrene sheet is brittle and would break or shatter. In addition, the biaxially oriented polystyrene shows a tenfold improvement in Izod impact strength, a twofold increase in tensile yield strength, substantial increases in yield elongation and in resistance to stress crazing. Similar important changes in the properties of plastic films can be obtained and they can be caused by accidental orientations resulting from the manufacturing process or intended uniaxial or biaxial orientation obtained by special stretching processes.

In this chapter the various methods of producing films with different types and amounts of orientation will be discussed, as well as the practical importance of orientation in several end-use applications. For a more elaborate description of film manufacturing techniques we refer to the literature.[4,5]

12.2 PRINCIPLES OF DRAWING, STRETCHING AND ANNEALING

Orientation in films and sheets can be produced in different ways most frequently by drawing and stretching. By the word 'drawing' we denote those operations which are carried out immediately after the main processing operation, e.g. extrusion or calendering, while the polymer is still fluid. In many cases the primary aim of drawing is to reduce the thickness of the article to be produced, but drawing may also result in molecular orientations.

The term 'stretching' is used to denote separate operations carried out after reheating of the already cooled extrudate or calendered sheet. It is mostly performed at lower temperatures than drawing and therefore requires higher stresses. Both drawing and stretching can be used to obtain monoaxial or biaxial orientations.

In all cases orientations in films and sheets are only obtained by applying a large deformation in one or in two perpendicular directions, at a suitable temperature. When the temperature is too low, e.g. when the polymer is in the glassy state, such deformations will mostly lead to rupture or if possible at all, require excessively high stresses. At too high temperatures, viz. in the purely viscous state deformations can easily be applied but will not result in molecular orientations because of rapid relaxation. Suitable conditions are those where the relaxation rate is exceeded by the rates of deformation and cooling; in that case rubber–elastic deformations occur and are frozen in as molecular orientations. For amorphous polymers this is the case in a temperature range between the glass–rubber transition point, T_g, and a somewhat higher value; with higher strain and cooling rates this range is extended to higher temperatures. Crystalline polymers, above the melting point, T_m, may have a temperature region where permanent orientations can be produced, again dependent on relaxation rate versus drawing rate and cooling rate. In both these cases molecular weight and distribution are important parameters since the relaxation time of a polymer melt strongly increases with increasing chain length and widening of the distribution.

If crystalline polymers are stretched below T_m the result will depend on the morphology of the polymer. At first it was believed that these polymers consisted of an amorphous continuous phase with dispersed crystalline regions, the so-called crystallites or fringed micelles.[6] During deformation the amorphous part would orient as an amorphous polymer and the crystalline regions would rotate and orient in the direction of stretching. Since it is known that the crystalline material in spherulites is continuous[7] and usually consists of folded molecules in lamellae, a different behaviour during stretching is assumed. At very low temperatures the crystalline material is too rigid and breakage will occur. Above a certain temperature slippage and twinning of the crystalline material can occur and the molecules orient by a rotation of the lamellae and finally by unfolding of the molecules.[8]

When oriented polymers are heated, they will try to reach their original high entropy, and shrinkage will occur as soon as the molecules can move sufficiently to recoil to their undisturbed dimensions. For amorphous polymers this will be the case when the glass transition point T_g is reached. For crystalline polymers the behaviour is more complicated.

At any temperature above T_g the oriented molecules in the amorphous regions will try to reach a higher entropy, but their coiling is hindered by the crystalline material. With increasing temperature the stress in the amorphous regions increases. Simultaneously the strength of the crystalline material decreases. Hence, more and more shrinkage due to deformation of the crystalline regions occurs accompanied by a twinning of crystals or slippage of crystal planes. Most of the shrinkage is observed near the melting point; above the melting point all orientation is lost. During this process, however, some slippage of the molecules may also occur and reversion measurements are therefore not an absolute yardstick for the original amount of orientation. That may be determined by the various methods discussed in the preceding chapters.

The resistance to shrinkage offered by the crystalline material depends on its quality and quantity. The number of lattice defects in a crystalline material increases if crystallisation is carried out at lower temperatures. Hence, the melting point of a polymer increases with increasing crystallisation temperature. Usually films to be stretched are quenched in order to obtain good stretchability and to prevent spherulite formation. Consequently, the amount of crystalline material is small and the melting point low. If such a material is heated, the amount of crystalline material will increase and the regularity of the lattice will improve, resulting in a higher melting point.

On this basis it is possible to understand what will happen if an oriented crystalline film is heated to near T_m and shrinkage is prevented by external stresses. The stress in the amorphous regions will increase but, as shrinkage is prevented, they will partly relax owing to molecular slippage. Simultaneously the strength of the crystalline material will increase.

This process—annealing or heat setting—is used to improve the thermal dimensional stability of stretched films and fibres. As the film which has undergone annealing, is heated freely, a strongly reduced shrinkage tendency will be observed up to the annealing temperature. By allowing a small predetermined amount of shrinkage to occur during annealing the shrinkage tendency can be reduced further but at the cost of the mechanical properties.

12.3 MANUFACTURE OF PLASTIC FILMS

Plastic films are manufactured by several completely different methods. Casting of films from solutions or dispersions is carried out on an appreciable scale. The solution or dispersion is cast on an endless belt of polished stainless steel or other suitable materials. The solvent is evaporated and the film is stripped from the belt, trimmed and wound up. This method is used for films from cellulose derivatives, rubber hydrochloride, etc., and can produce very uniform and clear films, virtually without orientation effects.

The other methods of film manufacture, calendering and several extrusion techniques, are much more interesting from our point of view.

12.3.1 Calendering of films

The calender, usually consisting of three or four heated steel rolls, is very simple in principle. However, in order to obtain wide thin films with narrow thickness tolerances and constant properties, numerous precautions must be taken, which make the designing and operating of a calender a skilled art.[9,10,11]

The calender is used for the manufacture of many products, sheets, floor coverings, artificial leather, etc. Many types of plastics and rubbers are suitable raw materials, but in the field of thin films its use is almost restricted to plasticised and rigid homopolymers and copolymers of vinyl chloride. The output of a calender decreases with decreasing thickness of the film. Hence, only films with a thickness of approx. 0·25 mm or more can be made economically.

If thinner films are required, an additional drawing or stretching process is necessary. Drawing is often carried out by taking the film from the last calender roll at a higher speed, while it is still very hot; the orientations relax rapidly so that not much anisotropy results from this operation. Another source of orientation effects is the shearing of the material in the nip of the calender. As this results in an unwanted anisotropy of the tear strength (*i.e.* a lower tear strength in the machine direction), raw materials and processing conditions are carefully selected to decrease the orientation effects. One remedy is, for example, an increase in the temperature settings of the rolls.

12.3.2 The stretching of rigid PVC films

The stretching of rigid PVC films has recently been discussed by Herner and Hatzmann.[11] They deal with the drawing and the stretching of rigid PVC film both in the HT (high-temperature) process and by the LT (low-temperature) or 'Luvitherm' process. The high-temperature process is carried out at a temperature of the material above approx. 150°C, when it is in the thermoviscous state, in order to reduce the anisotropy of the film. This type of film is used for less demanding applications in packaging or lamination and hence, more or less isotropic properties are required. Sometimes these films are stretched additionally at a lower temperature, for instance if film is to be made for shrink packaging.

The LT films are usually stretched with the aim of improving the mechanical properties in the machine direction. This is necessary for products like sound-recording tapes and adhesive tapes. The thickness and mechanical properties must be within very narrow tolerances and, in addition, the strength in the transverse direction must be high enough to prevent splitting. This can be achieved by various methods which will now be discussed.

Unidirectional stretching is always carried out in the tangential gaps between rolls or sets of rolls rotating at different circumferential velocities. Many variables—composition of the raw material, molecular weight and distribution, stretching rate and temperature, etc.—determine the result of the stretching operation. Another very important variable is the width of the tangential gap between the feeding and traction rolls not only for the stretching of PVC but for many stretching and drawing operations. This will be discussed in more detail.

If a round filament of plastic is stretched at a ratio of 1:4, the diameter decreases by a factor of 2, if minor variations in density due to, *e.g.*

crystallisation or void formation are disregarded. When film is stretched, however, the thickness decreases by a much higher factor than does the width. In addition, the decrease in thickness is not homogeneous and occurs more in the middle than at the edges.

This effect can be understood by the following simplified reasoning (see Fig. 2). The film is transported by the feeding rolls at a rate V_1 and stretched by the drawing rolls at a rate $V_2 > V_1$. The stretching ratio, λ, equals V_2/V_1. At the boundary of the film (*e.g.* the points on the curve AC) the material is only subjected to a tension in the direction of the tangent to the boundary. The stress is here monoaxial so that the material contracts freely both in the y-direction (transverse) and the z-direction (perpendicular to the film). This will result in monoaxial orientation.

In the middle of the sheet (points on the line EF) the same situation would occur if the material between C and D would be able to contract freely along the drawing rolls; as this is not the case the transverse contraction will not be complete and a stress will be present in the y-direction which reduces the film thickness and causes a transverse orientation. The way in which this stress increases as a function of y from the edge towards the middle, depends on the overall geometry: if l, the free length of the film between the rolls, is comparable to or greater than b,

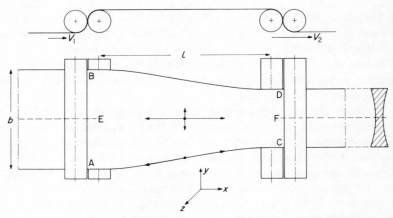

Fig. 2. Schematic representation of the apparatus for the stretching of films with a wide gap between feeding and stretching rolls and the cross-section of the film with exaggerated concavity.

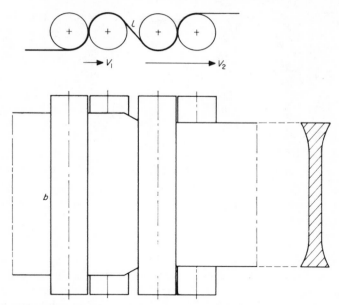

Fig. 3. Schematic representation of film stretching in a narrow tangential gap between feeding and stretching rolls.

the transverse stress will not be very high and moreover, will not reach an equilibrium value in the middle of the sheet so that the result is a concave cross-section indicated in Fig. 2 and inhomogeneous properties of the film in the transverse direction.

If, however, $l \ll b$ (see Fig. 3), the contraction of CD will be much less, the overall transverse stress and orientation will be greater and will virtually be constant over a considerable part of the cross-section. At the edges the material is still thicker and only monoaxially oriented, but this changes over a relatively short distance into a homogeneous thickness and transverse orientation. Only narrow strips need to be trimmed off afterwards to obtain an even film. Such a film can be considered as biaxially oriented with a stretch ratio in the transverse direction equal to the square root of the longitudinal stretch ratio. Owing to this the mechanical properties in the longitudinal direction are slightly inferior to those of a uniaxially oriented product, but the resistance to splitting is much improved.

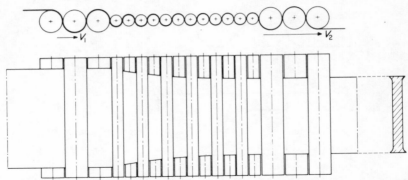

Fig. 4. Schematic representation of multiple roll-stretching equipment; the in-between rolls may rotate freely or may be driven at a predetermined, gradually increasing, speed.

It has appeared advantageous to carry out the stretching operation in a number of steps. This can be realised by arranging a large number of heated rolls closely together as indicated in Fig. 4. The transverse forces in each step are much lower than in a single-step operation so that transverse contraction is largely prevented by the friction between the film and the rolls. The result is a film of almost constant thickness.

At present most PVC films, both HT and LT, are stretched on this type of equipment. For the improvement of the mechanical properties by molecular orientation, the stretching temperature is in the range of 100–130°C in order to use the thermoelastic properties of the material.

12.3.3 Film extrusion

Many plastic films are manufactured by extrusion processes, which can be subdivided into three major categories:

(1) Flat film extrusion through a linear slit die into a water bath provided with submerged guiding rolls.

(2) Flat film extrusion through a linear slit die on a rotating chill roll.

(3) Tubular film extrusion through a circular die and film blowing.

Most films used for packaging have a thickness in the range 15–50 μm. In building, agriculture and heavy-duty packaging, thicker films are used. Thicknesses of more than approx. 200 μm are seldom required. On the other hand, the slits of extrusion dies are at least 500 μm wide in order to decrease flow resistance. Too narrow a slit would lead to excessive pressures in the extruder and the die, and hence to a decreased through-put and a risk of mechanical or thermal degradation of the material.

The molten film or web, when flowing freely, will have a thickness even larger than the slit width, owing to the so-called 'die-swell', the effect of the elastic recovery of the orientation in the die. Therefore a 'draw-down' is always applied; the material is removed from the die at a much higher

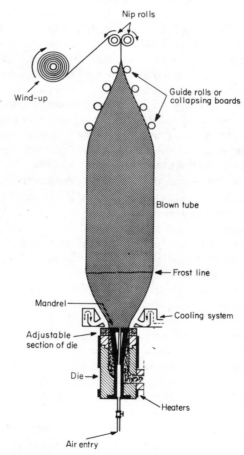

Fig. 5. Schematic representation of the tubular film extrusion process.

rate than the speed of the material in the die. The draw-down ratio varies between wide limits, according to the desired thickness of the film, and may be as high as 1:50.

The first extrusion method, water-bath quenching, has the advantage that a very rapid cooling hinders the growth of spherulites, leading to clear products. The drawing chiefly occurs in the gap between the die and the water surface. In the same way as discussed in Section 12.2.2, the edges will be thicker than the rest of the film. In order to decrease these edge effects, the distance between the die and the surface of the water bath is reduced as far as possible; usually it is only a few centimetres. Nevertheless, the thick edges must be trimmed before winding. The drawing prevents die-swell and causes even more orientation which is frozen-in in the water bath. Consequently, the properties of the film are always anisotropic. In order to reduce this, raw materials are selected with a low molecular weight and a narrow molecular-weight distribution; a high extrusion temperature is also used.

The principal drawback of water-bath quenching is that very thin films cannot be made from crystalline polymers such as polyethylene or polypropylene. Above a certain draw-down ratio there are thickness variations in the longitudinal direction. This phenomena is called 'draw resonance' and will be discussed in the next section.

Roll-quenching is also an important process because it gives a better gauge control. The web is extruded from a linear die on a rotating cold steel roll. The quenching is less sudden than in the water bath. In addition, especially at high speeds, an air cushion may be formed between the quenching roll and the web. In order to prevent this, a pressure is applied to the web, usually by means of an air knife. The quenching roll rotates at a much higher circumferential speed than the extrusion rate of the web in order to obtain sufficient draw-down; the distance between die and quenching roll is as short as possible to reduce edge thickening. Also in this method high melt temperatures are necessary to prevent anisotropy. The principal advantage of chill-roll quenching is that thin film can be made because 'draw resonance' effects do not occur.

The tubular process, film blowing, is very different and in some respects somewhat more complicated than flat-film manufacture. Hence, a schematic drawing is presented in Fig. 5 and a photograph of the actual balloon in Fig. 6.

After leaving the extruder, the melt flows through a circular die. The centre of the mandrel of the die is provided with an air inlet. The die is surrounded by an air-cooling ring, usually adjustable to provide for

Fig. 6. Photograph of the balloon in tubular-film manufacture. Note the clear visibility of the frost line where the material crystallises.

differences in blow-up ratio. At a certain distance from the die is the so-called frost line. The film solidifies here and the dimensions become fixed. The balloon is flattened between a pair of boards or sets of rolls and finally the flattened film passes a pair of nip rolls to prevent the air from escaping. It is then wound up or may be slit open to obtain flat film. It is a highly economic method, especially for the manufacture of sacks but also for flat film since there are no thick edges to be trimmed and reprocessed. Another advantage is that the die can be rotated completely or oscillated through a certain angle during the blowing process. In this way thickness variations in the film in the transverse direction are randomised, so that slightly thicker sections do not pile up on the roll.

The tubular-film process is only suitable for materials with sufficient melt strength to prevent blow-out of the balloon. Hence, grades with a higher molecular weight are sometimes used and the mass temperature is often 20–30°C lower than in flat-film extrusion. A low mass temperature is also required to decrease the height of the frost line, which results in a more stable balloon.

The molecular orientation in tubular films is chiefly obtained by that part of the drawing which occurs in the region just below the frost line, where the thermoelastic properties of the melt are most pronounced. Although the melt viscosity is higher here than near the die, the stresses are also much higher. The stress in the transverse direction increases with increasing diameter and with decreasing thickness of the balloon. In addition, the drawing in machine direction decreases the amount of material in a certain cross-section and hence increases the stress level both in the linear and transverse directions.

In this way states of biaxial orientation are produced; the ratio of the orientations in machine and transverse directions can be controlled within certain limits by adjusting the blow-up ratio with respect to the draw-down. Moreover, the overall level of orientation can be increased by decreasing the melt temperature and increasing the cooling rate.

In this respect the blow-extrusion technique is highly advantageous compared with flat film extrusion, where a similar combination of high molecular weight and low mass temperature would result in too much anisotropy.

The range over which the state of orientation can be changed in the film blowing operation and the resulting effects on mechanical properties have been demonstrated by Bates.[12] He investigated the effect of both transverse and longitudinal orientation on the impact strength of low density polyethylene.

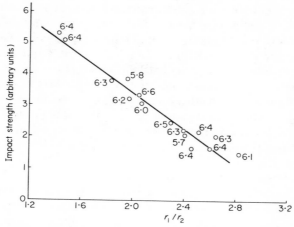

Fig. 7. Impact strength of low density polyethylene films with a varying ratio of machine-direction (r_1) versus transverse-direction orientation (r_2) as determined by reversion measurements. (The numbers on the graph denote r_1.)

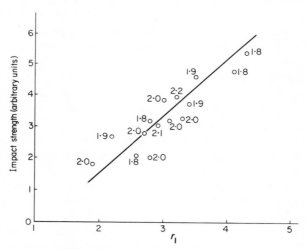

Fig. 8. Impact strength of low density polyethylene film as a function of the amount of orientation determined by reversion measurements. The ratio of longitudinal/transverse orientation (r_1/r_2) is kept constant at a value close to 2. (The numbers on the graph denote r_1/r_2.)

By carrying out reversion experiments he determined the actual orientations in both directions, which are, of course, lower than the corresponding draw ratios. His results are given in Figs. 7 and 8, which show that the impact strength decreases with increasing anisotropy, and increases with increasing orientation if equally balanced films are compared.

The tubular-film process is unsuitable for polymers with a very low melt strength such as polyethylene terephthalate. It is also not suitable for polypropylene films for packaging because the films are too crystalline, opaque and brittle due to too slow cooling. However, these films are often used as a precursor for fibrillated film fibres.

12.3.4 The 'draw resonance' phenomenon

Thickness variation in extrudates can be caused by many trivial factors such as inhomogeneous materials, fluctuation in the feeding rate of the extruder or in the melt temperature, surging of the extruder and lack of mechanical stability and stiffness of the equipment. A more fundamental reason for surface roughness or thickness variations is melt fracture, which appears suddenly as soon as a certain critical extrusion rate is exceeded.

The thickness variations due to 'draw resonance' have a different character. The thickness may vary by a factor of approx. 2, but there is no surface roughness. The transition from thick to thin regions is very gradual and occurs at regular distances; there is a certain 'wave-length', which may be as large as one metre or more. This defect has been called 'draw resonance' by Bergonzoni and Dicresce[13] because it is only observed when the draw-down ratio exceeds a certain value, usually approx. 1:4. Draw resonance also plays a role in paper coating and in the manufacture of fibres and monofilaments.[14] It is therefore remarkable that so far this technically important subject has only received scant attention in scientific and technical literature.

In older literature[15,16] the phenomenon is ascribed to circumstances such as improper extruder design or insufficient die pressure. Also Freeman and Coplan[17] consider it the effect of extruder surging, melt inhomogeneity or irregular cooling conditions. Bergonzoni and Dicresce[13] disagree with this view, because it would not explain the regularity or the dependence on the amount of draw-down. They try to explain it by melt thinning due to frictional energy losses in the melt between the die and the water bath, leading to a hot region of decreased viscosity in a similar way as the formation of hot necks during a stretching process. According to these authors, the hot region will stretch preferentially but, owing to its decreasing thickness, it will cool faster and stiffen. The tension, which decreased during the heating-up period, will increase again and a following hot region will form in a thick section. The difficulty of accepting this explanation is that the energy necessary to cause draw-down is several orders of magnitude smaller than the energy required for stretching a solid material. Hence, if all the energy applied during draw-down were converted adiabatically into heat in a hot region, the temperature rise would be of the order of $0.1-0.01°C$, which is obviously too small to explain the phenomenon.

Schuur[14] proposed another explanation. It is known that oriented polymers tend to lengthen spontaneously during crystallisation.[18] Hence, if sufficient draw-down is applied, the molecules will be oriented when entering the water-bath, and consequently the film will increase in length during crystallisation. As the flow rate of the melt and the speed of the take-off rolls are constant, this lengthening will result in a decreased draw-down and, hence, a region with a greater thickness and less molecular orientation will form. This region will show less spontaneous lengthening during crystallisation and the draw-down will increase. A more oriented region is formed, which will lengthen again during crystallisation and this process repeats itself with a regular 'wavelength'.

This explanation is in agreement with the fact that studies of shrinkage by reversion indicated the thin regions to be much more oriented. It has also been found that factors which increase the orientation of the melt also increase the tendency towards draw resonance. These factors are, for instance, a high draw-down ratio, a high production speed, a high molecular weight, a wide distribution and a low melt temperature. It has also been found that the 'wavelength' of the draw resonance increases with increasing distance between the die and the water bath.

Polymers with a low molecular weight and a high melt temperature, such as the polyamides and polyesters show much less tendency towards draw resonance than polyethylene or polypropylene.

It is also evident that draw resonance hardly occurs if at all during chill-roll quenching or film blowing. Chill-roll quenching prevents spontaneous lengthening of the melt mechanically, while film blowing results in a two-directional orientation, which prevents a preferential lengthening in one direction.

On the other hand, Bergonzoni and Dicresce[13] also observed that draw resonance can occur in polystyrene extrudates. If this observation is confirmed, it would be in disagreement with this explanation. Hence, the agreement of Schuur's explanation with a number of experimental facts does not prove that this hypothesis is complete or even correct, and further fundamental studies will be necessary. Whatever the true explanation may be, draw-resonance only occurs at high draw-down ratios indicating that orientation is involved.

12.4 TWO-DIMENSIONAL STRETCHED FILMS

12.4.1 Tubular-film stretching

Biaxial stretching can be carried out by tubular—or by flat—film processes. The first biaxially oriented films were manufactured from polystyrene in Germany in 1935. They were used for military applications, especially coaxial-cable insulation, and in capacitors. In this process a tube of polystyrene was extruded and drawn over a conically shaped extension of the mandrel of the die, as indicated in Fig. 9. Sufficient air pressure is used in the gap between the extruded tube and the mandrel to support the tube. If the pressure is increased, an air cushion may be formed between film and stretching cone. Additional air-cooling rings may also be used. The stretching cone may be internally cooled with water, but this is not shown in the drawing. The trouble with this process is the difficulty of starting-up. Recently a patent was applied for by Carlson.[19] to overcome

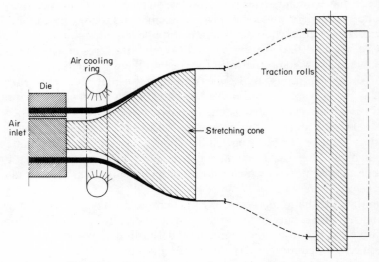

Fig. 9. Film stretching over a conical extension of the die mandrel.

this problem. During start-up, a nucleating and blowing agent is added to the polystyrene melt. The foaming causes the tube to expand spontaneously, so that it can be drawn over the cone without much difficulty. The addition of foaming agents is then stopped and simultaneously the temperature of the melt is adjusted. The clear polystyrene tube is biaxially oriented on the cone, slit open and wound up. Carlson applies his process to thicker sheets but it may also be suitable for thin films.

At present thin polystyrene film is made by blowing a film downwards. The difference with ordinary film blowing is that a cooled guide ring is used near the extrusion die in order to improve bubble strength and stability. In this process the amount of orientation depends on the internal air pressure, the diameter increasing with increasing pressure. The orientation also increases with increasing melt viscosity.

With polymers such as polyvinylidene chloride copolymers and polypropylene the bubble method is carried out in a different way. First a tube is extruded and quenched as rapidly as possible to keep the crystallinity as low as possible. As shown in Fig. 10 the collapsed tube passes through a pair of nip rolls (a) and is then heated to the desired temperature and blown up to a 4–5 times larger diameter before it reaches the next set of nip rolls (b) rotating at a higher speed to provide a longitudinal stretch ratio 1:3–1:4.

The film can then be flattened and wound up as a tube or slit to obtain a single sheet. A polyvinylidene chloride film obtained in this way is heat-shrinkable at temperatures above 80°C. If a heat-stabilised film is desired, an annealing step is applied by re-inflating the film and leading it through a radiant-heating tunnel (d). The third set of nip rolls (c) rotates at the same rate as the second set, and the air pressure in the bubble prevents the film from shrinking in the transverse direction. This is very effective

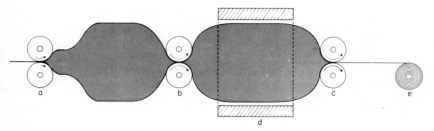

Fig. 10. Bubble stretching of tubular film.

and reduces the amount of heat shrinkage that occurs after 5 min immersion in boiling water from 30% in longitudinal and 40% in transverse direction to approx. 3% in both directions.

An advantage of this process is that the amorphous tube crystallises rapidly during biaxial stretching. This increases the modulus and therefore also the stress at a faster rate than the thickness decreases. Hence, the balloon will obtain a diameter where the developed tensile stress just balances the internal pressure. This diameter will, of course, depend on several factors, such as temperature, wall thickness and on the stretch ratio in the machine direction, but not on the amount of air. If more air is introduced, the balloon will become only longer, not wider and, hence, the balloon is self-stabilising.

More or less the same process as used for polyvinylidene chloride film can be applied to polypropylene, which also shows this self-stabilising property. In order to reduce the crystallinity of the unstretched film, internal cooling of the balloon by cold mandrels and (or) external cooling by water sprays is necessary.

An advantage of the bubble processes is that the stretching in the machine direction and transverse direction are carried out simultaneously, which makes it rather easy to obtain completely balanced films. In addition, there are no thick edges to be trimmed, which gives the bubble process an economic advantage.

Apart from the processes which were briefly discussed here, there are a number of other methods for stretching tubular film, such as octagonal film stretching and stretching over a horseshoe mandrel. These methods are based on similar principles and have been dealt with by Park and Conrad.[20]

12.4.2 Biaxial stretching of flat films

During the stretching of tubular films, as discussed in the previous section, longitudinal and transverse stretching are usually carried out simultaneously. This is also possible in flat-film stretching by means of a tenter frame.

The edges of the film are gripped by a series of clips. These clips diverge in the transverse direction and simultaneously accelerate in the machine direction; the increasing distance between the clips causes the longitudinal elongation as indicated in Fig. 11. The film is first heated to the desired temperature (a), then stretched (b), and finally passes through a cooling zone (c). The edges are subsequently trimmed and the film is wound up.

The heating of the film must be very accurately controlled in order to

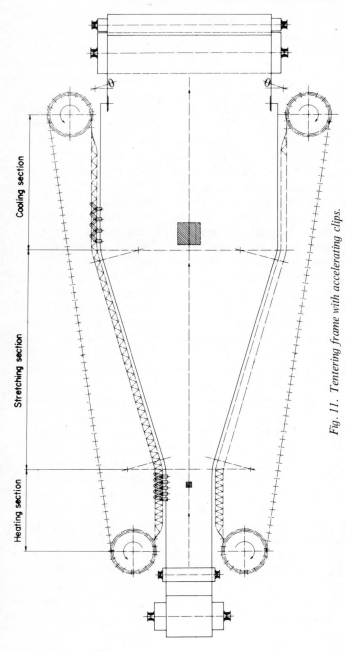

Fig. 11. Tentering frame with accelerating clips.

obtain exactly the same temperature over the whole surface. The temperature is always near the softening point if amorphous polymers are stretched to minimise relaxation. If crystalline polymers are stretched, the temperature is usually approx. 10–50°C below the melting point. The stretching ratios of amorphous polymers can be chosen at will up to a certain maximum and stretching may be balanced or unbalanced. If crystalline polymers are stretched, there is both a minimum and a maximum ratio because these polymers are stretched by cold drawing or necking-in. Hot regions[21,22] or mechanically unstable zones[23,24] are formed which travel through the film until it is completely stretched. If the stretch ratios are too low, the film contains unstretched regions.

The difficulty with these tenter frames is that mechanically they are very complicated and other systems have been proposed in patent literature, for instance the manufacture of films with beaded edges. The beads slide in diverging slots or are gripped by rubber strips; longitudinal orientation is realised by a difference in speed between feeding and take-off rolls.

Usually, however, simplification is achieved in a different way, viz. by carrying out longitudinal and transverse stretching in separate steps. The film is usually first stretched longitudinally in the tangential gap between two rolls or sets of rolls revolving at different speeds, and then transversely in a simplified tenter frame in which the clips only diverge without acceleration. The reverse is also possible: first stretching in the transverse direction by a tenter frame and then roll-stretching. This procedure has the advantage that the tenter frame is operated at a reduced speed.[25,26]

If the film is stretched too much in the first step, it may be too splitty for the second stretching process. This is especially a drawback for materials with stretch by necking-in since the first stretching cannot be reduced at will in order to prevent unstretched regions. In order to reduce the splitting tendency, the draw-down of the melt and the first stretching step are carried out in narrow gaps so that some orientation in the transverse direction is obtained.

The heating necessary to carry out the second stretching anneals the orientation obtained in the first stretching process. Hence, it is virtually impossible to obtain completely balanced films in a two-stage process. Not only is the mechanical strength anisotropic but also the heat shrinkage. Shrinkage will first occur in the direction of the last stretching process and shrinkage in the other direction will be observed at a higher temperature,[27] whereas a one-stage biaxially stretched film shows contraction in both directions simultaneously.

12.5 APPLICATIONS OF BIAXIALLY ORIENTED FILMS

A detailed discussion of the many types of oriented plastic films, their properties and applications falls outside the scope of this chapter. Instead, we shall discuss the reasons why biaxially oriented films are used and their advantages and disadvantages in more general terms.

Sometimes the prime reason for the biaxial stretching of plastic films is the cost reduction obtained by the decrease in thickness. Usually, however, this thickness reduction is possible only owing to the improvement of the tensile strength and modulus and other mechanical properties.

Polystyrene films and polyester (polyethylene terephthalate) films are only used in the biaxially stretched form, because unstretched films are very weak and brittle. To a lesser extent this is also true for polypropylene films. Roll-quenched flat film is used for many packaging applications, but below $0°C$ these films become brittle. Biaxially oriented polypropylene film, on the other hand, is not brittle at a temperature as low as $-50°C$ and is suitable for packaging frozen foods.

The optical properties, clarity and gloss, are also improved and biaxially oriented polystyrene and polypropylene films are used for envelope windows and overwraps. Biaxially oriented polyvinylidene chloride film is also used for food packaging in view of its good mechanical and optical properties, but for economic reasons it is only used if its low permeability to gases, especially oxygen, is required. The electrical properties are improved by biaxial orientation and the applications of polyester, polycarbonate and polypropylene films in capacitors are expected to show a rapid growth.[28] In these electrical applications crystalline films are always annealed in order to improve the dimensional heat stability.

Films for packaging are often annealed; isotropic or almost isotropic properties are usually preferred, but not if films are intended for use in which the principal stress is unidirectional. Films made from rigid PVC, polyester or polyolefin plastics for sound-recording, video-recording or adhesive tapes need a maximum strength and modulus in the longitudinal direction, but in the other direction the strength must only be high enough to prevent splitting.

The amount of biaxial orientation in films for packaging is not always increased to the maximum level. For blister packs, which are manufactured by deep drawing, a certain amount of stretchability is necessary. This can be obtained by a reduction of stretch ratios and avoidance of annealing. Blister packs from polystyrene must be manufactured from biaxially stretched films in order to reduce brittleness. Also blister packs from crystalline polymers are often made from biaxially oriented film to obtain good mechanical and optical properties and in addition to avoid variations in wall thickness. An unoriented film may show the necking-in phenomenon in highly strained regions, resulting in an article with thick unoriented and thin highly oriented parts. A biaxially oriented film does not show this necking and a more even wall thickness is obtained.

Annealing is avoided with films intended for shrink-packaging. In shrink-packaging the object or objects to be packed are surrounded by a loosely fitting film, sack or sleeve. The film is then heated in order to obtain sufficient shrinkage for a tight fitting. Sometimes machines, motors and other kinds of apparatus are packed in this way for protection against moisture and corrosion. The bulk of shrink-packaging is used for palletising and consumer multipacks for supermarket merchandising. For these applications biaxially stretched films are usually not necessary. Blown polyethylene films may show up to approx. 70% shrinkage on heating, which is quite sufficient. The shrinkage force is rather low, but the good strength and low price of these films make them very suitable for case wraps.[29,30]

An important property of plastic films is their heat sealability. During heat sealing the plastic is heated to above its glass transition point or melting point. In oriented films this heating causes shrinkage of the material and in addition, the molecular orientation in the seal is lost. Blown polyethylene film has a low shrinkage force and reasonable seals can be made by conventional techniques. Biaxially oriented films of polyesters, polyamides or polypropylene, however, show high shrinkage energy resulting in a poor appearance of the seal. In addition, the loss of biaxial orientation of the material and recrystallisation in spherulitic form cause weak and brittle seals. Figure 12 shows a seal in polypropylene film. Not only the weld itself, but also the bordering regions show a change in morphology. In order to overcome this difficulty several methods are used. In some cases heat sealing is not used and glues are used for overwraps. In other applications, for instance in the manufacture of bags from flat film, side welding or impulse welding may be applied.

The most universal method at present of improving heat sealability is coating with low-melting polymers which show good heat sealability at a temperature where the biaxial orientation is preserved and shrinkage is still absent or low. In order to manufacture heat-sealable, biaxially oriented polypropylene film, the film is first annealed, then corona-discharge treated in order to improve the adhesion and subsequently a

Fig. 12. Microphotograph of a heat-sealed biaxially oriented polypropylene film: black region (top) unheated part of film; middle region: recrystallised part adjacent to seal; bottom region: sealed material.

layer of a low-melting polyethylene copolymer is applied. Coatings with lattices or solutions of polyvinylidene chloride copolymers are more popular, owing to the excellent barrier properties of this material for gases. However, an adhesive intermediate layer, usually an isocyanate glue, is necessary between the corona-heated polypropylene film and the heat sealable coating in order to obtain a satisfactory seal strength.

12.6 UNIDIRECTIONALLY ORIENTED FILMS

Unidirectional orientation is often carried out to manufacture tapes or fibrillated fibres for various end uses. The films are manufactured by calendering or by flat film extrusion processes. Blown films are less suitable for tapes, owing to the fact that the cooling of the balloon occurs from the outside. This results in a slightly asymmetric structure of the film, as far as crystallinity and orientation are concerned. This asymmetric structure results in a tendency of the tapes to curl in the transverse direction. This curling is completely unacceptable for recording tapes and adhesive tapes. Tapes for weaving or knitting may have a slight curl but for maximum coverage flat tapes are better.

Stretching is always carried out between feeding rolls and faster rotating stretching rolls. If films are required for non-splitting tapes, *e.g.* for weaving or knitting, the smallest possible distance between the rolls is chosen in order to prevent a decrease in width of the film and to obtain some orientation in the transverse direction as described in Section 12.2.2. The splitting tendency of polypropylene tapes can be reduced further by the use of copolymers or additives which reduce splittiness, such as low-density polyethylene or certain kinds of rubbers. The tangential gap between the rolls is too small to install heating equipment; the desired stretching temperature of the film is obtained with hot feed rolls. The stretching rolls may also be heated to obtain some annealing. The method of stretching by means of a series of heated rolls as shown in Fig. 5 is only used for materials such as rigid PVC, which do not stretch by necking-in. Necking films stretch in a single gap anyway and a large number of gaps is superfluous.

After the stretching process the films may be annealed and are then slit into the desired number of tapes by rotating or stationary knives. Tapes for weaving or knitting polypropylene or high-density polyethylene are wound on bobbins, which are used in more or less conventional textile machinery.

If splitting is desirable the films are often slit into a number of tapes before stretching in order to minimise transverse orientation during stretching. The distance between the feeding and traction rolls is increased to a few metres and the film is heated in infra-red ovens or, more often, in hot air circulating ovens. Ovens with a symmetrical air circulation have a certain advantage over ovens with an asymmetrical circulation.[31]

The oven-stretched film is suitable for tapes for the weaving of backings for tufted carpets. The splitting tendency is then carefully controlled by the choice of proper stretching conditions and the proper polymer formulation. The splitting tendency should be low enough to prevent splitting during winding, beaming or weaving but, on the other hand, it must be high enough to prevent mechanical damage to the tapes by the tufting needles. As shown in Fig. 13, the needles will punch holes in tapes with insufficient splitting tendency, which is accompanied by a pronounced decrease in the strength of the backing, but if the tapes show sufficient splitting, only longitudinal slits are formed, which have a minor influence on the strength.

The oven-stretched films also always show a difference in thickness and orientation in the transverse direction due to the reduction in width during draw-down or stretching. Methods have been proposed to prevent this and to obtain a film which is only stretched in the machine direction over the whole width. The films can be extruded from an undulating die gap instead of the normal straight one; draw-down does not cause a decrease in the width of the web but only the undulations disappear.[32] In a similar manner the film can be undulated in the transverse direction before stretching.[33] These methods are too complicated for general use but may be applied if a maximum stretch ratio and hence maximum strength and splitting tendency of the stretched films or the tapes is required.

A fairly recent development is the manufacture of cross-laminated unidirectionally oriented films which can be used for the manufacture of sacks. Rasmussen[34] invented this process in which films are spirally cut from unidirectionally stretched tubes and laminated by means of an adhesive. The orientation directions of both layers are perpendicular to each other and at an angle of approx. 45° to the longitudinal direction of the laminate. Sacks manufactured from cross-laminated high-density polyethylene film have very high strength, tear strength and puncture resistance and are sold under the trade name 'Valeron'.

Another development is the manufacture of 'over-stretched' products. If films are stretched at too high a ratio and especially at too low a temperature, numerous microscopic cracks form and the film has an

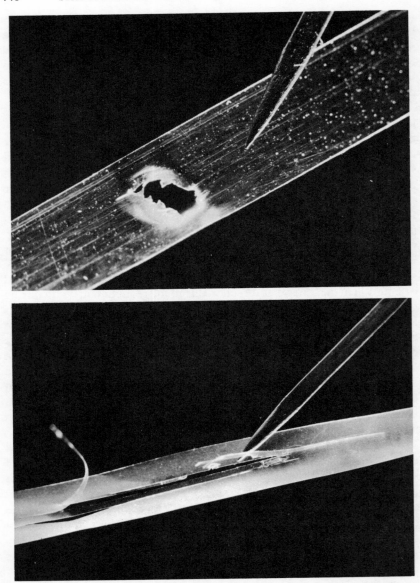

Fig. 13. The effect of needle punching on tapes in carpet backing (Top: non-splitting tapes with some transverse orientation stretched in a narrow tangential gap. Bottom: splitting tapes stretched in an air oven).

attractive pearly appearance. Polypropylene especially, if stretched at temperatures at or below approximately 100°C shows this 'overstretching'. Polypropylene copolymers show it under a wide range of conditions. The strength is not reduced appreciably by these cracks and the film is being used for the manufacture of binding tapes and twines with a silvery appearance for fancy packaging.

Recently, however, this overstretching has been used for the manufacture of polypropylene film with very fine pores, sold under the trade name 'Celgard'. A polypropylene film is extruded, cold-stretched in the transverse direction and then hot-stretched in the machine direction to open the cracks. Finally, the film is annealed under tension.[35] Hot stretching is more effective and consequently the tensile strength in the longitudinal direction is 10 times higher than in the transverse direction. The pore size is less than 0·1 μm. Air and water vapour pass the film readily but it prevents the passage of water unless specially treated. It is intended for battery separations, ultrafiltration and several new engineering and industrial uses.[36]

12.7 FIBRILLATED FILMS

12.7.1 Manufacture of fibres from plane films

The splittiness of unidirectionally oriented plastics has been known for a very long time and the first patent to make use of this property was applied by Jacqué[37] before the second world war. Jacqué developed a process for fine fibres by grating, turning, twisting or brushing unidirectionally oriented films from PVC, polystyrene and cellulose derivatives. He indicated that the fibrillation could be achieved by techniques of the textile industry like reeling, combing or carding or by beating the films in the direction of orientation with wedge- or knife-shaped hammers. Other early attempts were made by Costa et al.[38] who manufactured thread by shredding and twisting oriented films of vinyl-chloride and vinylidenechloride homo- and copolymers. Also Rasmussen[39] showed that oriented films can be split or fibrillated to obtain fibres, yarns or net-like products.[40]

None of those earlier developments was used on an appreciable technical scale. The reason is that the polymers which were available were not easily fibrillated or not very suitable for textile uses.

In order to be splittable the film must have a high strength in the orientation direction. The absolute level of the cohesive strength in the transverse direction is not of prime importance as long as it is low enough to allow the cracks to propagate without plastic flow at the tip. This condition can be fulfilled for those polymers for which the forces between the chains are low, *e.g.* the polyolefins. Polymers with strong polar forces or hydrogen bonding between the chains, such as cellulose derivatives, polyamides, polyesters or PVC are less suitable.

In unoriented polymers the splitting tendency is not only lowered by the presence of strong primary chemical bonds in all directions but to a much higher degree by the fact that a plastically deformed layer of material near the tip of the crack absorbs large amounts of irreversible work during crack propagation. Hence, Griffith's well-known equation $\tau = (2\gamma_s E/\pi c)^{1/2}$, where τ is the tensile strength, γ_s is surface energy, E is Young's modulus and c is half the length of the crack must be modified into $\tau = [2E(\gamma_s + \gamma_f)/\pi c]^{1/2}$, where γ_f is the work of plastic deformation at the tip of the crack.

This viscous flow in the tip of the crack results in molecular orientation of the surface which can be seen on fresh fracture surfaces of polymethylmethacrylate as bright interference colours due to the difference in refraction index of oriented and bulk polymer.[41−43] The amount of energy absorbed by this plastic deformation is orders of magnitude larger than the surface energy. This amount of plastic flow strongly depends on the amount of orientation and in well-oriented polymers it is very low or absent if the crack propagates in the orientation direction. Unidirectional orientation may also elongate weak regions resulting in an increase in the effective dimension in that direction and consequently in more effective crack initiation. Hence, in order to manufacture easily splittable films, a high stretch ratio must be applied and the decrease in film width during stretching must not be prevented to avoid transverse orientation. This implies that very wide films must be slit into a plurality of strips before stretching in order to prevent a decreased splittability in the middle of the film.

Numerous methods have been proposed to obtain splitting by using turbulent air or water, air jets, ultrasonics, etc., but only mechanical fibrillation is of practical importance through means already indicated by Jacqué.[37]

Coarse products suitable for the manufacture of baler twines, ropes and hawsers are made by flat film extrusion, water-bath quenching, air-oven stretching and optionally annealing thick film strips. The oriented strips are twisted and coarse fibrillation then occurs spontaneously. These polypropylene twines and ropes are stronger than ropes of sisal or manilla of equal weight. They do not become stiff and heavy when wet, do not rot and they float on water. These practical advantages in combination with low cost, explain why these ropes and twines are preferred by the majority of end users.

Fine fibres or yarns cannot be obtained in this way because thin films cannot be made by water-bath quenching and the degree of fibrillation obtained by twisting is not high enough. Quench-roll film can be thin enough but usually blown film is used as raw material. The blow-up ratio is kept low to prevent transverse orientation. After extrusion the film can be slit and stretched in line or subjected to a two-stage process. In the latter case care must be exercised because the crystallinity of the film may increase during storage and thus influence the stretching process and product properties.

Slitting is necessary when making continuous yarns, but for staple fibre it is sometimes possible to avoid slitting and to stretch the full width of the sheet or tube because the product is afterwards cut to the desired staple length and the cards or garnets will disintegrate the product to a more or less individual fibre state.

After stretching and, optionally, heat setting, the films are fibrillated. In general, rotating rolls provided with projections such as needles or knives are used. These rolls rotate in the direction of film transport but at a higher circumferential velocity. The ratio of the film speed to that of the needles is the fibrillation ratio. The final product depends on the construction of the fibrillator roll, *e.g.* the diameter and number of pins, the fibrillation ratio, the angle of contact between fibrillator and film, the film tension and film properties. Very important is the degree of orientation of the film, which has a great influence on splittiness.

A large number of patents has been taken out relating to the construction of fibrillators. In one patent, for instance, rolls are described with rows of pins in staggered position, which can be used to obtain regular networks.[44] Regular networks have no advantage over irregular structures for textile applications but a certain coherence and strength of the fibrillated product is necessary in order to give the film the required tension and to remove the fibrillated web from the fibrillator. Such an irregular structure is shown in Fig. 14. Fibrillation can also be obtained using pinned rolls, with the rows of needles wide apart to prevent a large number of needle rows touching the film simultaneously and lifting the film instead of penetrating it (fakir effect) and to ensure that each needle gives its own slit instead of penetrating existing slits.[45] Other fibrillators have a surface appearance reminiscent of that of a card cloth, but the

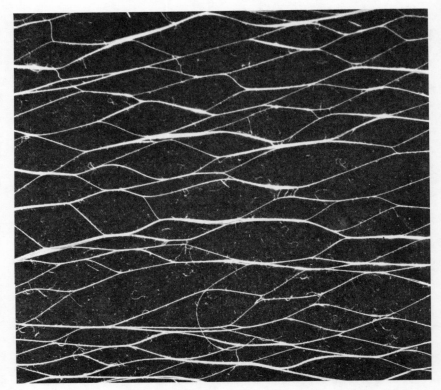

Fig. 14. Fibrillate obtained by passing an oriented film over a pinned-roll fibrillator.

needles are finer and at a much smaller distance as shown in Fig. 15.

Fibrillation decreases the strength appreciably. Unfibrillated tapes vary in strength between approx. 4 and 9 grams/denier but after fibrillation the strength is between approx. 1 and 3 grams/denier. However, the original strength can almost be restored by the introduction of twist.

Whatever the type of fibrillator, a distribution of fibre width is always obtained. This distribution and the rectangular cross-section as shown in Fig. 16 gives fabrics a somewhat coarse handle and may also introduce a certain shine or gloss, which is not always desirable. Therefore, attempts to improve the fibrillators continue and products with a much

Fig. 15. The needled surface of a fibrillator roll.

finer fibrillation and narrower width distribution than shown in Fig. 17 have already been made on laboratory scale.

Continuous fibrillated yarns are suitable for technical fabrics and for the backings of woven carpets. Staple fibres are used in large quantities at present. Sometimes these fibres are texturised to increase the bulkiness and to reduce the gloss by the same methods used by the synthetic fibre industry, *e.g.* the stuffing box. The fibres are used in blends with other fibres for the manufacture of knitting yarns and upholstery fabrics. The rectangular cross-section and somewhat frayed edges give the fibres a high coefficient of friction. As a consequence they are very suitable for non-wovens. Needle-punched products tend to be stronger and more dimensionally stable than those of spun staple fibre.[46] These needle-

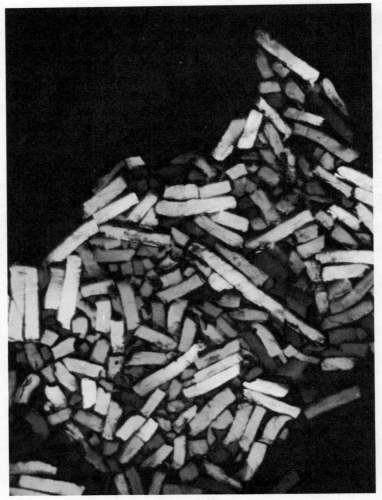

Fig. 16. Microphotograph of a cross-section of fibrillated yarns.

punched products are used as technical felts for many applications and for non-woven carpets and floor tiles.

12.7.2 Manufacture of fibres from profiled films
In order to overcome the difficulty of a broad width distribution and to

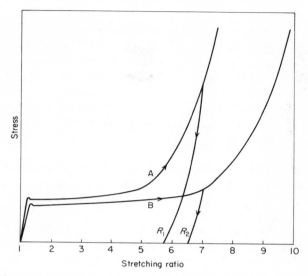

Fig. 17. Schematic representation of the stress–strain curves of a pre-oriented high molecular weight layer (A) and a less oriented low molecular weight layer (B) in a laminated film during stretching.

obtain fibres with regular and more circular cross-sections various methods have been proposed, with the common feature that profiled films, with thin and thick sections in the transverse direction, are used as raw material. These films may be obtained by extrusion of the melt through a flat-film die of which one or both lips have a fixed geometrical pattern.[47,48] The profiled melt, with the ridges connected by thin webs of film, may be extruded over the full width of the die but can also be obtained in the form of a number of tapes. These films or film strips are then cooled, stretched in hot-air ovens and annealed as described in the previous section. Some splitting may be obtained during the stretching process but for complete splitting into individual continuous fibres an additional fibrillation step is necessary. Owing to the weakness of the webs between the ridges, this fibrillation can be carried out by less severe treatments such as twisting, flexing and compression. Therefore the loss of tensile strength is negligible and the cross-section of the fibres approaches that of spun fibres or filaments in shape and regularity. They can be made in a wide thickness range but approx. 25 denier filaments find the largest use. They are used for cordage, carpet backings and industrial fabrics.

Another way of obtaining the profiled films is by melt-embossing.[49] The hot melt is extruded from a linear die into the nip formed by a plain roll and a V-grooved embossing roll. Cooling of the melt occurs on the cool roller and the embossed film is stretched and heat-set in the conventional way. The cross-section of the yarns is uniform but the shape depends not only on the size and shape of the grooves but also on operating conditions[50] and may approach the circular cross-section of spun filaments.

The third method of obtaining a profiled film is roll-embossing an extruded film from a linear die.[51] The embossing is carried out below the melting point of the polymers between an embossed roller and a pressure roller. The rest of the process is as usual. The yarns obtained by this roll-embossed film (REF) process have an excellent strength and knot strength. They are suitable for cordage and a wide variety of non-apparel fabrics including canvases and tarpaulins, wall coverings, upholstery and carpet backings.

12.8 BICOMPONENT FIBRES FROM FIBRILLATED FILMS

Natural fibres, such as wool or cotton, have a curled or irregular shape. Yarns and fabrics made from these fibres are bulky and have a high thermal insulation and a pleasant grip and appearance. Many methods have been developed to give synthetic yarns and fibres similar properties. This is called texturisation. In texturisation processes unannealed yarns or fibres are usually deformed by twisting, stuffing or knitting. They are then heat-set in the deformed state, which makes the deformation more or less permanent. Most of those processes can be applied to split fibres but we shall only deal with techniques in which orientation effects play an important role.

One of those methods is sometimes applied to spun yarns from staple fibres. Instead of a single type of fibre, a blend of two or more fibres is used which differ in shrinkage properties. The yarns are then heated in the wet or dry state, depending on the type of fibres. Part of the fibres shrink and the other fibres must form loops or curls, resulting in an increased bulkiness of the yarn. Sometimes fibres are blended from polymers which differ in melting point; the temperature is so selected that only the lower melting fibres will show an appreciable shrinkage. It is also possible to use fibres from the same polymer which differ in the amount of orientation or, more often, to use a blend of annealed and unannealed fibres. Heating to near the annealing temperature will only shrink the unannealed fibres. In this way texturised yarns can be made but the individual fibres are not crimped.

In order to achieve crimping of individual fibres by a similar technique, one side of the fibre must have different shrinkage properties than the other. This has been achieved in spun-fibre technology by the use of complicated spinnerets in which each orifice is fed by two different channels. Usually a polymer with a high melting point and consequently a low heat shrinkage tendency is fed through one channel and another polymer or copolymer with a lower melting point through the other. In this way a bicomponent fibre is made and, after stretching, a fine curling is obtained by heating the fibres at a temperature where the low-melting half will only show an appreciable shrinkage.

In a similar way, but more simply because of the less complicated construction of a laminated film die, bicomponent split fibres can be made. In Japan[52] a process was developed in which a laminated film is manufactured from two polymers differing in thermoshrinkage properties. The films are then stretched and fibrillated in the usual way and the split fibres are heated to produce crimps. The laminated films are made by gluing two different films together, by coating a film with another polymer or, and most important from a practical point of view, by extruding two different polymers through double-slit flat or double annular dies.

It is also possible to obtain crimped fibres by the split-fibre technology in a different way. If a laminated film is produced by extruding the raw material or raw materials in such a way that the layers display during stretching stress–strain curves which do not coincide, then a spontaneous crimp can be obtained immediately after fibrillation without additional thermal treatments of the fibres.[53] It is, of course, necessary that the layers adhere well to each other. If both layers consist of polypropylene or copolymers of propylene, this is no problem. However, if different polymers are used, the adhesion is usually insufficient.

One way of achieving non-coinciding stress–strain curves is to use two polymers which differ appreciably in molecular weight. During extrusion and draw-down the high-molecular weight material will have more orientation in the machine direction than the other. During stretching the high molecular layer will show a slightly higher yield stress and stretching will be completed at a lower stretch ratio than the other layer as shown in Fig. 17. The stress increases rapidly as soon as a certain stretch ratio is exceeded. The other layer, however, will still be at a much lower stress level. Hence, if a stretching ratio is applied of, for instance, 1:7 and the stress is relieved, the highly oriented layer will show more elastic recovery (R1) than the lower-oriented layer (R2). This difference in length will lead to an internal stress and as soon as the film is fibrillated and the fibres are free from external stresses, they will show a spiral crimp, with the high-molecular weight layer on the inside and the low-molecular weight layer on the outside. This structure is shown in Fig. 18.

A difference in average molecular weight is only one way of achieving these non-coinciding stress–strain curves. A difference in the width of the molecular-weight distribution may also lead to crimping, the layer with the

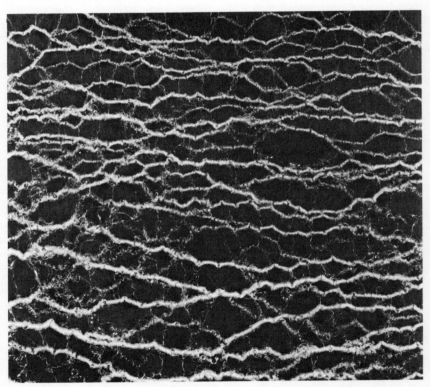

Fig. 18. Microphotograph of a stretched and fibrillated bicomponent film.

widest distribution being more oriented than the other. Manufacturing conditions play an important role. If the film is made by blowing, for instance, the high-molecular weight or wide-distribution layer must be on the outside of the balloon. This side will cool more rapidly and, hence, will show less relaxation than the slower-cooling interior layer. Many other factors such as extrusion temperature, draw-down ratio, orientation rate and temperature, the presence of pigments or fillers and the method of fibrillation have a certain influence. A maximum crimp is only obtained if these factors are carefully selected and co-operate. On the other hand, if they work in opposite directions insufficient crimping or no crimping at all will be observed.

This method is not easily applied to spun fibres because the spinning process requires molecular weights and distributions within rather narrow limits to prevent melt fracture and other deficiencies. The split-film bicomponent fibres also differ from spun fibres in another respect. Spun fibres usually show a homogeneous crimp but in split film the crimp depends on the degree of fibrillation, the fine fibres showing a higher crimp than the coarse fibres. Therefore, it is difficult to express the crimp level as the number of turns per centimetre because the fibres have both a width distribution and a crimp distribution. If the bicomponent fibres are made from profiled films, however, the width and the degree of crimping can be more or less uniform.

The bicomponent fibres are suitable for felts, needle-punched carpets and tiles and upholstery fabrics; the appearance of these products is more pleasant than that of uncrimped fibres owing to the absence of gloss and they have a bulkier, softer handle. Continuous crimped yarns could be suitable for certain types of stretch fabrics or knitted products but so far these products have not reached the commercial stage. Attempts to use the bicomponent fibres for tufted carpets have not been satisfactory so far for two reasons. One reason is that the resilience of the piles is less than that of wool or nylon. Another reason is that fibrillated fibres may show a tendency towards a further fibrillation under severe wear or abuse.

Bicomponent films can also be made from two different polymers if sufficient adhesion between the layers can be achieved to prevent delamination during stretching and fibrillation. Mehta[54] has pointed out that bicomponent films can be made from polypropylene and low density polyethylene. After stretching and fibrillation the fibres look like ordinary split fibres but on heating they develop a spiral crimp with the LDPE on the inside. These fibres have the important advantage that a carded fleece, when needle-punched and exposed to temperatures above the melting point of LDPE (120°C), undergoes heat-bonding owing to the adhesion of the molten LDPE at the cross points of the fibres while leaving the polypropylene unaltered. The use of these heat-bondable fibres in non-woven products obviates the need for external bonding media. Depending on pressure and temperature, the products obtained from these fibres possess characteristics ranging from a paper-like to a textile-like feel and appearance.

REFERENCES

1. Kranz, M. R., Newman, H. D. and Stigale, F. H. (1972). *J. Appl. Polymer Sci.*, **16**, 1249.
2. Broutman, L. J. and Kalpakjian, S. (1969). *S.P.E. Journal*, **25**, 46 October.
3. Thomas, L. S. and Cleereman, K. G. (1972). *S.P.E. Journal*, **28**, 62 April.
4. Park, W. R. R. (1969). *Plastic Film Technology*, Van Nostrand-Reinhold Co., New York, London.
5. Elden, R. A., Swan, A. D. and Griffin, G. J. L. (1971). *Calendering of Plastics*, The Plastics Institute, London, Iliffe Books.
6. Hermann, K., Gerngross, O. and Abitz, W. (1932). *Z. Phys. Chem.*, **B10**, 371.
7. Schuur, G. (1955). *Some Aspects of the Crystallisation of High Polymers*, Thesis, Delft.
8. Frank, F. C., Keller, A. and O'Connor, A. (1958). *Phil. Mag.*, **3**, 64.
9. Kopsch, H. (1971). *Kunststoffe*, **61**, 18.
10. Schuller, R. (1971). *Kunststoffe*, **61**, 89.
11. Herner, M. and Hatzmann, G. (1972). *Kunststoffe*, **62**, 81.
12. Bates, T. W. Private communication.
13. Bergonzoni, A. and Dicresce, A. J. (1965). *S.P.E. Techn. Papers*, **11**, VI 1/2.
14. Schuur, G. (1966). *Kolloid-Z/Z, f.P.*, **208**, 123.
15. Kennaway, A. and Weeks, D. J. (1957). *Polyethylene* (ed. A. Renfrew and P. Morgan), Iliffe, London, p. 304.
16. Kresser, T. O. J. (1960). *Polypropylene*, Reinhold, New York, p. 121.
17. Freeman, H. S. and Coplan, M. J. (1964). *J. Appl. Polymer Sci.*, **8**, 2389.
18. Brenschede, W. (1950). *Z. Elektrochem.*, **54**, 191.
19. Carlson, F. A. Jr. Mobil Oil. Corp. U.S.P. 3.619.445.
20. Park, W. R. R. and Conrad, J. (1965). *Encyclopedia of Polymer Science and Technology*, Vol. 2, Interscience, New York, p. 339.
21. Marshall, I. and Thompson, A. B. (1953). *Nature*, **171**, 38.
22. Müller, F. H. and Jäckel, K. (1953). *Makromol. Chem.*, **9**, 97.
23. Vincent, P. I. (1960). *Polymer*, **1**, 7.
24. Hansen, D. and Rusnock, J. A. (1965). *J. Appl. Phys.*, **36**, 332.
25. Lange, M. Brückner–Maschinenbau Germ. Ols 2.042.741 (28.8.1970); Gosper, D. L. and Moffitt, M. L. Mobil Corp. U.S.P. 3.551.546 (5.7.1968).
26. Mobil U.S.P. 3.551.546 (5.7.1968).
27. Tanaka, H., Kurihara, K., Morita, M., Mori, K. and Okajima, S. (1971). *J. Polymer Sci., Part B*, **9**, 723.
28. Wellenhofer, P. (1972). *Kunststoffe*, **62**, 373.
29. Editorial Mod. Plast. Int., Jan. 1972, p. 59.
30. Editorial Mod. Plast. Int., July 1972, p. 20.
31. Schuur, G. and Gouw, L. H. 'Fourth Shirley Internation Seminar', The Hague 6–8 Oct. 1971, pp. 71–73.

32. Rasmussen, O. B. U.K. Pat. 1.229.793 (30.5.67).
33. Polymer Processing Research Institute, O.L.S. 2.138.328 (Jap. 12.8.70).
34. Rasmussen, O. B., U.K. Pat. 792.976 (4.6.1954).
35. Celanese Plastics Co., OLS 2.055.193 (U.S. 13.11.69).
36. *Edit. SPE Journal,* **27,** 17 (Nov. 1971).
37. Jacqué, H., I. G. Farben U.S.P. 2.185.789 (30.6.1936).
38. Costa, J. E., Leboeuf, E. W. and Lefevre, L. A. DOW Chem. Comp. U.S.P. 2.853.741 (27.5.1954).
39. Rasmussen, O. B., U.S.P. 2.954.587 (23.5.1955); U.S.P. 2.948.927 (24.4.1957) U.S.P. 3.233.029 (6.8.1962).
40. Rasmussen, O. B. (1964). *Text. Inst. Industr.,* **2,** 258.
41. Higuchi, M. (1958). *Rep. Res. Inst. Appl. Mech., Kyushu Univ.,* **6,** 173.
42. Wolock, I., Kies, J. A. and Newman, S. B. (1959). *Fracture* (Swampscot Symposium 1959), Wiley, New York, p. 251.
43. Newman, S. B. and Wolock, I. (1962). *Adhesion and Cohesion,* Elsevier, Amsterdam, p. 218.
44. Slack, P. T., Plasticisers Ltd., U.K. 1.073.741.
45. Dekker, J. and Schuur, G. Shell, U.K. 1.167.631.
46. Schuur, G. (1969). *Textile Month,* Feb., 45.
47. Courtaulds Ltd., U.K. 1.035.657 (1.3.1963).
48. Barmag, U.K. 1.176.357 (1.1.1970).
49. Smith and Nephew, U.K. 914489 (2.1.1963).
50. Dow, J., Patchell, A. G. and Piskozub, Z. T. (1972). *Plastics and Polymers,* April, p. 80.
51. Nichols, G. J. (1972). *Plastics and Polymers, April,* p. 84.
52. Polymer Processing Research Institute Ltd., U.K. 1.132.641 (24.8.1965).
53. Schuur, G. Shell, U.K. 1.181.249 (31.8.1966).
54. Mehta, H. Textiles from Film II Conference, London, 6–7 July, 1971. Paper No. 15.

CHAPTER 13

ULTRA-HIGH MODULUS ORGANIC FIBRES

G. B. CARTER and V. T. J. SCHENK

13.1 INTRODUCTION

Fibres, whether man-made like nylon and polyester, or natural like wool and cotton, represent some of the most highly oriented forms of organic polymers. The uses of such conventional organic fibres fall into two main categories. Woven, knitted and felted materials are used for apparel or furnishings, and cords or heavy fabrics of various kinds are used in industrial products such as ropes, motor vehicle tyres, drive belts and conveyor belts. Recently, however, there has appeared in the patent literature a range of novel fibres with physical properties, particularly elastic modulus, which are greatly superior to those familiar to most polymer or fibre scientists. It is these new fibres which form the subject matter of this chapter. We shall call them ultra-high modulus fibres to distinguish them from those conventionally considered to be of high modulus such as the rayon and polyester used in tyre cords. It is not our intention to discuss conventional fibres in any detail; this has been done elsewhere.[1] To set the scene, however, and to provide a background against which to view the new ultra-high modulus fibres, it is worth briefly considering the range of tensile properties exhibited by a selection of the common organic fibres in relation to their chemical structures. Table 1, which is far from comprehensive, contains a variety of the more common fibres and also includes a small number of experimental fibres. In the case of the man-made fibres variation in properties is possible according to the degree to which they are 'drawn' during or after spinning. Generally the initial modulus and tenacity (breaking strength) of a fibre are increased when the degree of alignment of its component polymer molecules is increased by drawing while the amount by which it can be extended before breaking is reduced.

A word is necessary here about the units employed in this chapter. The unit g dtex^{-1} is one familiar to organic fibre scientists. It incorporates the density of the polymer and is based on the definition that fibre or yarn of one tex weighs 1 g per 1000 m length. Since the densities of many of the new fibres to be described are only approximately known (they lie in the region of 1·3–1·5 g cm^{-3}) while the decitex value of a fibre can be determined by weighing a measured length, we have not attempted the conversion to SI units but offer the conversion equations:

$$1 \text{ g dtex}^{-1} = 9\cdot81 \times 10^7 \rho \text{ N m}^{-2}$$
$$= 9\cdot81 \times 10^8 \rho \text{ dyn cm}^{-2}$$
$$= 1\cdot42 \times 10^4 \rho \text{ lb in}^{-2}$$
$$= 1\cdot11 \text{ g denier}^{-1}$$

(where ρ = fibre density in g cm^{-3})

The unit frequently employed in the literature, g denier^{-1}, is based on a 9000 m length of fibre in the same way.

All the fibres in Table 1 are made up of inherently flexible molecules. Many are wholly or predominantly open-chain aliphatic in character so that the polymer chains can readily fold. Cellulose in spite of its rigid anhydroglucose rings contains ether linkages which can rotate and even Nomex, which consists of rigid *m*-phenylene and amide units, is so arranged geometrically that the chains are probably able to adopt irregular conformations. Roughly speaking those polymers which contain rigid rings tend toward the higher end of the stiffness range as far as practical fibres are concerned, although in the laboratory at least, polyethylene can be drawn until a surprisingly high modulus is achieved.[2]

All conventional organic fibres then are made from polymers whose molecules can chain-fold and consequently are morphologically complex. They consist of folded regions interspersed with regions of varying degrees of order. The practical fibre stiffnesses are governed by the more flexible parts of the structure and bear no relation to the extremely high stiffness values which have been obtained by techniques which only measure the properties of the crystalline regions. Values obtained in this way are found to correlate fairly well with those calculated for polymer chains on theoretical grounds as described in Chapter 7.

TABLE 1

TENSILE PROPERTIES OF COMMON ORGANIC FIBRES INCLUDING SOME LESS WELL KNOWN EXAMPLES

Polymer	Initial modulus $(g\ dtex^{-1})^a$	Tenacity $(g\ dtex^{-1})$	Ext. to break(%)	Density $(g\ cm^{-3})$
Cellulose (cotton)	36–72	2·3–4·5	5–10	1·54–1·56
Cellulose (hemp)	198	5·2–6·1	1·8	1·48
Cellulose (high tenacity viscose rayon)	113–153	5·4	6–8	1·52
Cellulose diacetate	26–38	1·2–5·4	5–30	1·32
Polypeptide (wool)	22–36	0·9–1·5	30–45	1·30–1·32
Polypeptide (silk)	54–72	2·7–4·5	20–25	1·34
Segmented polyurethane (Spandex elastomeric fibres)	0·03–0·1	0·5–0·9	500–700	1·21
Polyethylene (high-density)	45–68	4·1–7·2	10–20	0·95
Polyethylene (experimental fibre[2])	230	—	—	—
Polypropylene	~81	4·5–8·6	15–25	0·91
Polyacrylonitrile (acrylic fibres)	36–45	1·8–3·2	25–45	1·14–1·19
Polyvinyl alcohol (high tenacity)	36–76	6·5	10–18	1·26–1·32
Polyvinyl alcohol[3]	225	9·0	9	—
Poly(hexamethylene adipamide) (high tenacity nylon 66)	41–72	5·4–8·6	12–26	1·14
Poly(metaphenylene isophthalamide) (Nomex[b] high temperature resistant fibre)	126–135	5·0	14–17	1·38
Poly(ethylene terephthalate) (high tenacity Terylene[b])	99–117	5·4–7·7	6–14	1·38
Poly(ethylene-1,2-diphenoxy ethane-p,p'-dicarboxylate)[4]	206[c]	7·9	6·1	—

[a] For a density of 1 g cm^{-3}, 1 g dtex^{-1} is approximately equivalent to 10^8 N m^{-2} (see text).
[b] Trademark.
[c] 2% secant modulus. The initial modulus is greater.

13.2 THE KNOWN ULTRA-HIGH MODULUS ORGANIC FIBRES

Until recently ultra-high moduli had only been attained in macroscopic structures such as fibres by using inorganic materials like E-glass (325 g dtex^{-1}; $8·3 \times 10^{10}$ N m^{-2}), chrysotile asbestos (650 g dtex^{-1}; $1·6 \times 10^{11}$ N m^{-2}), boron nitride (480 g dtex^{-1}; 9×10^{10} N m^{-2}) or carbon (2200 g dtex^{-1}; $4·0 \times 10^{11}$ N m^{-2}). However, one of the results of the quest during the last decade for organic high performance polymers and fibres has been the accumulation of a great deal of evidence linking thermal stability and stiffness with the presence in the polymer chain of high proportions of ring systems which are themselves thermally stable and extremely rigid. In many cases this thermal stability and stiffness can be associated with resonance stabilisation. Such rings often occur in the so-called ladder polymers in many of which the polymer chain can only be broken by rupturing two chemical bonds rather than one as is the case for most polymers.

This account is concerned with organic polymeric materials which, in fibre form, lie in the range of initial modulus between steel or glass and the lower levels of carbon. We have taken an initial modulus of approximately 300 g dtex^{-1} as our lower limit for ultra-high modulus fibres which are thus set well clear of all conventional organic fibres. Many experimental high-temperature fibres have been reported in recent years exhibiting initial moduli in the range 130–230 g dtex^{-1} (Ref. 5). It can now be seen that, in many cases, their chemical structures, often very rigid and containing high proportions of aromatic and other ring systems, lacked only the linear geometry which can increase stiffness by an order of magnitude.

A number of ultra-high modulus fibres have, however, been appearing in the patent literature over the last few years and one well publicised example was patented as long ago as 1961.[6] These materials represent a major step toward practical realisation of the levels of mechanical properties which have been predicted theoretically for organic polymers and may constitute the first examples in a new era of organic engineering fibres. Chemically they all belong to the class of condensation polymers and are all in some degree similar in molecular structure to conventional fibre formers such as the nylons. They consist of aromatic or other rigid rings linearly connected by amide groups which occur either singly or in pairs as hydrazide or oxamide groups.

Table 2 lists all the organic ultra-high modulus fibre types known to the present authors together with typical physical properties. The fibres listed range from a laboratory curiosity (D in the table) to one which threatens to make a large impact on the market for tyre cords and composite reinforcements (A and B in the table). The mechanical properties of these fibres can often be modified over a wide range by heat treatment and/or drawing subsequent to spinning and, in many cases are only fully developed by high temperature treatments. Table 2 therefore contains examples of these effects.

13.3 SPINNING METHODS

Having mentioned fibre spinning it is appropriate to describe briefly the techniques of wet spinning and dry spinning. For detailed accounts the

TABLE 2

TYPICAL TENSILE PROPERTIES OF ULTRA-HIGH MODULUS ORGANIC FIBRES

Polymer	Temperature of heat treatment (°C)	Initial modulus (g dtex⁻¹)	Tenacity (g dtex⁻¹)	Ext. to break (%)	Remarks	Ref.
A	550	927	13·9	1·6	Dry spun from T.M.U./LiCl	7, Ex. VI
	as spun	458	7·4	3·1		
A	500	1116	10·7	1·1	Wet spun from T.M.U./LiCl	Author's results
	as spun	374	4·1	3·0		
A	200 for 64 h	386	11·4	7·8	Wet spun from D.M.Ac., with carbamate additive (see p. 473)	8, Ex. II
	as spun	372	8·2	7·6		
A	—	1260	16·4	1·6	Preparation not described	7, Ex. XXIII
B	500	799	12·3	1·6	Wet spun from oleum	9a, Ex. II
	as spun	156	6·3	9·1		
Mixture of A + B	640 (drawn 1·08 ×)	949	12·9	1·4	Wet spun from oleum	9a, Ex. XV
	as spun	198	6·4	10·0		
B	350	828	28·8	2·1	Dry jet Wet spun from sulphuric acid	9b
	as spun	639	27·9	3·2		
B	525	855	21·0	2·2	Dry jet Wet spun from sulphuric acid	9c, Ex. I
	as spun	675	26·0	3·7		
C	536 (drawn 1·2 ×)	546	11·2	2·3	Wet spun from D.M.Ac./LiCl into water	10, Ex. X
	536 (drawn 1·1 ×)	607	10·6	1·9		
	480 (drawn 1·05 ×)	544	10·0	2·1		
	450 (drawn 1·05 ×)	412	8·0	2·4		
	as spun	67	2·0	25·0		
D	hot drawn	739	8·5	1·4	Spun from D.M.Ac./LiCl	11
	—	528	8·7	2·4		
	as spun	177	5·9	18·6		
E	300 (drawn 1·65 ×)	567	11·3	2·4	Wet spun from D.M.Ac.	12, Ex. II
	300	256	7·0	13·5		
E	300 (drawn 2 ×)	544	5·8	1·3	Wet spun from D.M.Ac.	Author's results
	as spun	87	2·0	18·0		
E	350 (drawn 1·41 ×)	—	17·4	4·3	1% sorbitan monopalmitate included	13, Ex. VII
F	300 (drawn)	546	12·3	2·8	Wet spun from D.M.Ac./LiCl	14, Ex. II
	300 (drawn 1·47 ×)	470	9·6	2·6		
	300 (drawn 1·25 ×)	372	10·8	4·2		
	drawn in hot water	134	6·0	28·3		
	as spun	79	3·9	55·9		
G	400	441	5·4	1·3	Dry spun from D.M.S.O./LiCl	15, Ex. II
	as spun	225	2·3	2·1		
H	351 (drawn 2·6 ×)	373	10·9	3·3	Dry spun from D.M.Ac./LiCl	6, Ex. VI
	as spun	53	1·7	48·0		
H	273 (drawn 2·3 ×)	403	5·8	1·6	Wet spun from D.M.Ac./LiCl	Author's results
	300	243	4·6	2·2		
	as spun	96	1·6	23·4		
J	500 after hot stretch at 250	767	11·8	1·4	Dry jet wet spun from D.M.Ac./LiCl	59, Ex. II
	as spun	73	2·4	45·0		
K	275 (drawn 2·5 ×) then 350 (drawn 1·2 ×)	435	14·1	5·1	Dry jet wet spun from D.M.Ac./LiCl	60, Ex. V
L	320 (drawn 2·2 ×) then 380 (drawn 1·05 ×)	774	20·9	3·0	Wet spun from D.M.Ac./CaCl₂	58, Ex. X
	as spun	158	3·7	20·5		

Polymer A is poly (p-benzamide).
Polymer B is poly (p-phenylene teraphthalamide).
Polymer C is an aromatic–aliphatic ordered copolyamide containing repeating units of the general formula:

$$\{(HN \bigcirc CO)_a\!-\!NH \bigcirc NH(-CO \bigcirc NH)_b\!-\!CO(CH_2)_cCO\}_n$$

and where $a + b = 23$, $c = 4$.

Polymer D is an ordered azoaromatic polyamide of formula:

$$-[NH-\langle\bigcirc\rangle-CONH-\langle\overset{CH_3}{\bigcirc}\rangle-N{=}N-\langle\overset{CH_3}{\bigcirc}\rangle-NHCO-\langle\bigcirc\rangle-NHCO-\langle\bigcirc\rangle-CO]_n$$

Polymer E is the copolymer produced by reaction of p-aminobenzhydrazide with terephthaloyl chloride:

$$-[(NH-\langle\bigcirc\rangle-CONHNH){-}CO-\langle\bigcirc\rangle-CO]_n$$

Polymer F is the ordered copolymer produced by the reaction of the diamine below and terephthaloyl chloride:

$$H_2N-\langle\bigcirc\rangle-CONHNHCO-\langle\bigcirc\rangle-CONHNHCO-\langle\bigcirc\rangle-NH_2$$

Polymer G is the aromatic polyhydrazide from terephthalic dihydrazide and terephthaloyl chloride:

$$-[NHNHCO-\langle\bigcirc\rangle-CONHNHCO-\langle\bigcirc\rangle-CO]_n$$

Polymer H is the aromatic copolymer from 2,5-dimethylpiperazine and 3,3′-dimethyl-4,4′-diisocyanato-diphenylene:

$$-[N\underset{CH_3}{\overset{CH_3}{\diagdown}}NCONH-\langle\overset{CH_3}{\bigcirc}\rangle-\langle\overset{CH_3}{\bigcirc}\rangle-NHCO]_n$$

Polymer J is a copolyoxamide-terephthalamide of p-phenylene diamine, *i.e.* it contains units of B and of

$$-[NH-\langle\bigcirc\rangle-NHCOCONH-\langle\bigcirc\rangle-NH]-$$

Polymer K is a copolyterephthalamide of oxalic dihydrazide and terephthalic dihydrazide, *i.e.* it contains units of G and of

$$-[NHNHCOCONHNHCO-\langle\bigcirc\rangle-CO]-$$

Polymer L is a copolymer containing units of A and B together with units of

$$-[CO-\langle\bigcirc\rangle-CONHNH]-$$

T.M.U. is tetramethyl urea, D.M.Ac is dimethyl acetamide, D.M.S.O. is dimethylsulphoxide.

reader is referred to standard works such as that by Mark *et al.*[16] The usual process of converting a polymer into a fibre consists of preparing the polymer in a liquid form such as a melt or viscous solution, extruding this through a small orifice (the spinneret) and solidifying the emergent stream by cooling the melt or removing the solvent from the solution. We are not concerned here with melt spinning; the polymers which form ultra-high modulus fibres cannot be melted in the usual way but simply decompose at extremely high temperatures. Powerful hydrogen bonding solvents or strong acids are therefore used to prepare spinning solutions or dopes. In wet spinning the dope is extruded into a bath of liquid which dissolves the dope solvent but not the polymer whereas in dry spinning the dope is extruded into a chamber at a sufficiently high temperature to evaporate the dope solvent.

13.4 PREPARATION AND PHYSICAL PROPERTIES OF THE POLYMERS AND FIBRES

13.4.1 Aromatic Polyamides

The synthesis of this type of polymer and indeed of nearly every ultrahigh modulus fibre forming polymer, in a sufficiently pure state and at high enough molecular weight to allow the best advantage to be taken of its stiffness potential is well illustrated by the preparation of poly-(p-phenylene terephthalamide)[17-19] and poly(p-benzamide).[7,20-23] The method of choice is solution polycondensation, preferably at low temperatures, of a diacid chloride and a diamine or of an amino-acid chloride. The substituted amide solvents employed are necessarily powerful and can maintain the rigid polymer chains in solution by virtue of strong hydrogen bonding with the polymer amide groups. This effect is enhanced in a way not fully understood by the inclusion in the solvent system of alkali metal halides, the preferred examples being lithium and calcium chlorides.[24,25]

Poly(p-phenylene terephthalamide) (I) is prepared for instance by the reaction of stoichiometric quantities of terephthaloyl chloride and p-phenylene diamine in a mixture of hexamethyl phosphoramide and N-methyl pyrrolidone, the temperature being maintained between about $-10°C$ and $30°C$.

I

TABLE 2 NOTES

The hydrogen chloride eliminated in the reaction is neutralised by salt formation with some of the amide solvent.

Variations on the diamine–diacid chloride theme have provided other rigid linear aromatic polyamides which are ultra-high modulus fibre formers, structures II–V being notable examples and numerous 'co-monomers' have been described.[9c] Polymer II is said only to give good fibre properties if its structure is deliberately disordered slightly by inclusion of small quantities of co-reactants, the effect being attributed to better solubility and consequent improved spinnability.[26] However, this same polymer is claimed elsewhere to give excellent fibre moduli with or without a co-reactant.[27]

II   (Ref. 27)

III   (Ref. 28)

IV   (Ref. 11)

V   (Ref. 29)

Preparation of poly(p-benzamide) (VI) is a little more complex and allows some variation in approach. The basic starting material, p-aminobenzoic acid, can be converted to the polymer by two routes as exemplified below.

p-Thionylaminobenzoyl chloride, the intermediate common to both routes, can be purified by vacuum distillation. In route A judicious hydrolysis of the thionyl compound leads, via the amino acid chloride, directly to polymer formation[23,30] while in route B the amino acid chloride is prepared as its stable hydrochloride which may be isolated and stored until required.[7] Two recent variations which have been described[31] are preparation of the thionyl intermediate in the amide solvent and preparation of a polymerisable 'dimer' and oligomer mixture similar to the amino acid chloride hydrochloride.

Typically tetramethyl urea or dimethylacetamide are employed as amide solvents. Lithium chloride may be added to stabilise the resulting polymer solution or may be produced *in situ* by addition of lithium hydroxide or carbonate. The degree of polymerisation can be controlled by adding a chain stopper such as *p*-aminobenzoic acid.[32]

The general method apparently works just as well for amino acids such as

$$H_2N\langle\rangle CONH\langle\rangle COOH \quad \text{(Refs. 22, 33) and}$$

$$H_2N\langle\rangle\langle\rangle COOH \quad \text{(Ref. 34)}$$

Numerous 'co-monomers', mostly of similar chemical type, have been described and, of course, random copolymers of poly(*p*-benzamide) and poly(*p*-phenylene terephthalamide)[35] which may be seen to be less ordered versions of II and III.

A particularly interesting copolymer type (VII) which deserves special mention because its structure seems slightly at odds with the usual rule for ultra-high modulus fibre formation is a kind of block copolymer in which rigid segments are 'hinged' together by flexible aliphatic diacid units.[10]

$$\text{VII} \quad \left[(NH\langle\rangle CO)_a NH\langle\rangle NH(CO\langle\rangle NH)_b CO(CH_2)_c CO\right]_n$$

Benzene ring substituents have been included in variations on many of the above examples and the interested reader should refer to the original literature.

The polymers described above may be isolated by precipitation into water with stirring when they are obtained as white powders which exhibit no melting points and no detectable glass transition temperatures by DSC (Fig. 1). The DSC endotherms in Fig. 1 which are centred at about 580°C are not reversible and are the result of decomposition rather than melting. As precipitated the polymers are not especially crystalline judging by X-ray diffraction but nevertheless are often quite difficult to re-dissolve in amide solvent–lithium chloride systems.[36] They are, however, readily soluble in strong acids such as concentrated sulphuric acid, oleum, hydrofluoric acid and methane sulphonic acid[17] and in mixtures of these with other powerful polar solvents.[9c] The density of the best studied example, poly(p-benzamide) is approximately 1·45 g cm^{-3} in fibre form.

The solution properties of these materials are unusual. They form optically anisotropic solutions in both amide and acid solvent systems over quite wide ranges of concentration and polymer molecular weight. In other words they are among the few known examples of synthetic polymers which can form lyotropic liquid crystals. (That is to say liquid crystals formed by the action of a solvent.) The usual example quoted in this context is poly(γ-benzyl-L-glutamate) which forms cholesteric meso-morphic solutions in certain organic solvents. The helical structure adopted by the polypeptide in these solvents behaves as a rigid rod and it is

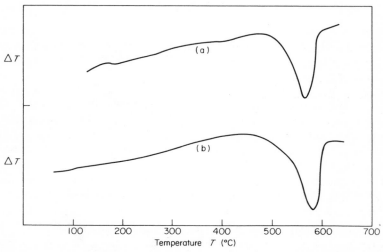

Fig. 1. D.S.C. thermograms for (a) poly(p-benzamide) and (b) poly(p-phenylene terephthalamide).

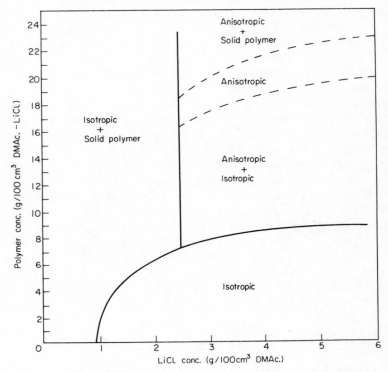

Fig. 2. Phase diagram for poly(p-benzamide) [$\eta_{inh} = 1·18$] in D.M.Ac./LiCl relating anisotropy to LiCl content. (From B.P. 1,262,002.)

this rigidity of structure which seems to be the prerequisite for spontaneous formation of the mesomorphic phase.[37] In the case of the rigid aromatic polyamides the liquid crystal type involved is thought to be nematic. This is the simplest form of liquid crystal and involves parallel orientation of large groups of rodlike molecules but no other constraints.

The particular conditions defining an optically anisotropic aromatic polyamide solution are polymer molecular weight, solvent, temperature, polymer concentration and alkali-metal halide concentration. Variation of some of these parameters while maintaining others constant allows the construction of phase diagrams such as Figs. 2 and 3. Solutions having compositions which place them in the anisotropic regions in these figures exhibit a characteristic opalescence when stirred which fades in a few

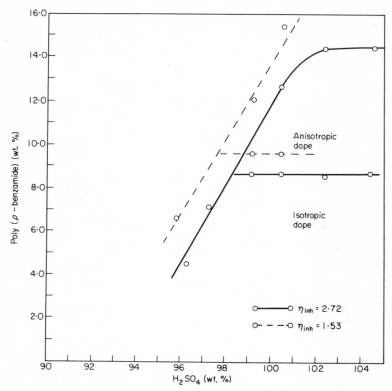

Fig. 3. Phase diagram for poly(p-benzamide) in oleum relating anisotropy to SO_3
content and polymer inherent viscosity. (From B.P. 1,283,064.)

seconds when the shearing action stops. Viewed through a polarising microscope an anisotropic solution of poly(p-benzamide) in tetramethylurea–lithium chloride has the appearance shown in Fig. 4, while Fig. 5 clearly shows the effect of shearing the liquid on the microscope slide by drawing a pin point through it. Russian workers appear to have independently observed the formation of lyotropic liquid crystals of poly(p-benzamide).[38]

The molecular weights of polymers of this type are difficult to measure because of their unusual solution properties and limited range of solvents. Attempts have been made, however, and the relationship between molecular weight and solution viscosity (the Mark–Houwink equation,

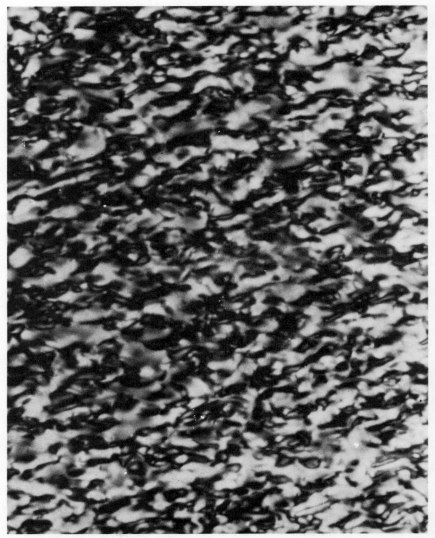

Fig. 4. Anisotropic solution of poly(p-benzamide) [$\eta_{inh} = 1 \cdot 2$] in T.M.U. viewed through crossed polars (authors' results).

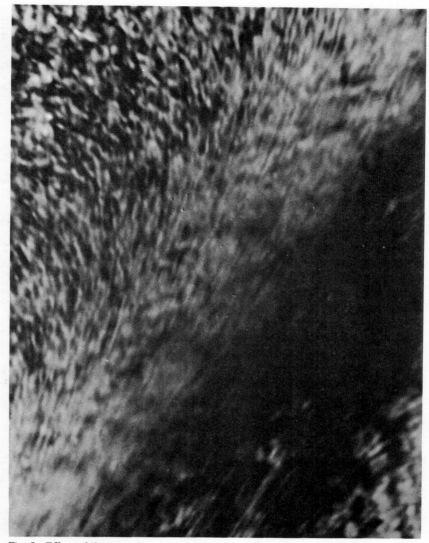

Fig. 5. Effect of shear on the anisotropic solution shown in Fig. 4. (authors' results).

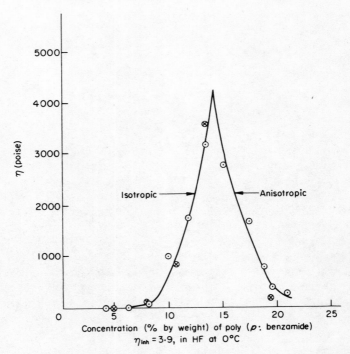

Fig. 6. Effect of anisotropy on the relationship between viscosity η and concentration for poly(p-benzamide) in HF at 0°C (B.P. 1,273,564).

$[\eta]$† $= KM^{\alpha}$) has been determined for poly(p-benzamide) in sulphuric acid although only the exponent α has been published.[39] It is reported to be abnormally high, $\simeq 1.6$, reflecting the extreme rigidity of the polymer chains in the chosen solvent. A similar result is described for poly(γ-benzyl-L-glutamate).[40]

A final point of similarity between these two polymers is the occurrence of a 'critical' concentration for a given polymer–solvent combination. Above this point the viscosity falls with further increase in concentration as illustrated in Fig. 6 for poly(p-benzamide) in hydrofluoric acid. This is a consequence of the transition to a lyotropic liquid crystalline state.

† The authors have adopted the conventional symbols used to express polymer solution viscosity. These are defined in standard texts on polymer science.

The molecules in such a solution are more readily oriented by shearing than are the molecules in an isotropic solution with the result that the effective viscosity is lower. A more detailed discussion will be found in Ref. 37.

The usual indicator of molecular weight or degree of polymerisation reported for these polymers is the conveniently measured inherent viscosity (η_{inh}) in sulphuric acid or other suitable solvent. At inherent viscosities of 0·7–0·8 and above these polymers are fibre forming by wet or dry spinning.

In wet spinning warm water is a convenient and effective coagulant while dry spinning requires extrusion into a chamber heated at about 200°C owing to the high boiling points of the amide solvents. An important variation which has been described is dry-jet wet spinning [9b,9c] in which the dope is extruded into the air for a short distance and then passes into a liquid coagulant such as water. Dope preparation is straightforward in some cases, merely requiring the polymer to be stirred with the solvent, but elaborate techniques involving alternate heating and freezing have often been employed.[36] The most elegant method is the direct use of freshly prepared polymer dissolved in its amide reaction medium as the spinning dope.[41] Anisotropy of the spinning dope appears to be necessary for the greatest development of the fibre properties during spinning. Table 3 shows the effect on fibre modulus and tenacity of the presence or absence of liquid crystals in typical spinning dopes. The fibres in this table have not been treated after spinning except to wash out residual solvent. From the available data the general rule seems to be that fibres spun from anisotropic dopes exhibit higher moduli in the as-spun state than those spun from isotropic dopes. Because of the importance of the optical anisotropy of these solutions optical methods have been worked out for convenient estimation of the degree of anisotropy. These rely on the change in light transmission of the solution as viewed through crossed polars.[17]

It appears that for these rigid aromatic polyamides a major breakthrough has been achieved in the technique of producing oriented fibres. The molecules in the liquid crystal state presumably have the short range order typical of nematic mesophases and are themselves readily oriented in the extensional flow[37] produced by passage through a spinneret. The rate of relaxation to a disordered state is evidently slower than the rate of coagulation of the polymer by dry spinning or wet spinning and so a fibre consisting of highly oriented rigid chains is produced. The fibres produced in this manner, once dry, are generally not capable of being

TABLE 3

EFFECT OF DOPE ANISOTROPY ON AS-SPUN PROPERTIES OF AROMATIC POLYAMIDE FIBRES

Polymer	η_{inh}	Solvent	Solution	Initial mod. $g\ dtex^{-1}$	Tenacity $g\ dtex^{-1}$	Ext. to break %	Ref.
Poly(p-benzamide)	2·36	N,N dimethyl acetamide	Anisotropic[a]	255	6·5	8·1	20, Ex. XXV
Poly(p-benzamide)	2·36	N,N dimethyl acetamide	Only slightly anisotropic[a]	58	1·1	90·0	9a, Exs. XL and XLII
Poly(p-phenylene terephthalamide)	4·00	HF	Anisotropic	234	7·8	5·4	
Poly(p-phenylene terephthalamide)	3·06	HF	Isotropic	71	3·1	550·0	17, Exs. XXXIX and XLI

[a] These solutions were obtained by centrifuging a dope containing anisotropic and isotropic components, thereby substantially separating them.

drawn in the usual way. Only at the lower end of the as-spun modulus range can a significant draw be applied. Because of the extreme thermal stability of the aromatic polyamides, however, they can be given heat treatments at up to 700°C for brief periods or at 400–500°C for longer periods. In some cases it is necessary to apply enough tension during heat treatment to take up a very slight elongation which occurs.[42] The effect of heat treatment is shown in Table 4. The fibre modulus, whether initially very high or not, is always increased, frequently by a factor of two or three, and occasionally by an order of magnitude when starting with poorly oriented fibre. Heat treatment can often outweigh differences due to the degree of anisotropy of the spinning dope and the increase in modulus is usually accompanied by a similar increase in tenacity. As can be seen in Tables 2 and 4 the ultimate mechanical property levels are only attained by very high temperature treatment.

A typical heat treatment consists in passing the fibre continuously over a hot bar or through a hot tube and a nitrogen atmosphere may be used to minimise oxidation. The contact time need only be about one second, *i.e.* enough to bring the polymer filament up to the required temperature. Longer contact times are not necessarily disastrous since these polymers do not decompose rapidly even at 500°C but experiments in the authors' laboratory have shown that poly(p-benzamide) does give off very small quantities of the expected decomposition products, aniline for instance, at these temperatures. Some chain extension by water elimination is to be expected and Table 4 contains examples showing how η_{inh} is increased on heat treatment.

TABLE 4

EFFECT OF HEAT TREATMENT ON POLY(p-BENZAMIDE) FIBRE PROPERTIES

Nominal[a] heat treatment (°C)	Initial modulus (g dtex^{-1})	Tenacity (g dtex^{-1})	Ext. to break %	Fibre η_{inh}
Room temp.	458	7·4	3·1	1·67
100	437	7·2	2·8	1·69
200	491	8·4	2·5	1·74
304	544	9·2	2·4	1·78
400	604	11·9	2·5	1·92
500	775	14·4	2·1	2·32
525	936	15·2	1·9	2·40
536	926	14·1	1·7	2·61
550	927	13·9	1·6	2·75

Data taken from Ref. 7, Ex. VI.
[a] Temperature at the centre of a steel tube through which the fibre was passed.

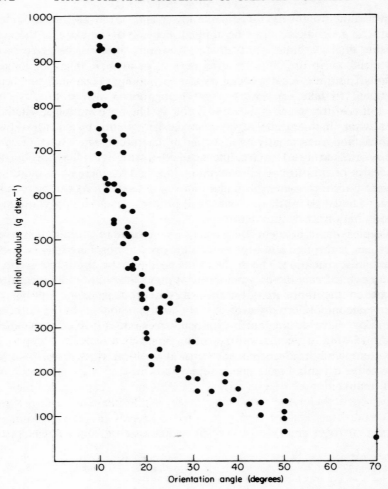

Fig. 7. Plot of initial modulus against orientation angle for poly(p-benzamide) and poly(p-phenylene terephthalamide) fibres. (Data compiled from Refs. 7, 9, 19, 20, 23.)

Potentially, at least, the mechanical properties of these fibres can be varied over a wide range, simply by varying the post-spinning heat treatment to which they are subjected. Unfortunately the increases in fibre initial moduli and tenacities brought about by heat treatment are normally accompanied by reduction in the extensibilities, as shown in Tables 2 and

4, although, even at the highest modulus levels, the extensibility (1–2%) is still as good as that of glass fibre. In certain forseeable applications of this type of fibre, such as motor vehicle tyre reinforcement or industrial conveyor belting, it is desirable to have much higher extension to break, and the use of polyfunctional carbamic acid esters as polymer additives has been reported to reduce the loss of extensibility on heat treatment in poly(p-benzamide) fibres.[8] The additive probably decomposes to a di- or tri-isocyanate on heating which then cross-links and chain-extends the aromatic polyamide.

The levels attained in mechanical properties are clearly related to the degree of orientation of the polymer chains and several techniques are available for estimating the degree of orientation in a fibre. Two of the most widely used for these fibres are sonic velocity and orientation angle determinations.[9a] Sonic velocity is claimed to be a measure of total molecular orientation[43] while the determination of the orientation angle, being an X-ray technique, gives a measure of crystalline orientation.[44] The latter specifies an angle about the fibre axis in which a given percentage of crystallites are aligned. A plot of initial modulus against measured orientation angle for a range of poly(p-benzamide) and poly(p-phenylene terephthalamide) fibres is shown in Fig. 7 and clearly shows the effect of crystallite orientation on fibre modulus. The orientation angles plotted in this figure have been determined in such a way as to represent twice the angle about the fibre axis within which approximately 77% of the crystallites are oriented.

A limited amount of data only is available on the crystal structures of poly(p-benzamide) and poly(p-phenylene terephthalamide). From X-ray diffraction diagrams of highly oriented examples of both fibres there appears to be a great deal of similarity between the crystal structures.

Indeed the fibre repeat distance[45] for both polymers is identical ($\simeq 13$ Å). In poly(p-benzamide) this corresponds to two 'monomer units' while in poly(p-phenylene terephthalamide) it corresponds to one. An analogous situation is found in nylon 6 and nylon 6·6. As has been reported recently in the case of poly(m-phenylene isophthalamide)[46] it seems unlikely that the aromatic rings and the amide groups are coplanar for steric reasons. This will be discussed later.

A possible crystal structure for poly(p-benzamide) which is in accord with the measured repeat distance is shown in Fig. 8. It has been estimated that about 36 repeat units would have to be involved in one chain fold in this polymer. No obvious spacing corresponding to chain folds was found by low angle X-ray measurements on highly oriented poly(p-benzamide) fibre. However, a strong equatorial streak was obtained which is consistent either with a microfibrillar structure having axially oriented fibrils or with axially oriented voids or regions of low density.[47]

The authors have made an approximate estimate of the dimensions of the unit cell for poly(p-benzamide) from X-ray diffraction photographs (Fig. 9) and find the cell to be orthorhombic with $a = 5\cdot2$ Å, $b = 7\cdot7$ Å, $c = 12\cdot95$ Å. While the values of a and b will probably need refining as better diffraction photographs become available the cross-sectional area

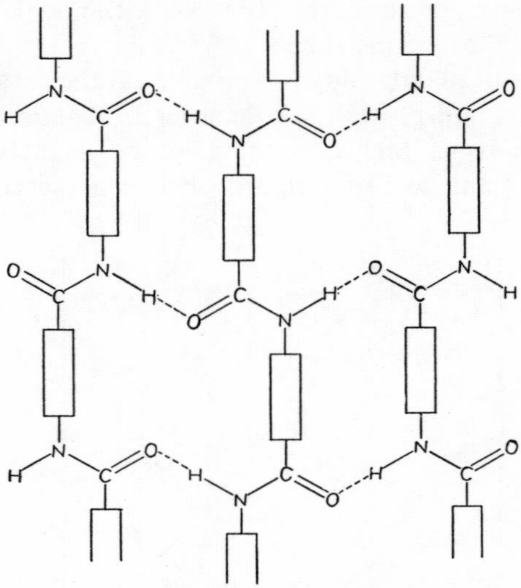

Fig. 8. Possible crystal structure for poly(p-benzamide).

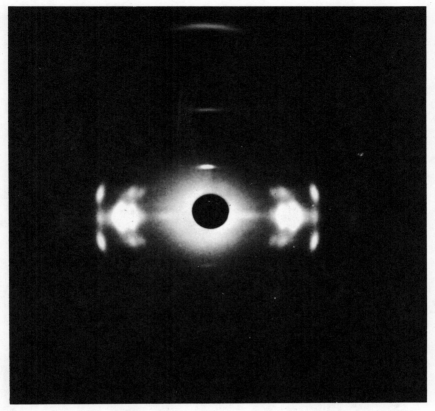

Fig. 9. X-ray diffraction diagram for poly(p-benzamide) fibre (initial modulus 1100 g dtex⁻¹) (authors' results).

($ab = 40$ Å²) is almost certainly of the correct order. The unit cell contains two polymer chains, each therefore of average cross-sectional area 20 Å² which agrees well with assumptions made elsewhere about the cross-sections of these molecules.[48] On this basis the crystal density is about $1·52$ g cm⁻³ which can be compared to measured fibre densities of $1·45$ g cm⁻³. A similar examination of poly(p-phenylene terephthalamide) fibres which has recently been reported[49] gave unit cell dimensions of $5·18$ Å, $7·87$ Å and $12·9$ Å and a crystallographic density of $1·48$ g cm⁻³.

A number of attempts have been made to measure the birefringence of a heat treated poly(p-benzamide) fibre sample prepared by the authors.

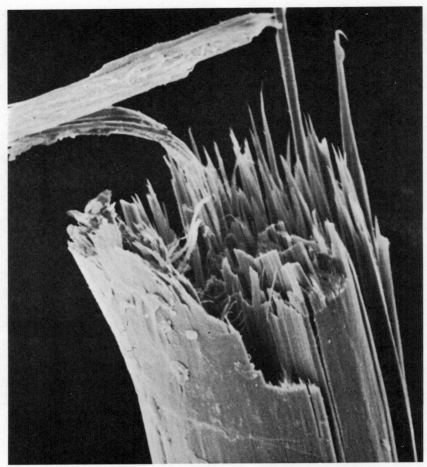

Fig. 10. Poly(p-benzamide) fibre (initial modulus 1100 g dtex⁻¹) broken in tension (× 3600). (Courtesy of S. Simmens, Shirley Institute.)

Using a Berek compensator a value of 0·65 was found. This was felt to be too high and an attempt was made to determine the individual refractive indices using a Becke line assessment with immersion liquids of different refractive indices.[50] The refractive index $\lambda n_\perp$ perpendicular to the fibre axis, was found to be in the range 1·63–1·64. The refractive index $\lambda n_\parallel$ parallel to the fibre axis, was very much more difficult to measure but

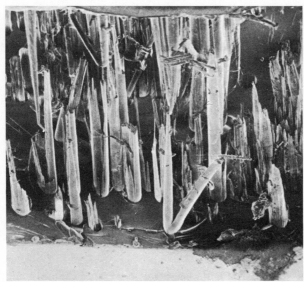

Fig. 11. Fracture surfaces (× 200) of fibre–epoxy composites using (a) glass fibre, (b) heat treated poly(p-benzamide) fibre.

values in the range 2·0–2·1 were recorded. These results agree well with calculations made for the theoretical refractive indices of this polymer assuming a density of 1·45 g cm^{-3} which gave $n_\perp = 1·64$ and $n_\parallel = 1·99$.[51] On the basis of these measurements the birefringence ($n_\parallel - n_\perp$) appears to be of the order of 0·37–0·47. Other aromatic polyamides such as poly(p-aminobenzhydrazide terephthalamide) show similar high orders of birefringence.[12] These will be discussed in more detail later.

Finally, a word about the fracture properties of poly(p-benzamide) fibre. Not unexpectedly it appears that the extreme axial properties of the fibre have to be paid for in terms of very poor transverse properties. Indeed the higher the modulus the greater is the tendency for the fibre to exhibit a failure mode consisting of fibrillation and splintering. Words cannot describe this as well as the Stereoscan photograph (Fig. 10) and Fig. 11 which compares fracture surfaces in a glass fibre–epoxy composite (a) and a poly(p-benzamide)–epoxy composite (b) prepared in the authors' laboratories.

13.4.2 Aromatic polyhydrazides and polyamide-hydrazides

The basic chemistry of the preparation of these polymers needs no elaboration here since it is strictly analogous to the diamine–diacid chloride polycondensations already described; the hydrazide group reacts in the same manner as the amino group. The examples which to the authors' knowledge have been shown to lead to ultra-high modulus organic fibres are made by condensing terephthaloyl chloride with p-aminobenzhydrazide (VIII), the all-para substituted diaminodihydrazide (IX) and terephthalic dihydrazide (X) (polymers E, F and G in Table 2).

VIII   H₂N⟨◯⟩CONH NH₂

IX   H₂N⟨◯⟩CONH NHCO⟨◯⟩CONH NHCO⟨◯⟩NH₂

X   H₂N NHCO⟨◯⟩CONH NH₂

The usual amide solvent systems are used for the polymers from VIII and IX while the all-hydrazide polymer from X and terephthaloyl chloride has been reported[52] to be insoluble in the usual organic solvents.

More recently it has been found to dissolve in dimethyl sulphoxide–lithium chloride[15] from which fibres can be spun. In the authors' experience the polyamide-hydrazide from VIII and terephthaloyl chloride dissolves readily enough in sulphuric acid but the inherent viscosity of the polymer quickly falls, presumably owing to degradation of the polymer chains.

In the case of the amino hydrazide VIII consideration has to be given to the way in which this compound enters the polymer chain during the solution polycondensation. At first sight it might be supposed that, in a purely random way, VIII could link up by either of its amino groups. However, one of these forms part of a hydrazide group and is considerably more reactive toward acid chlorides than the other. It is found therefore that there is always a tendency for diamine IX to form first leading to some order in the final polymer. The degree of order can be controlled by varying the reaction conditions[53] and monitored by electron diffraction[53] and NMR.[54] The less ordered the polymer the more readily it may be fabricated but this has to be paid for by a corresponding decrease in thermal stability.

The polymers do not melt or exhibit glass transitions but decompose at about 400–525°C (Ref. 14) and the spinning methods used are essentially the same as for the aromatic polyamides including dry-jet wet spinning.[12,13]

As in the case of the aromatic polyamides minimum values of η_{inh} for the production of good fibres are quoted and the useful number average molecular weight range for the polymers based on VIII and IX is said[12] to be 50 000–100 000. Spin stretch appears to be a factor in the development of the as-spun properties.[12,53]

Heat treatment of these fibres in the relaxed state at about 300°C produces a marked improvement in their mechanical properties.[55] Even greater improvements, however, are obtained on hot drawing as shown in Table 2, although only a slight draw is possible and the addition of the wax sorbitan monopalmitate to the spinning solutions is reported to allow greater drawing of the filaments.[13] Apparently the wax reduces the coagulation rate and helps to prevent the rapid formation on the fibre surface of a 'skin' which inhibits the diffusion of dope solvent and inorganic salts from the fibre interior. The more homogeneous fibre thus produced is drawn more readily. The hydrazide group is relatively unstable and on heating will lose water and cyclise to an oxadiazole ring producing an amide–oxadiazole polymer such as XI.

XI $\left[\text{CO}\middle\langle\bigcirc\middle\rangle\text{CONHNHCO}\middle\langle\bigcirc\middle\rangle\text{NH}\right]_n$

$$\xrightarrow[-H_2O]{\Delta} \left[\text{CO}\middle\langle\bigcirc\middle\rangle - \text{C} \overset{O}{\underset{N-N}{\diagup\diagdown}} \text{C} - \middle\langle\bigcirc\middle\rangle\text{NH}\right]_n$$

Apparently this does not occur at the moderate temperature (300°C) and during the very short time involved in heat treating or hot drawing the fibres.[14]

The hot drawn fibres are pale yellow in colour and have densities in the same range as the aromatic polyamides; reported figures are 1·425, 1·44 and 1·47 g cm^{-3}. Birefringence values are also high, ranging up to 0·45.[12]

X-ray diffraction measurements suggest a chain extended structure in that many orders of the repeat unit are found in the meridional direction indicating a highly regular arrangement of atoms over great distances along the chain axis.[56] Using highly crystalline films obtained from dilute solution the unit cell dimensions were found to be: a, 8·5 Å; b, 4·9 Å; c, 29·6 Å by transmission electron diffraction, giving a density of 1·59 g cm^{-3} (Ref. 57). This would correspond to the ordered structure designated polymer F in Table 2. As for the aromatic polyamides the assumption is made that the unit cell contains two molecules.

The range of fibres obtainable by varying the chemical order of the polymer and the various conditions of fibre preparation and treatment is said to be defined by the equation:

$$\text{Initial modulus (g denier}^{-1}) = 900\ E_b^{-0.58}$$

where E_b is the extension to break (%) for a single filament.[12]

A recent patent[58] describes copolymers (L in Table 2) consisting of the repeat units of poly(p-benzamide) and poly(p-phenylene terephthalamide) together with the hydrazide unit

$$\left[\text{CO}\middle\langle\bigcirc\middle\rangle\text{CONHNH}\right]$$

It is claimed that ultra-high modulus fibres can be made from those compositions which fall within a defined area when plotted on a triangular composition diagram for the three types of repeat unit. These workers have, in effect, mixed together various combinations of the bifunctional compounds which give rise to polymers A, B, E and G. They used the usual amide–LiCl solvents but also added acid-acceptors such as N,N-dimethylaniline. Fibre formation was by wet spinning from the same solvents and the moduli were developed by slight drawing at up to 380°.

13.4.3 Aromatic polyamides and polyhydrazides containing the oxalyl group

Fibres containing the oxalyl unit in the polymer chain are also new-comers to the scene. For convenience we describe them under a separate heading although formally they belong in the two classes already discussed. Two types of structure have been reported, one (XII) belonging to the polyamide class[59] while the other (XIII) is a polyhydrazide.[60]

XII $\left[\text{NH} \bigcirc \text{NHCOCONH} \bigcirc \text{NHCO} \bigcirc \text{CONH} \bigcirc \text{NHCO} \bigcirc \text{CO} \right]_n$

XIII $\left[\text{NHNHCOCONHNHCO} \bigcirc \text{CONHNHCO} \bigcirc \text{CONHNHCO} \bigcirc \text{CO} \right]_n$

No doubt both the structures shown above are idealised since some block formation is to be expected in these polymers. Both polymers (J and K in Table 2) are prepared in the familiar way from terephthaloyl chloride and mixtures of the requisite diamines or diahydrazides in amide–LiCl solvents and wet spun from the same solvents.

Like the polyhydrazide fibres described in the previous section these oxalyl-containing fibres develop their highest moduli on heat treatment and are capable of a limited amount of drawing. The polyoxamide XII seems to be able to stand the higher treatment temperature. After a pre-drawing at a moderate temperature it increases its modulus almost without drawing when heated to 500° in a way reminiscent of polymers A and B.

13.4.4 Copolyureas

Finally we come to the only example in the literature so far of an ultra-high modulus fibre which does not contain exclusively aromatic rings in the polymer chain. Instead the chain contains both aromatic and aliphatic rings but the latter are almost certainly stiff as will be seen. The possibilities for preparing ultra-high modulus fibre forming polymers of this type do not seem to have been as thoroughly investigated as have those for the polyamides and polyhydrazides already described.

Essentially we find in the literature just two polymers which have been spun to such fibres. Their preparation is simplicity itself merely requiring the bringing together of 2,5-dimethyl piperazine (XIV) and a bis-isocyanate of the type (XV) (where R = H or CH_3) in a suitable solvent such as dimethylacetamide–lithium chloride.[6,61] The resulting viscous polymer solution may be employed as the spinning dope.

The best figure is obtained when R = CH_3 and the tensile properties of this fibre are adequately exemplified in Table 2 (polymer H).

The polymer can be isolated as a white solid by precipitation into water and, in the authors' experience, is not as stable at high temperatures as the aromatic polyamides and possibly the polyamide-hydrazides as well. Nevertheless the fibre can be heat-treated at 350°C and exhibits a marked increase in tensile properties which is greatest when slight draw is simultaneously applied. The glass-transition temperature of this polymer is reported to be 205°C (Ref. 6). Although it was originally produced by dry spinning, the authors have found that equally good fibre results from wet spinning into acetone. The initial modulus was in fact greater than that reported for dry spun fibre (Table 2).

13.5 THE STRUCTURAL ORIGINS OF THE FIBRE STIFFNESS

It must be stated at the outset in any discussion of these remarkable fibres that little information is available at the time of writing about their detailed physico-chemical structures. Their properties seem to the authors to admit of only one explanation in general terms but, as so often occurs, one or two awkward facts must be accommodated. We shall attempt here to draw together the evidence of which we are aware, most of which has been laid before the reader in the foregoing descriptive sections. Much of the information available relates to the aromatic polyamide fibres and, in particular, to poly(p-benzamide) and poly(p-phenylene terephthalamide). These are the best studied examples and also exhibit the most striking properties but it seems probable that the other fibres, such as those containing the hydrazide group, are to be understood in similar terms.

What the evidence points to is that the fibres we have described consist of polymer molecules which are extremely inflexible and are largely oriented along the fibre axis. Other workers have also commented on the rigidity of these chains.[39,57] The folded crystallite concept is difficult to envisage in the context of these fibres since chain folding would seem to be unlikely unless extremely long folds are postulated incorporating many repeat units. Nevertheless the fibres are highly crystalline and some of the 'earlier' morphological concepts of fibre structure such as the fringed fibrillar or fringed micellar models[62] would appear to give a more realistic picture. Essentially what is required is a morphological model showing crystallites, high alignment of chains along the fibre axis and high axial polymer chain continuity.

13.6 FIBRE PROPERTIES

A model of the kind suggested above is supported strongly by the observed fibre properties. The fibre moduli, at best, come close to the theoretical levels calculated for many extended polymer chains (see Chapter 7). In particular the theoretical modulus has recently been calculated[49] for both poly(p-benzamide) and poly(p-phenylene terephthalamide) to be 2.00×10^{12} dyn cm^{-2} corresponding to 1350 g dtex^{-1} assuming a density of 1·5 g cm^{-3}. This may be compared to the best value so far reported for poly(p-benzamide) of 1260 g dtex^{-1} (Ref. 63). The assumption made for this calculation is that the polymer chain responds to axial stress by bond stretching and bond angle deformation rather than bond rotation. The fact that the calculated modulus and the experimentally determined figure are of the same order would seem to suggest that this is indeed the actual response to applied stress. Bond rotation and the straightening out of disordered conformations, therefore, do not contribute greatly to the response mechanism. The model is also in agreement with orientation angle and sonic velocity measurements which indicate that very high degrees of orientation are involved at the highest moduli levels.

The authors have not found convincing evidence from low angle X-ray diffraction photographs of heat-treated poly(p-benzamide) fibre for appreciable long range order which might be identified as chain folding. Observations of multiple orders of the repeat unit in the polyamide-hydrazide fibre X-ray diffraction photographs, as already stated, have been interpreted[56] as arising from regular structure over great distances along the fibre axis rather than from chain folding. Clearly more detailed and painstaking X-ray diffraction work would give some valuable answers.

Further evidence for a rigid linear chain model is provided by the spinning behaviour of the aromatic polyamides and by the remarkable behaviour of most of these fibres on heat treatment. The attainment of an initial modulus of some 400 g dtex^{-1} in the aromatic polyamides merely by solution spinning from a liquid crystal state clearly demonstrates that, in this state at least, the polymer chains are unusually easy to orientate under shear. Moreover, as illustrated in Table 2, not only the aromatic polyamides but one of the polyamide-hydrazides (E) and the copolyurea (H) exhibit very large increases in modulus on high temperature treatment, even in the relaxed state. The fibres behave as though their constituent molecules are essentially oriented on passing through the spinneret and do not relax to a disordered state once coagulated but are locked in a non-ideal condition which relaxes to a preferred extended condition on heat treatment.

The most simple picture is of a fibre consisting mainly of rigid rod-like molecules held together by intermolecular polar attractions, especially hydrogen-bonding between the carbonyl and amino moieties of the amide groups. One may imagine that heat treatment permits the chains to slip over one another until hydrogen-bonding groups are brought into register and regions of crystalline order are formed.

A striking feature of poly(p-benzamide) fibre is the extremely splintered nature of the broken end of the heat treated material (Fig. 10). The tendency of the fibres to break up into fibrillar or high aspect ratio particles and to show very poor mechanical properties in a direction perpendicular to the fibre axis must clearly be related to the extremely high axial orientation of the polymer molecules. One can envisage these fibrillar particles as built up of submicroscopic fibrils, themselves built up of the ultimate linear crystallites.

13.7 SOLUTION PROPERTIES

Solutions of poly(p-benzamide) of quite low concentration (2–3%) were found by the authors to be highly viscous, which is thought to reflect the stiffness of the polymer chains rather than unusually high molecular weight. This is confirmed by the high value ($\simeq 1 \cdot 6$) of the exponent α in the Mark–Houwink equation which was reported for poly(p-benzamide) in sulphuric acid.[39] The magnitude of α is often considered to be a measure of chain stiffness.[64] High resolution NMR spectra obtained in the authors' laboratory from such solutions showed unusually broad spectral lines as compared with the spectra of other aromatic condensation polymers. Such a result is consistent with a low degree of chain flexibility although it might also be explained by some form of aggregation in solution.

13.8 CHEMICAL CONSIDERATIONS

The amide group probably has a certain amount of double bond character as a result of resonance.[65] It may therefore be regarded as a fairly stiff unit in its own right. The barrier to rotation about the central

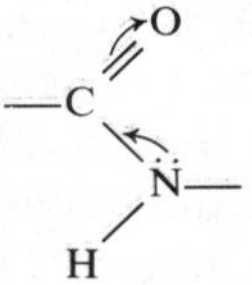

C–N bond in many systems of the type $R_1CONR_2R_3$ is a significant one, ΔG usually being in the region of 15–20 kcal mol^{-1} (Ref. 66). Thus if we consider a portion of an all-para oriented aromatic polyamide chain as shown below we see that the only bond rotation which could give rise to chain folding is inhibited. If a space-filling molecular model is constructed it can be seen that free rotation about the bonds joining the amide

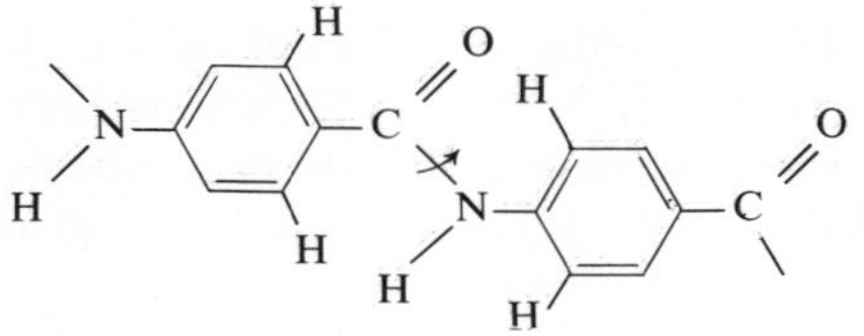

groups to the aromatic rings is probably hindered by the ortho-hydrogens on the rings. Such rotation could in any case not lead to chain folding. In spite of the expected tendency of the amide group to resonate with the adjacent aromatic ring, a more probable configuration seems to be one in which the amide group lies in a plane perpendicular to that of the ring. This is supported by a recent crystal structure determination for poly(m-phenylene isophthalamide).[46] As far as bond rotation is concerned similar considerations apply to the azo-group $-N=N-$ which appears in one of the polymers (D in Table 2).

One of the ultra-high modulus fibre forming polymers, designated H, contains non-aromatic piperazine rings which in principle could be in either of the two interconvertible conformations known as 'boat' and 'chair'. A further complication is introduced by the possibility of 'cis' and

'trans' configurations at the nitrogen atoms. In view of the fibre properties it seems most probable that we are dealing with the linearly oriented and relatively rigid trans-chair form. It is known[67] that the 'chair' form of piperazine is $3·8$ kcal mol^{-1} more stable than the 'boat' form. The authors' high resolution NMR measurements of *trans*-2,5-dimethylpiperazine in dichloromethane solution were consistent with the 'chair' form and remained unchanged between $-55°C$ and $130°C$. The 'chair' form is therefore probably the stable one over a wide temperature range.[68] The barrier to ring inversion in N,N-dimethylpiperazine has been found to be $13·3$ kcal mol^{-1} by low temperature NMR measurements.[69]

All the fibres discussed in this review are of the amide or hydrazide type. This seems to be significant in that, provided we assume the amide group to be a rigid planar unit, it is only possible to have 'offsets' in the polymer chains leaving the aromatic ring axes still parallel. Obviously many linear chemical structures exist. For instance the Carborundum Company's aromatic polyester 'Ekonol', the condensation polymer of p-hydroxybenzoic acid, is structurally very similar to poly(p-benzamide). Unfortunately, it is quite insoluble although ways around this problem are suggested by the aromatic copolyester–amides which have been reported.[70] Of course the stiff linear polymer *par excellence* would be poly(p-phenylene). Developments in the fabrication of this and other linear rigid polymers which are currently on the 'insoluble list' may appear in due course.

If, however, the chain contains inherently 'bent' units, are ultra-properties precluded? It seems probable to the authors that, in due course, research will show that such structures can be coaxed into a high modulus form. Maybe some of the stiff polymers with less linear chemical structures can, with sufficient attention to purity, spinning conditions, molecular weight, etc., be fabricated into extended-chain fibres. We see the beginnings of this even in flexible polymers like polyethylene and polypropylene while a great many high temperature fibres have moduli bordering on the 'ultra-levels'. Now that the attainment of these properties is shown to be possible for polymers which naturally seem to take up an extended chain conformation, we may reasonably expect to hear more of the geometrically less linear but still inherently rigid structures.

In all of the foregoing discussion we have not mentioned the curious aromatic–aliphatic copolyamide designated polymer C in Table 2. It seems to be in a class of its own at present and many of the arguments advanced in favour of extended chains do not take proper account of it. Judging by its mechanical properties in fibre form, however, its morphology is not as disordered as one might expect from the inclusion of sequences of $-CH_2-$ units. One can only conclude therefore that it can still be described by some such model as we have assumed above. Perhaps, in such an environment, the short polymethylene chain segments are constrained to adopt a fully extended conformation. Clearly many points still need to be resolved before a truly comprehensive explanation of the structural origins of the stiffness of these fibres can be given.

13.9 COMMERCIAL APPLICATIONS

Fibres having such high stiffnesses and low extensibilities are unlikely to find uses in apparel or furnishings except perhaps for very specialised applications. They have obvious possibilities, however, in the field of 'industrial' yarns and fabrics such as are used in motor vehicle tyres, drive belts, conveyor belts and various high performance composites covering a very wide range from aerospace to sporting goods. The fibres are still in the development or early commercial stages and costs limit their use to materials which command high prices. Good examples are, of course, aerospace components for which the ratios of strength and stiffness to weight have to be maximised.

Du Pont and Monsanto, the two U.S. chemical companies responsible for the discovery of the known ultra-high modulus organic fibres, are actively testing their fibres in conjunction with potential users. The former company is now apparently committed to production of quite large quantities of tyre cord. Very little information is available about the chemical natures of the proprietary fibres but it seems reasonably certain that the information in Table 5 is correct.

TABLE 5

ULTRA-HIGH MODULUS FIBRES AVAILABLE IN DEVELOPMENT OR COMMERCIAL QUANTITIES

Company	Designation of fibre[a]	Typical initial modulus (g dtex^{-1})	Typical tenacity (g dtex^{-1})	Ext. to break (%)	Density (g cm^{-3})	Ref.	Apparent chemical constitution
Du Pont	Fibre B — Yarn	405	10·8	7·0		71	Poly(p-benzamide or poly(p-phenylene terephthalamide) or possibly both as copolymer or blend[72,77]
	Fibre B — Dipped tyre cord	{306, 315	10·0, 16·2	4·2, 4·0	1·44	71, 72	
Du Pont	PRD-49-1	1000–1100	16·5	1·8	1·45	73, 74	
	PRD-49-3	875–970	19·5	2·0	1·45	73, 75, 76	
	PRD-49-4	580	21·0	3·3	1·45	75, 76	
Monsanto	X500 type 1	630	14·5	3·2	1·45	78, 79	Probably based on the aromatic polyamide-hydrazides with varying degrees of chemical order[78]
	X500 type 2	390	10·0	8·0	1·45	78, 79	
	X500 type 3	200	7·2	23·0	1·45	78, 79	

[a] Fibre B has now been trade-named by Du Pont as Kevlar and PRD-49 as Kevlar 49 (Refs. 80, 81). The earlier names are used in the references cited in the table.

13.9.1 Tyre cord
Although there are signs that X-500 type fibres are under development for this application[78] Fibre B has pride of place at present. Originally based on poly(p-benzamide), its breaking strength and competitive edge with respect to glass and steel cords were markedly improved by a change to poly(p-phenylene terephthalamide). In radial tyre belts it is theoretically as cost-effective as four or five times its weight of steel. Fibres based on aromatic polymers of this type are far superior to conventional organic tyre cords in terms of their retention of physical properties at the elevated temperatures produced under running conditions.[72,82] It is intended to increase production of Fibre B from some 500 000 lb/annum in 1972 to 6 000 000 lb/annum in 1973 with the aim of reaching full output by 1975 when the fibre will probably cost $2·00–2·50 per lb.[82] The reactions of tyre manufacturers who have tested development quantities are said to be very favourable.[83]

13.9.2 Composites for military and aerospace uses
The preparation and properties of composites using aromatic polyamide-hydrazide fibres have been described,[12,78] and one area in which the X-500 fibres have clearly been tested as composite reinforcements outside Monsanto is that of personnel armour for ballistic protection. Type 3 seems to have given the most promising results, the stiffer types tending to fail by longitudinal splitting.[78,79]

Again it is Du Pont's PRD-49 group of aromatic polyamides which have now appeared from Boeing and Lockheed detailing the results of their which is presumably a heat-treated version of Fibre B, was called PRD-49-1 and details were supplied to the Boeing Co. in 1968.[84] Reports have now appeared from Boeing and Lockheed detailing the results of their tests on PRD-49 reinforced parts for the BO-105 helicopter,[85] the L1011 Tristar Airbus[73] and the 737 transport[76] while Boeing Aerospace have described the use of the fibre in filament wound pressure vessels.[86]

The fibre itself has undergone modification and the version which is being actually put into production, PRD-49-3, has somewhat lower stiffness (Table 5) and higher strength. A third version, PRD-49-4, is about half as stiff but has a higher extensibility and is intended for special fabrics (e.g. for ballistic armour) and for cables. Type 3 has significantly higher shear strength and impact resistance in epoxy composites than type 1.[75] Readers interested in composites will find a great deal of information in the Du Pont commercial literature and in the various aerospace reports.[73–76,84–86] One obvious weakness of PRD-49 as a composite reinforcement seems to be the low compressive strength of its composites and this is being studied.

Fibre production began in 1972 at 100 000 lb/annum with plans to increase as the market develops. Large price reductions for the fibre were timed by Du Pont to coincide with the 1972 Farnborough air show bringing epoxy 'pre-preg' fabric for instance down to $20–30 per lb,[87] and projected costs are expected to be lower than for carbon and boron fibres.[85,87] It has been reported that the fibre will be available as collimated tape, chopped fibre, yarns, rovings and as woven fabric.[88]

It would seem that Russian workers are developing a very similar type of fibre which they call Terlon. The name is faintly evocative of poly(p-phenylene terephthalamide) and reported properties[89] support this. No information is available, however, on the chemical nature of the fibre.

13.9.3 Other possible uses

Likely outlets for PRD-49 composites are reported to be in high-speed machine components because of their combination of low weight and high stiffness and in electronics where dielectric properties and dimensional stability are important. The high priced sports equipment market has not escaped Du Pont and PRD-49, either along or combined with other reinforcements, is apparently being evaluated by professional athletes in various equipment from tennis racquets to kayaks.[87]

ACKNOWLEDGEMENTS

The authors would like to thank the following people for their contributions in terms of experimental work and useful discussions: D. J. Adam, J. M. Bell, C. G. Cannon, J. N. K. Hyland, J. E. McIntyre, G. W. Meacock, S. Simmens and G. J. Tyler.

REFERENCES

1. (a) Goodman, I. (1967). *Synthetic Fibre-forming Polymers* (RIC Lecture Series, No. 3).
   (b) Hearle, J. W. S. and Peters, R. H. (1963). *Fibre Structure*, Textile Institute and Butterworths, London.
   (c) Peters, R. H. (1963). *Textile Chemistry*, Vol. 1, Elsevier, Amsterdam.
2. Andrews, J. M. and Ward, I. M. (1970). *J. Mater. Sci.*, **5**(5), 411; Capaccio, G. and Ward, I. M. (1973). *Nature Phys.-Sci.*, **243**, 143.
3. de Winter, W. (1970). *Textile Industrie*, **72**(11), 833.
4. Imperial Chemical Industries Ltd, B.P. 1,047,978.
5. Carter, G. B. (1970). *Reports on the Progress of Applied Chemistry*, **55**, 92.
6. Du Pont de Nemours & Co., E.I., B.P. 876,491. See also Ref. 1a, p. 40.
7. Du Pont de Nemours & Co., E.I., U.S.P. 3,600,350.

8. Du Pont de Nemours & Co., E.I., U.S.P. 3,595,951.
9. (a) Du Pont de Nemours & Co., E.I., B.P. 1,283,065.
   (b) Du Pont de Nemours & Co., E.I., Ger. Offen. 2,219,646.
   (c) Du Pont de Nemours & Co., E.I., Ger. Offen. 2,219,703.
10. Du Pont de Nemours & Co., E.I., B.P. 1,328,680.
11. Bach, H. C. and Hinderer, H. E. (1970). *Polymer Preprints,* **11**(1), 334.
12. Monsanto Co., U.S.P. 3,600,269.
13. Monsanto Co., U.S.P. 3,642,706.
14. Monsanto Co., B.P. 1,220,455.
15. Du Pont de Nemours & Co., E.I., U.S.P. 3,642,707. See also U.S.P. 3,536,651.
16. Mark, H. F., Atlas, S. M. and Cernia, E. (1967). *Man-made Fibres, Science and Technology,* Vol. I, Wiley, New York.
17. Du Pont de Nemours & Co., E.I., B.P. 1,283,064.
18. Du Pont de Nemours & Co., E.I., U.S.P. Appln. 713304 Ex. XIX.
19. Du Pont de Nemours & Co., E.I., B.P. 1,259,788.
20. Du Pont de Nemours & Co., E.I., B.P. 1,262,002.
21. Du Pont de Nemours & Co., E.I., B.P. 1,283,066. Ex. III.
22. Du Pont de Nemours & Co., E.I., U.S.P. 3,541,056.
23. Du Pont de Nemours & Co., E.I., Neth. P. Appln. 68/18740.
24. Du Pont de Nemours & Co., E.I., B.P. 871,580.
25. Du Pont de Nemours & Co., E.I., U.S.P. 3,360,598.
26. Preston, J., Morgan, H. S. and Black, W. B. (1973). *J. Macromol Sci. Chem.,* **A7**(1), 325.
27. Ref. 9a Ex. XXIX.
28. Ref. 9a Ex. VII.
29. Ref. 9a Ex. IV.
30. Imperial Chemical Industries Ltd., B.P. 1,170,171.
31. Du Pont de Nemours & Co., E.I., U.S.P. 3,699,085 and 3,719,642.
32. Du Pont de Nemours & Co., E.I., U.S.P. 3,637,606.
33. Ref. 23 Exs. XIV and XXI.
34. Ref. 17 Ex. XXV.
35. Ref. 17 Exs. XI–XIII. Ref. 9c Ex. IV.
36. See for example Ref. 7, column 6.
37. Hermans, J. Jr. (1967). In *Ordered Fluids and Liquid Crystals* (ed. R. S. Porter and J. F. Johnson), Amer. Chem. Soc. Advances in Chemistry Series, Vol. 63, p. 282.
38. Kalmykova, V. D. *et al* (1971). *Vysokomol. Soed.,* Ser. B, **13**(10), 707.
39. See Ref. 17, p. 5.
40. Tsvetkov, V. N. *et al.* (1968). *J. Polymer Sci. C,* **16,** 3205.
41. Ref. 20 Exs. XI–XVII.
42. Authors' results. Increases in length of about 2% were obtained with poorly oriented poly(*p*-phenylene terephthalamide) on heat treating at 500°C in the relaxed state. See also Refs. 9b and 9c.
43. Charch, W. H. and Moseley, W. W. (1959). *Textile Research J.,* **29,** 525.
44. Krimm, S. and Tobolsky, A. V. (1951). *Textile Research J.,* **21,** 805.
45. Statten, W. O. (1967). *Handbook of X-Rays* (ed. E. F. Kaelble), McGraw-Hill, New York, pp. 21–27.
46. Herlinger, H. and Knoell, H. (1972). Paper presented at the 164th ACS Meeting, New York, Aug.–Sept.
47. Private communication, P. G. Owston (formerly ICI Corporate Lab).
   Private communication, C. G. Cannon (ICI Fibres Ltd).
48. Fielding-Russell, G. S. (1971). *Textile Research J.,* **41** (10), 861.
49. Northolt, M. G. and van Aartsen, J. J. (1973). *Polymer Letters,* **11,** 333.
50. Chamot, E. M. and Mason, C. W. (1946). *Handbook of Chemical Microscopy,* Wiley, New York, Vol. 1, p. 362.
51. Private communication, G. Longman (ICI Corporate Lab).

52. Frazer, A. H. and Wallenberger, F. T. (1964). *J. Polymer Sci. A*, **2**, 1150.
53. Monsanto Co., U.S.P. 3,632,548.
54. Morrison, R. W., Preston, J., Randall, J. C. and Black, W. B. (1973). *J. Macromol. Sci.— Chem.*, **A7**(1), 99.
55. Authors' results.
56. See Ref. 12, column 8.
57. Holland, V. F. (1973). *J. Macromol. Sci.—Chem.*, **A7**(1), 173.
58. Toray Industries Inc., Ger. Offen. 2,232,504.
59. Monsanto Co., U.S.P. 3,738,964.
60. Monsanto Co., Ger. Offen. 2,247,567.
61. Du Pont de Nemours, E.I., U.S.P. 3,318,849.
62. Hearle, J. W. S. in Ref. 1b, p. 209.
63. Ref. 7, Ex. XXIIIC.
64. Burchard, W. (1963). *Makromol. Chem.*, **67**, 182.
65. Cannon, C. G. (1955). *Mikrochimica Acta*, No. 2–3, 555; (1956). *J. Chem. Phys.*, **24**, 491; (1958). *Disc. Farad. Soc.*, **25**, 59.
66. Kessler, H. and Rieker, A. (1967). *Z. Naturforsch.*, **22B**, 456; Mannschreck, A. (1965). *Tetrahedron Lett.*, 1341; Jackman, L. M. *et al.* (1969). *Org. Mag Res.*, **1**, 109; Matsubayashi, G.-E. and Tanaka, T. (1969). *J. Inorg. Nucl. Chem.*, **31**, 1963; Gehring, D. G. *et al.* (1966). *J. Org. Chem.*, **31**, 3436.
67. Eliel, E. L., Allinger, N. L., Angyal, S. J. and Morrison, G. A. (1965). *Conformational Analysis*, Wiley, New York, p. 250.
68. Spragg, R. A. (1968). *J. Chem. Soc. B*, 1128.
69. Reeves, L. W. and Strømm, K. O. (1961). *J. Chem. Phys.*, **34**, 1711.
70. Giori, C. (1971). *Polymer Preprints*, **12**(1), 606.
71. *Rubber World*, 1970, March, 59.
72. *Rubber World*, 1972, **166**(4), 40–41, 44, 46–47, 50.
73. Stone, R. H. (1972). Paper presented at 17th National SAMPE Symposium, L.A., California, April 11–13.
74. U.S. Government Rept. P, 177,697.
75. Sturgeon, D. L., Wolffe, R. A., Miner, L. H. and Wagle, D. G. (1972). *PRD-49 Fibre and Composite Performance*, Du-Pont Technical Literature.
76. Moore, J. W. (1972). Paper presented at 27th Annual SPI Conf., Washington D.C., Feb. 8–11.
77. Hamm, G. (1972). *Plastica*, **25**(10), 437.
78. *J. Macromol. Sci.—Chem.*, 1973, **A7**(1). The whole of this volume is devoted to a symposium on the Monsanto Company's fibres and is recommended reading.
79. Lilyquist, M. R. (1971). U.S. Govt. Rept., A.D. 730775, Jan.
80. *Eur. Chem. News*, 1973, Aug. 10, 14.
81. *Fin. Times*, 1973, July 19, 10.
82. See for example: *Rubber World*, 1972, **161**(1), 56; *Chemical Age*, 1972, June 2, 18; *Rubber J.*, 1972, Feb., 23; *Chem. Week*, 1972, March 22, 37; Ref. 69; *Fin. Times*, 1973, Aug. 2, 8.
83. See for example: *Chem. Week*, 1972, 23rd Aug., **111**, 28; *Chem. Week*, 1972, 22nd Mar., 37; *Daily News Record*, 1972, April 21st, 23; *Text. Inst. Ind.*, 1973, June, 142.
84. Stratton, W. K. (1971). Paper presented at the 16th National SAMPE Symposium, April 22nd, Anaheim, California.
85. Hooker, D. M. (1972). Paper presented at the 31st Annual Conf. of the Society of Aeronautical Weight Engineers, Inc., 22–25th May, Atlanta, Georgia.
86. Hoggatt, J. T. (1971). Paper presented at 1971 National SAMPE Technical Conf. on Space Shuttle Materials, Huntsville, Alabama, Oct. 5.
87. *The Engineer*, 1972, 14th Sept., 60.
88. *Non Wovens Report*, 1971, Oct., 6.
89. *Khim. Volokna*, 1972 6, 20.

INDEX